HANDBOOK OF ATOMIC, MOLECULAR, AND OPTICAL PHYSICS

Handbook of Atomic, Molecular, and Optical Physics

Editor:

Maria Emilova Velinova, Ph.D.

ArclerPress

www.arclerpress.com

Handbook of Atomic, Molecular, and Optical Physics

Editor: Maria Emilova Velinova, Ph.D.

708 3rd Avenue, 6th Floor
New York
NY 10017
United States of America
www.arclerpress.com

ISBN: 978-1-68094-440-2

Library of Congress Control Number: 2016951130

About the Editor

Maria Emilova Velinova, Ph.D.

Maria Velinova is Ph.D. holder in Quantum chemistry at the University of Sofia since April 2012. Her major research experience is in the field of Computational Chemistry, especially in statistical mechanics methods applied to different sorts of biomolecules. Member of the Laboratory of Quantum and Computational Chemistry at the University of Sofia.

Contents

List of Contributors

Norah Alonizan
Department of Physics and Astronomy, College of Science, King Saud University, Riyadh 11451, Saudi Arabia

Rabia Qindeel
Department of Physics and Astronomy, College of Science, King Saud University, Riyadh 11451, Saudi Arabia

Nabil Ben Nessib
Department of Physics and Astronomy, College of Science, King Saud University, Riyadh 11451, Saudi Arabia

W. Nelson
Department of Physics, University of York, Heslington, York YO10 5DD, United Kingdom

P. Bokes
Department of Physics, Faculty of Electrical Engineering and Information Technology, Slovak University of Technology, Ilkovičova 3, 841 04 Bratislava, Slovak Republic
European Theoretical Spectroscopy Facility (ETSF)

Patrick Rinke
European Theoretical Spectroscopy Facility (ETSF)
Fritz-Haber-Institut der Max-Planck-Gesellschaft, Faradayweg 4-6, 14195 Berlin, Germany

R. W. Godby
Department of Physics, University of York, Heslington, York YO10 5DD, United Kingdom
European Theoretical Spectroscopy Facility (ETSF)

A A Svinarenko
Odessa State Environmental University, L'vovskaya str.15, Odessa-16, 65016, Ukraine

A V Ignatenko
Odessa State Environmental University, L'vovskaya str.15, Odessa-16, 65016, Ukraine

V B Ternovsky
Odessa State Environmental University, L'vovskaya str.15, Odessa-16, 65016, Ukraine

V V Nikola
Odessa State Environmental University, L'vovskaya str.15, Odessa-16, 65016, Ukraine

S S Seredenko
Odessa State Environmental University, L'vovskaya str.15, Odessa-16, 65016, Ukraine

T B Tkach
Odessa State Environmental University, L'vovskaya str.15, Odessa-16, 65016, Ukraine

Mahmood Payehghadr
Department of Chemistry, Payame Noor University, Tehran 19395-4697, Iran

Sousan Esmaeilpour
Department of Chemistry, Payame Noor University, Tehran 19395-4697, Iran
Standard Research Institute, Karaj 31585-163, Iran

Mohammad Kazem Rofouei
Faculty of Chemistry, Tarbiat Moallem University, Tehran 15815-3587, Iran

Laleh Adlnasab
Department of Chemistry, Shahid Beheshti University, Evin, Tehran 1983963113, Iran

Farid Shakerian
Department of Chemistry, Faculty of Science, Yazd University, Yazd 89195-741, Iran

Ali Mohammad Haji Shabani
Department of Chemistry, Faculty of Science, Yazd University, Yazd 89195-741, Iran

Shayessteh Dadfarnia
Department of Chemistry, Faculty of Science, Yazd University, Yazd 89195-741, Iran

Mahdieh Shabani
Department of Chemistry, Faculty of Science, Yazd University, Yazd 89195-741, Iran

Wei Gao
School of Information Science and Technology, YunnanNormalUniversity, Kunming 650500, China

Weifan Wang
Department of Mathematics, Zhejiang Normal University, Jinhua 321004, China

Mohammad Reza Farahani
Department of Applied Mathematics, Iran University of Science and Technology, Narmak, Tehran 16844, Iran

Anoop Kumar Pandey
Department of Physics, University of Lucknow, Lucknow, India

Shamoon Ahmad Siddiqui
Department of Physics, Centre for Advanced Materials and Nanoengineering, Najran University, Najran, Kingdom of Saudi Arabia

Apoorva Dwivedi
Department of Physics, University of Lucknow, Lucknow, India

Kanwal Raj
Medicinal and Process Chemistry Division, CDRI, Lucknow, India

Neeraj Misra
Department of Physics, University of Lucknow, Lucknow, India
B-981, Sector-A, Mahanagar, Lucknow 226006, India

Jesus Baldenebro-Lopez
NANOCOSMOS Virtual Lab, Centro de Investigación en Materiales Avanzados, S.C., Miguel de Cervantes 120, Complejo Industrial Chihuahua, 31190 Chihuahua, CHIH, Mexico
Facultad de Ingeniería Mochis, Universidad Autónoma de Sinaloa, Prol. Ángel Flores y Fuente de Poseidón, S.N., 81223 Los Mochis, SIN, Mexico

Norma Flores-Holguin
NANOCOSMOS Virtual Lab, Centro de Investigación en Materiales Avanzados, S.C., Miguel de Cervantes 120, Complejo Industrial Chihuahua, 31190 Chihuahua, CHIH, Mexico

Jose Castorena-Gonzalez
Facultad de Ingeniería Mochis, Universidad Autónoma de Sinaloa, Prol. Ángel Flores y Fuente de Poseidón, S.N., 81223 Los Mochis, SIN, Mexico

Jorge Almaral-Sanchez
Facultad de Ingeniería Mochis, Universidad Autónoma de Sinaloa, Prol. Ángel Flores y Fuente de Poseidón, S.N., 81223 Los Mochis, SIN, Mexico

Daniel Glossman-Mitnik
NANOCOSMOS Virtual Lab, Centro de Investigación en Materiales Avanzados, S.C., Miguel de Cervantes 120, Complejo Industrial Chihuahua, 31190 Chihuahua, CHIH, Mexico

A. Farcaş
Department of Mathematics and Computer Science, "Iuliu Hatieganu" University of Medicine and Pharmacy, L. Pasteur 6, 400349 Cluj-Napoca, Romania

C. Iacoviţă
Department of Pharmaceutical Physics-Biophysics, "Iuliu Hatieganu" University of Medicine and Pharmacy, L. Pasteur 6, 400349 Cluj-Napoca, Romania

E. Vinţeler
Faculty of Physics, "Babes-Bolyai" University, Kogalniceanu 1, 400084 Cluj-Napoca, Romania

V. Chiş
Faculty of Physics, "Babes-Bolyai" University, Kogalniceanu 1, 400084 Cluj-Napoca, Romania

R. Ştiufiuc
Department of Pharmaceutical Physics-Biophysics, "Iuliu Hatieganu" University of Medicine and Pharmacy, L. Pasteur 6, 400349 Cluj-Napoca, Romania

C. M. Lucaciu
Department of Pharmaceutical Physics-Biophysics, "Iuliu Hatieganu" University of Medicine and Pharmacy, L. Pasteur 6, 400349 Cluj-Napoca, Romania

Har Lal Singh
Department of Chemistry, Faculty of Engineering & Technology, Mody University of Science and Technology, Laxmangarh, Sikar, Rajasthan 332311, India

Jangbhadur Singh
Department of Chemistry, Faculty of Engineering & Technology, Mody University of Science and Technology, Laxmangarh, Sikar, Rajasthan 332311, India

Jolanta Kumirska
Faculty of Chemistry, University of Gdansk, Sobieskiego 18/19, PL-80-952 Gdansk, Poland

Małgorzata Czerwicka
Faculty of Chemistry, University of Gdansk, Sobieskiego 18/19, PL-80-952 Gdansk, Poland

Zbigniew Kaczyński
Faculty of Chemistry, University of Gdansk, Sobieskiego 18/19, PL-80-952 Gdansk, Poland

Anna Bychowska
Faculty of Chemistry, University of Gdansk, Sobieskiego 18/19, PL-80-952 Gdansk, Poland

Krzysztof Brzozowski
Faculty of Chemistry, University of Gdansk, Sobieskiego 18/19, PL-80-952 Gdansk, Poland

Jorg Thöming
UFT-Centre for Environmental Research and Sustainable Technology, University of Bremen, Leobener Straße UFT, D-28359 Bremen, Germany

Piotr Stepnowski
Faculty of Chemistry, University of Gdansk, Sobieskiego 18/19, PL-80-952 Gdansk, Poland

Ø. Langangen
Institute of Theoretical Astrophysics, University of Oslo, Blindern, 0315 Oslo, Norway
Department of Biology, Centre for Ecological and Evolutionary Synthesis (CEES), University of Oslo, Blindern, 0316 Oslo, Norway

A. Vaskinn
Department of Physics, The Norwegian University of Science and Technology, 7491 Trondheim, Norway

B.-S. Skagerstam
Department of Physics, The Norwegian University of Science and Technology, 7491 Trondheim, Norway
Centre for Advanced Study (CAS), Drammensveien 78, 0271 Oslo, Norway

J. F. Goodwin
Quantum Optics and Laser Science, Blackett Laboratory, Imperial College London, Prince Consort Road, London SW7 2AZ, United Kingdom

G. Stutter
Quantum Optics and Laser Science, Blackett Laboratory, Imperial College London, Prince Consort Road, London SW7 2AZ, United Kingdom

R. C. Thompson
Quantum Optics and Laser Science, Blackett Laboratory, Imperial College London, Prince Consort Road, London SW7 2AZ, United Kingdom

D. M. Segal
Quantum Optics and Laser Science, Blackett Laboratory, Imperial College London, Prince Consort Road, London SW7 2AZ, United Kingdom

Jianqiang Zhu
National Laboratory on High Power Laser and Physics, Shanghai Institute of Optics and Fine Mechanics, Chinese Academy of Sciences, Number 390, Qinghe Road, Jiading, Shanghai 201800, China

Xinglong Xie
National Laboratory on High Power Laser and Physics, Shanghai Institute of Optics and Fine Mechanics, Chinese Academy of Sciences, Number 390, Qinghe Road, Jiading, Shanghai 201800, China

Meizhi Sun
National Laboratory on High Power Laser and Physics, Shanghai Institute of Optics and Fine Mechanics, Chinese Academy of Sciences, Number 390, Qinghe Road, Jiading, Shanghai 201800, China

Qunyu Bi
National Laboratory on High Power Laser and Physics, Shanghai Institute of Optics and Fine Mechanics, Chinese Academy of Sciences, Number 390, Qinghe Road, Jiading, Shanghai 201800, China

Jun Kang
National Laboratory on High Power Laser and Physics, Shanghai Institute of Optics and Fine Mechanics, Chinese Academy of Sciences, Number 390, Qinghe Road, Jiading, Shanghai 201800, China

Sonia Boscolo
Photonics Research Group, School of Engineering and Applied Science, Aston University, Birmingham B4 7ET, UK

Christophe Finot
Laboratoire Interdisciplinaire Carnot de Bourgogne, UMR 6303 CNRS, Université de Bourgogne, 21078 Dijon, France

T. Hanke
Department of Physics and Center for Applied Photonics, University of Konstanz, D-78457 Konstanz, Germany

G. Krauss
Department of Physics and Center for Applied Photonics, University of Konstanz, D-78457 Konstanz, Germany

D. Tra¨utlein
Department of Physics and Center for Applied Photonics, University of Konstanz, D-78457 Konstanz, Germany

B. Wild
Department of Physics and Center for Applied Photonics, University of Konstanz, D-78457 Konstanz, Germany

R. Bratschitsch
Department of Physics and Center for Applied Photonics, University of Konstanz, D-78457 Konstanz, Germany

A. Leitenstorfer
Department of Physics and Center for Applied Photonics, University of Konstanz, D-78457 Konstanz, Germany

Preface

Atomic, molecular and optical physics (AMO Physics) is the study of the interaction between light and matter. These interactions are investigated on various scales (from the atomic to molecular level) with the purpose to find answers to important scientific questions. All that is the driving reason to extent the research opportunities in atomic, molecular and optical physics to a broad range of topics.

The first part of the book presents the leading investigations in the theoretical and experimental study of the atomic systems. The Hartree-Fock and multiconfiguration Hartree-Fock methods are discussed on the example of neutral oxygen. The Green's function technique is used for the calculation of self-interaction part of the GW self-energy on atomic hydrogen. A relativistic model for valuating of radiation transition parameters in spectra of multicharged ions is described. The experimental aspect of the atomic physics by describing in details the flame atomic absorption spectrometry for determining of trace amount of metals is also included in the book. The second part starts with a discussion of molecular structures from computational and theoretical perspective. A larger space is given to the application of the DFT methods to systems at different levels of complexity and nature. A theoretical study of topological indices of a molecular structure, which can provide us a basis for the manufacturing of drugs and chemical materials without using of experiments, is also discussed. The last articles of this book section look at the experimental aspects of the molecular physics involving various types of spectroscopy. A brief description of the basic ideas behind these techniques is given, with emphasis on the Raman, Electronic, Infrared, NMR, X-Ray and Mass spectroscopy. The last part of the book presents a collection of articles whose main topic is the optical physics, which studies the dynamical interactions of atoms and molecules with electromagnetic fields. Michelson-Morley Interferometer (MMI) is an example of an optical device where one, in terms of intensity measurements, can exhibit the difference between classical and quantum-mechanical light. Other optical devices that have wide applications are the lasers. Moreover, nonlinear phenomena have many attractive features for modifications of optical properties, but its potential has not been yet fully explored. Here are discussed examples of nonlinear pulse shaping in fibres both for pulse generation and for nonlinear light emission of single gold optical antennas.

Editor
Maria Emilova Velinova, Ph.D.

CHAPTER 1

ATOMIC STRUCTURE CALCULATIONS FOR NEUTRAL OXYGEN

Norah Alonizan, Rabia Qindeel, and Nabil Ben Nessib

Department of Physics and Astronomy, College of Science, King Saud University, Riyadh 11451, Saudi Arabia

ABSTRACT

Energy levels and oscillator strengths for neutral oxygen have been calculated using the Cowan (CW), SUPERSTRUCTURE (SS), and AUTOSTRUCTURE (AS) atomic structure codes. The results obtained with these atomic codes have been compared with MCHF calculations and experimental values from the National Institute of Standards and Technology (NIST) database.

INTRODUCTION

Oxygen atom (O I) is the most abundant element after hydrogen and helium in the Universe. Its spectroscopic study is very important for the knowledge of the structure of stars, galaxies and in general the whole Universe. It is also important for studying the life on the earth and the possibility of life on other planets or exoplanets. The studies of earth's atmosphere and its radiative properties need these data. Industrial and technical applications need the characteristics of this element.

Pradhan and Saraph [1] calculated oscillator strengths for dipole transitions in O I using the SUPERSTRUCTURE (SS) code [2] with spectroscopic type orbitals for 1s, 2s, and 2p and correlation type orbitals for the $\overline{3d}$. Tayal and Henry [3] calculated oscillator strengths and electron collisional excitation cross sections for O I. They used the Hibbert CIV3 atomic structure code [4] with the eight orthogonal one-electron orbitals 1s, 2s, 2p, 3s, 3p, 3d, 4s, and 4p. Using the same CIV3 atomic structure code, Bell and Hibbert [5] calculated oscillator strengths for allowed transitions in O I with more single electron orbitals. Hibbert et al. (HBGV) [6] used the CIV3 code to calculate E1 transitions connecting the $n = 3$ and $n = 4$ energy levels in O I. Biémont et al. [7] calculate oscillator strengths of astrophysical interest for O I using the CIV3 configuration interaction code and the Hartree-Fock pseudorelativistic (HFR) suite of Cowan (CW) codes [8]. Using the SS code, Biémont and Zeippen [9] calculated oscillator strengths for $2p^4$-3s and 3s-3p allowed or spin-forbidden transitions in O I. Zheng and Wang [10] used the Weakest Bound Electron Potential Model (WBEPM) theory to calculate radiative lifetime, transition probabilities, and oscillator strengths for atomic carbon and oxygen. Using the Multiconfiguration Hartree-Fock (MCHF) method [11], Tachiev and Froese Fischer (TFF) [12] calculated ab initio Breit-Pauli energy levels and transition rates for nitrogen-like and oxygen-like sequences. Froese Fischer and Tachiev (FFT) [13] calculated Breit-Pauli energy levels, lifetimes, and transition probabilities for the beryllium-like to neon-like sequences in the adjusted with experimental values. Fan et al.

Citation: Norah Alonizan, Rabia Qindeel, and Nabil Ben Nessib, "Atomic Structure Calculations for Neutral Oxygen," International Journal of Spectroscopy, vol. 2016, Article ID 1697561, 7 pages, 2016. doi:10.1155/2016/1697561

[14] used the WBEPMT theory to calculate energy levels of high states in O I. Çelik and Ateş [15] employed the WBEPMT theory to calculate radial transition matrix elements and then atomic transition probabilities for O I.

Using CW or SS or AS codes, we did atomic structure calculations for several atoms and ions [16–18] that are needed for ab initio Stark broadening calculations [19, 20] and for emission line ratio calculations [21], but we never compare results obtained by the three codes for the same element.

About O I atomic data in databases, we used the National Institute of Standards and Technology (NIST) data [22] for fine structure energy levels and oscillator strengths. There are energy levels and oscillator strengths of O I without fine structure in the Opacity Project TOPbase [23] and NORAD-Atomic-Data [24] atomic structure databases. TIPbase database [25] of the Opacity Project used NIST data for the fine structure energy levels and Galavis et al. [26] data for the oscillator strengths fine structure data. Galavis et al. [26] used the SS atomic structure code with spectroscopic type orbitals for 1s, 2s, and 2p and correlation type orbitals for $\overline{3s}$, $\overline{3p}$, $\overline{3d}$, and $\overline{4f}$.

In the Chianti project [27], they used the NIST database for experimental energy levels and oscillator strengths. For theoretical energy levels they used the Zatsarinny and Tayal [28] and FFT [13] for the theoretical oscillator strengths.

In this work, we will calculate atomic data for transitions with fine structure in O I using CW and SS and AS codes. Comparison with other theoretical and experimental data available in the literature will be presented.

METHODS FOR CALCULATION

Hartree-Fock Pseudorelativistic (HFR) Method

In this method a set of orbitals are obtained for each electron configuration by solving the Hartree-Fock equations [8]. A totally antisymmetric wave-function is a combination of single electron solution of the hydrogen atom Ψ (Slater determinant):

$$\Psi = \frac{1}{\sqrt{N}} \begin{vmatrix} \psi_1(1) & \psi_1(2) & \cdots & \psi_1(N) \\ \psi_2(1) & \psi_2(2) & \cdots & \psi_2(N) \\ \cdots & \cdots & \cdots & \cdots \\ \psi_N(1) & \psi_N(2) & \cdots & \psi_N(N) \end{vmatrix} \quad (1)$$

$\psi_j(i)$ means that the ith electron's space and spin are in the one-electron state ψ_j. This will automatically satisfy the Pauli principle, because a determinant vanishes if two columns are the same.

Relativistic corrections are introduced by a Breit-Pauli Hamiltonian and treated by the perturbation theory. The relativistic corrections include the Blume-Watson spin-orbit, mass-variation, and one-body Darwin terms. The Blume-Watson spin-orbit term contains the part of the Breit interaction that can be reduced to a one-body operator.

The Cowan (CW) atomic structure suite of codes (RCN, RNC2, RCG, and RCE) uses this HFR method. The three first codes are for ab initio atomic structure calculations and the fourth one (RCE) is used to make least-squares fit calculations using an iterative procedure.

Thomas-Fermi-Dirac-Amaldi (TFDA) Method

In this method and to have atomic parameters of an atom or ion, a statistical TFDA potential is used. For an atom or ion having Z protons and N electrons, this potential is in the following form [29]:

$$V(r) = \frac{Z_{\text{eff}}(\lambda_{n\ell}, r)}{r} = -\frac{Z}{r}\phi(x), \quad (2)$$

where

$$\phi(x) = e^{-Zr/2} + \lambda_{n\ell}\left(1 - e^{-Zr/2}\right), \quad x = \frac{r}{\mu} \quad (3)$$

with μ the constant:

$$\mu = \frac{1}{4}\left(\frac{N}{N-1}\right)^{2/3}\left(\frac{9\pi^2}{2Z}\right)^{1/3} \approx 0.8853\left(\frac{N}{N-1}\right)^{2/3} Z^{-1/3} \quad (4)$$

and $\lambda_{n\ell}$ are the orbital scaling parameters.

The function ϕ verifies the following equation:

$$\frac{d^2\phi}{dx^2} = \frac{1}{\sqrt{x}}\phi(x)^{3/2} \quad (5)$$

with the boundary conditions:

$$\phi(0) = 1, \qquad \phi(\infty) = -\frac{Z - N + 1}{Z}. \quad (6)$$

The SUPERSTRUCTURE (SS) and AUTOSTRUCTURE (AS) atomic structure codes use this method. Relativistic corrections are also done by a perturbation method using the Breit-Pauli Hamiltonian. The SS atomic structure code used in this work [30] is an updated version of the original one of 1974 [2]. Some relativistic corrections are introduced in this version and orbital scaling parameters $\lambda_{n\ell}$ are dependent on n and ℓ [31] and not like the original SS version of 1974, where scaling parameters were depending only on ℓ (λ_ℓ). The AS code [32, 33] is an extension of the SS code incorporating various improvements and new capabilities like two-body non-fine-structure operators of the Breit-Pauli Hamiltonian and polarization model potentials. Comparing the two atomic structure codes SS and AS we can see that even they used the same techniques in general; they gave different results mainly because they incorporated different relativistic corrections of the Hamiltonian. For the comparison between the two codes, we can refer to the work of Elabidi and Sahal-Bréchot [34] where they studied excitation cross section by electron impact for O V and O VI levels. They showed that the incorporation of the two-body non-fine-structure operators (contact spin-spin, two-body Darwin, and orbit-orbit) in AS and not in the initial SS code is the main reason of the different results obtained by the two codes.

RESULTS AND DISCUSSION

Energy Levels

We performed ab initio calculations of energy levels for O I using the three atomic structure codes CW, SS, and AS with the 5 configurations expansion $2p^4$, $2p^3$ 3s, $2p^3$ 3p, $2p^3$ 3d, and $2p^3$ 4s. This same set of configurations expansion was used by Tachiev and Froese Fischer (TFF) in the ab initio calculations [12] and by Froese Fischer and Tachiev in the adjusted with experimental values calculations [13]. For the SS and AS atomic codes, the scaling parameters are determined variationally by minimizing the sum of all the nonrelativistic term energies (Table 1).

Table 1: The scaling parameters used after minimization for the SS and AS atomic structure codes

Scaling parameters	1s	2s	2p	3s	3p	3d	4s
$\lambda_{n\ell}$ (SS)	1.23103	1.23103	1.15865	1.23103	1.15865	0.92431	1.23103
$\lambda_{n\ell}$ (AS)	1.55498	1.16135	1.20894	1.07693	2.38871	0.91505	2.33806

In Tables 2–6, calculated fine structure energy levels are presented. The obtained values are compared with the NIST atomic database [22] and with Tachiev and Froese Fischer ab initio calculations [12] using the Multiconfiguration Hartree-Fock (MCHF) method [11].

Table 2: Energy levels for configuration $2s^2\,2p^4$. E(NIST) are from NIST database [22]. E(CW), E(SS), and E(AS) are energy levels calculated using, respectively, the Cowan, SUPERSTRUCTURE, and AUTOSTRUCTURE atomic codes. E(TFF) are energy levels values from Tachiev and Froese Fisher [12]. All energies are in cm^{-1}

Configuration	Term	*J*	E(NIST)	E(CW)	E(SS)	E(AS)	E(TFF)
$2s^22p^4$	^3P	2	0	0	0	0	0
$2s^22p^4$	^3P	1	158	141	171	204	156
$2s^22p^4$	^3P	0	227	210	255	303	223
$2s^22p^4$	^1D	2	15868	14941	18881	17364	16122
$2s^22p^4$	^1S	0	33793	37013	45746	42140	33844

Table 3: The same as Table 2 for configuration $2s^2\,2p^3$ 3s

Configuration	Term	*J*	E(NIST)	E(CW)	E(SS)	E(AS)	E(TFF)
$2s^22p^3(^4S^o)3s$	$^5S^o$	2	73768	70285	50342	67409	74012
$2s^22p^3(^4S^o)3s$	$^3S^o$	1	76795	72704	54546	70365	76910
$2s^22p^3(^2D^o)3s$	$^3D^o$	3	101135	96475	80769	99037	—
$2s^22p^3(^2D^o)3s$	$^3D^o$	2	101148	96474	80768	99036	—
$2s^22p^3(^2D^o)3s$	$^3D^o$	1	101155	96473	80767	99035	—
$2s^22p^3(^2D^o)3s$	$^1D^o$	2	102662	97684	82871	100502	—
$2s^22p^3(^2P^o)3s$	$^3P^o$	2	113911	113471	100024	119170	—
$2s^22p^3(^2P^o)3s$	$^3P^o$	1	113921	113469	100023	119169	—
$2s^22p^3(^2P^o)3s$	$^3P^o$	0	113928	113468	100022	119168	—
$2s^22p^3(^2P^o)3s$	$^1P^o$	1	115918	114666	101453	121545	—

Table 4: The same as Table 2 for configuration $2s^2\ 2p^3\ 3p$ except that we did not put the AS data because they are too far from the other results

Configuration	Term	J	E(NIST)	E(CW)	E(SS)	E(TFF)
$2s^22p^3(^4S^o)3p$	^5P	1	86626	83148	62276	86645
$2s^22p^3(^4S^o)3p$	^5P	2	86628	83149	62279	86647
$2s^22p^3(^4S^o)3p$	^5P	3	86631	83152	62284	86650
$2s^22p^3(^4S^o)3p$	^3P	1	88631	85516	67495	88590
$2s^22p^3(^4S^o)3p$	^3P	2	88631	85519	67492	88591
$2s^22p^3(^4S^o)3p$	^3P	0	88631	85514	67497	88591
$2s^22p^3(^2D^o)3p$	^1P	1	113204	108935	91748	—
$2s^22p^3(^2D^o)3p$	^3D	3	113295	108877	91674	—
$2s^22p^3(^2D^o)3p$	^3D	2	113295	108871	91665	—
$2s^22p^3(^2D^o)3p$	^3D	1	113298	108864	91655	—
$2s^22p^3(^2D^o)3p$	^3F	4	113714	109075	92012	—
$2s^22p^3(^2D^o)3p$	^3F	3	113721	109072	92007	—
$2s^22p^3(^2D^o)3p$	^3F	2	113727	109069	92003	—
$2s^22p^3(^2D^o)3p$	^1F	3	113996	109318	92419	—
$2s^22p^3(^2D^o)3p$	^1D	2	116631	113447	101312	—
$2s^22p^3(^2P^o)3p$	^3D	3	127283	126148	111485	—
$2s^22p^3(^2P^o)3p$	^3D	2	127288	126149	111486	—
$2s^22p^3(^2P^o)3p$	^3D	1	127292	126149	111485	—
$2s^22p^3(^2P^o)3p$	^1P	1	127668	126335	111820	—
$2s^22p^3(^2P^o)3p$	^1D	2	128595	127944	115611	—
$2s^22p^3(^2P^o)3p$	^1S	0	130943	132423	111145	—
$2s^22p^3(^2D^o)3p$	^3P	1	—	111274	96720	—
$2s^22p^3(^2D^o)3p$	^3P	2	—	111280	96730	—
$2s^22p^3(^2D^o)3p$	^3P	0	—	111272	96715	—
$2s^22p^3(^2P^o)3p$	^3P	1	—	127427	113836	—
$2s^22p^3(^2P^o)3p$	^3P	2	—	127418	113812	—
$2s^22p^3(^2P^o)3p$	^3P	0	—	127432	113848	—

Table 5: The same as Table 2 for configuration $2s^2\ 2p^3\ 3d$

Configuration	Term	J	E(NIST)	E(CW)	E(SS)	E(AS)	E(TFF)
$2s^22p^3(^4S^o)3d$	$^5D^o$	4	97421	94227	72721	89755	97147
$2s^22p^3(^4S^o)3d$	$^5D^o$	3	97421	94227	72721	89755	97147
$2s^22p^3(^4S^o)3d$	$^5D^o$	2	97421	94227	72721	89755	97147
$2s^22p^3(^4S^o)3d$	$^5D^o$	1	97421	94227	72721	89755	97147
$2s^22p^3(^4S^o)3d$	$^5D^o$	0	97421	94227	72721	89754	97148
$2s^22p^3(^4S^o)3d$	$^3D^o$	1	97488	94299	72820	89836	97205
$2s^22p^3(^4S^o)3d$	$^3D^o$	2	97488	94299	72820	89836	97205
$2s^22p^3(^4S^o)3d$	$^3D^o$	3	97489	94299	72821	89836	97205
$2s^22p^3(^2D^o_{3/2})3d$	$^3P^o$	2	123297	120045	102401	120957	—
$2s^22p^3(^2D^o_{3/2})3d$	$^3P^o$	1	123356	120049	102405	120959	—
$2s^22p^3(^2D^o_{3/2})3d$	$^3P^o$	0	123387	120050	102406	120961	—
$2s^22p^3(^2D^o_{5/2})3d$	$^3F^o$	4	124214	119938	102042	120523	—
$2s^22p^3(^2D^o_{5/2})3d$	$^3F^o$	3	124219	119935	102039	120520	—
$2s^22p^3(^2D^o_{5/2})3d$	$^3F^o$	2	124224	119934	102037	120518	—
$2s^22p^3(^2D^o)3d$	$^3G^o$	4	124239	119945	102051	120530	—
$2s^22p^3(^2D^o)3d$	$^3G^o$	5	124240	119946	102053	120531	—
$2s^22p^3(^2D^o_{5/2})3d$	$^1S^o$	0	124243	119935	102037	120519	—

$2s^22p^3(^2D^\circ_{5/2})3d$	$^3D^\circ$	3	124247	119957	102070	120544	—
$2s^22p^3(^2D^\circ)3d$	$^3G^\circ$	3	124253	119944	102051	120529	—
$2s^22p^3(^2D^\circ_{5/2})3d$	$^3D^\circ$	2	124258	119958	102072	120545	—
$2s^22p^3(^2D^\circ_{3/2})3d$	$^1G^\circ$	4	124259	119952	102061	120538	—
$2s^22p^3(^2D^\circ_{5/2})3d$	$^3D^\circ$	1	124264	119959	102072	120546	—
$2s^22p^3(^2D^\circ_{5/2})3d$	$^1P^\circ$	1	124274	120043	103055	120030	—
$2s^22p^3(^2D^\circ_{3/2})3d$	$^1D^\circ$	2	124319	120017	102181	120609	—
$2s^22p^3(^2D^\circ_{3/2})3d$	$^1F^\circ$	3	124327	120033	102174	120629	—
$2s^22p^3(^2D^\circ_{3/2})3d$	$^3S^\circ$	1	124336	120002	102133	120577	—
$2s^22p^3(^2P^\circ)3d$	$^1D^\circ$	2	137928	137105	121625	141057	—
$2s^22p^3(^2P^\circ)3d$	$^3P^\circ$	0	137947	137093	121609	141050	—
$2s^22p^3(^2P^\circ)3d$	$^3P^\circ$	1	137947	137094	121611	141052	—
$2s^22p^3(^2P^\circ)3d$	$^3P^\circ$	2	137947	137097	121615	141054	—
$2s^22p^3(^2P^\circ)3d$	$^3D^\circ$	1	137963	137117	121642	141063	—
$2s^22p^3(^2P^\circ)3d$	$^3D^\circ$	2	137963	137119	121645	141066	—
$2s^22p^3(^2P^\circ)3d$	$^3D^\circ$	3	137963	137121	121647	141067	—
$2s^22p^3(^2P^\circ)3d$	$^1P^\circ$	1	137981	137186	121755	141153	—
$2s^22p^3(^2P^\circ)3d$	$^3F^\circ$	4	—	137084	121596	141041	—
$2s^22p^3(^2P^\circ)3d$	$^3F^\circ$	3	—	137085	121597	141041	—
$2s^22p^3(^2P^\circ)3d$	$^3F^\circ$	2	—	137085	121597	141041	—
$2s^22p^3(^2P^\circ)3d$	$^1F^\circ$	3	—	137136	121667	141098	—

Table 6: The same as Table 2 for configuration $2s^2\ 2p^3\ 4s$

Configuration	Term	*J*	E(NIST)	E(CW)	E(SS)	E(AS)	E(TFF)
$2s^22p^3(^4S^\circ)4s$	$^5S^\circ$	2	95477	92425	71070	110558	95576
$2s^22p^3(^4S^\circ)4s$	$^3S^\circ$	1	96225	93053	72757	122242	96264
$2s^22p^3(^2D^\circ)4s$	$^3D^\circ$	3	122420	118248	100741	144123	—
$2s^22p^3(^2D^\circ)4s$	$^3D^\circ$	2	122433	118247	100740	144121	—
$2s^22p^3(^2D^\circ)4s$	$^3D^\circ$	1	122441	118246	100739	144121	—
$2s^22p^3(^2D^\circ_{3/2})4s$	$^1D^\circ$	2	122798	118560	101552	149959	—
$2s^22p^3(^2P^\circ)4s$	$^3P^\circ$	0	*135682*	135369	120288	164620	—
$2s^22p^3(^2P^\circ)4s$	$^3P^\circ$	1	135682	135370	120289	164621	—
$2s^22p^3(^2P^\circ)4s$	$^3P^\circ$	2	135682	135372	120291	164623	—
$2s^22p^3(^2P^\circ)4s$	$^1P^\circ$	1	136353	135684	121113	170467	—

For the fundamental configuration $2s^2\ 2p^4$, CW code gives 7% difference from NIST values, while SS and AS codes give, respectively, 15% and 19% difference from NIST values. With CW code and for the other four excited configurations ($2p^3$, 3ℓ, $\ell = 0, 1, 2$, and $2p^3\ 4s$), the agreement with NIST values is about 3%, while SS and AS give an agreement of about 20% with the NIST values except for $2p^3\ 3s$ where the agreement of AS with NIST is 4% and 7% for $2p^3\ 3d$. AS gives bad energy levels (one hundred greater values) for the $2p^3\ 3p$ configuration comparing to the other data and the calculated values are not reported in Table 4. TFF energy levels are less than 1% near the NIST values but there are many missed values (they give energy levels for only 24% of the NIST data for the 4 excited configurations $2p^3\ n\ell$).

We obtain six energy levels for the $2p^3\ 3p$ configuration ($2p^3(^2D^\circ)3p\ ^3P_{0,1,2}$ and $2p^3(^2P^\circ)3p\ ^3P_{0,1,2}$) and four new energy levels for the $2p^3\ 3d$ configuration ($2p^3(^2P^\circ)3d\ ^3F_{2,3,4}$ and $2p^3(^2P^\circ)3d\ ^1F_3$) which are not in the NIST atomic database (see the end of Tables 4 and 5).

Oscillator Strengths

We have also computed oscillator strengths for three multiplets of the transition $2p^4$-$2p^3\ 3s$, four multiplets of the transition

$2p^3$ 3s-$2p^3$ 3p, one multiplet of the transition $2p^4$-3d, and four multiplets of the transition $2p^3$ 3p-$2p^3$ 3d using the atomic structure codes CW, SS, and AS (see Tables 7–10). They are compared with those of FFT [13] and HBGV [6] and tabulated in NIST [22]. Our calculations with CW code give an agreement of about 2% with NIST values while FFT gives 8% with the NIST ones. The HBGV values have an agreement of 2% with the NIST ones but many data are missing.

Table 7: Oscillator strengths for 3 multiplets $2p^4$-$2p^3(^4S)$ 3s ^3S. and are, respectively, the statistical weights of initial and final levels of the transition. gf(NIST) are from NIST database [22]. gf(CW), gf(SS), and gf(AS) are calculated using, respectively, the Cowan, SUPERSTRUCTURE, and AUTOSTRUCTURE atomic codes. gf(FFT) are values from Froese Fisher and Tachiev [13] and gf(HBGV) are values from Hibbert et al. [6]

Terms	g_i	g_k	λ (Å)	gf(NIST)	gf(CW)	gf(SS)	gf(AS)	gf(FFT)	gf(HBGV)
^3P-$^3S^o$	1	3	1306.03	$5.19E-02$	$7.26E-02$	$1.39E-01$	$5.13E-02$	$4.78E-02$	$5.29E-02$
^3P-$^3S^o$	3	3	1304.86	$1.56E-01$	$2.18E-01$	$4.17E-01$	$1.54E-01$	$1.43E-01$	$1.59E-01$
^3P-$^3S^o$	5	3	1302.17	$2.60E-01$	$3.66E-01$	$6.95E-01$	$2.59E-01$	$2.40E-01$	$2.64E-01$
^1D-$^3S^o$	5	3	1641.31	$2.22E-06$	$6.04E-06$	$1.31E-05$	$6.95E-06$	$2.13E-06$	—
^1S-$^3S^o$	1	3	2324.74	$1.12E-08$	$3.08E-09$	$1.60E-08$	$5.50E-08$	$1.59E-10$	—

Table 8: The same as Table 7 for 4 multiplets $2p^3(^4S)3s$ S°-$2p^3(^4S)$ 3p P°

Terms	g_i	g_k	λ (Å)	gf(NIST)	gf(CW)	gf(SS)	gf(AS)	gf(FFT)	gf(HBGV)
$^5S^o$-$^5P^o$	5	3	7775.39	$1.00E+00$	$1.09E+00$	$9.50E-01$	$2.00E+00$	$9.67E-01$	$1.01E+00$
$^5S^o$-$^5P^o$	5	5	7774.17	$1.67E+00$	$1.81E+00$	$1.58E+00$	$3.33E+00$	$1.61E+00$	$1.68E+00$
$^5S^o$-$^5P^o$	5	7	7771.94	$2.34E+00$	$2.54E+00$	$2.22E+00$	$4.67E+00$	$2.26E+00$	$2.35E+00$
$^5S^o$-$^3P^o$	5	3	6726.54	$1.31E-05$	$5.73E-07$	$9.21E-06$	$7.10E-05$	$1.02E-05$	—
$^5S^o$-$^3P^o$	5	5	6726.28	$4.00E-05$	$1.77E-06$	$2.81E-05$	$2.18E-04$	$3.51E-05$	—
$^3S^o$-$^5P^o$	3	3	10169.35	$9.16E-06$	$2.86E-06$	$1.68E-05$	$4.43E-05$	$7.91E-06$	—
$^3S^o$-$^5P^o$	3	5	10167.26	$2.80E-05$	$4.65E-08$	$5.51E-06$	$1.36E-04$	$2.74E-05$	—
$^3S^o$-$^3P^o$	3	1	8446.25	$3.44E-01$	$3.88E-01$	$3.81E-01$	$4.20E-01$	$3.39E-01$	$3.60E-01$
$^3S^o$-$^3P^o$	3	3	8446.76	$1.03E+00$	$1.16E+00$	$1.14E+00$	$1.26E+00$	$1.02E+00$	$1.08E+00$
$^3S^o$-$^3P^o$	3	5	8446.36	$1.72E+00$	$1.94E+00$	$1.91E+00$	$2.10E+00$	$1.70E+00$	$1.80E+00$

Table 9: The same as Table 7 for the multiplet 2p P-$2p^3(^4S)$ 3d $^3D^o$

Terms	g_i	g_k	λ (Å)	gf(NIST)	gf(CW)	gf(SS)	gf(AS)	gf(FFT)	gf(HBGV)
^3P-$^3D^o$	5	7	1025.76	$8.45E-02$	$3.5E-01$	$1.29E+00$	$1.05E-01$	$8.17E-02$	$8.93E-02$
^3P-$^3D^o$	5	5	1025.76	$1.51E-02$	$6.24E-02$	$2.31E-01$	$1.87E-02$	$1.46E-02$	$1.59E-02$
^3P-$^3D^o$	5	3	1025.76	$9.98E-04$	$4.16E-03$	$1.54E-02$	$1.25E-03$	$9.69E-04$	$1.05E-03$
^3P-$^3D^o$	3	5	1027.43	$4.52E-02$	$1.87E-01$	$6.94E-01$	$5.60E-02$	$4.37E-02$	$4.10E-02$
^3P-$^3D^o$	3	3	1027.43	$1.51E-02$	$6.22E-02$	$2.31E-01$	$1.86E-02$	$1.45E-02$	$1.36E-02$
^3P-$^3D^o$	1	3	1028.16	$2.00E-02$	$8.30E-02$	$3.09E-01$	$2.48E-02$	$1.94E-02$	$2.12E-02$
^3P-$^5D^o$	5	7	1026.47*	—	$8.45E-09$	$1.00E-06$	$5.79E-09$	$4.35E-07$	—
^3P-$^5D^o$	5	5	1026.47*	—	$1.13E-07$	$3.25E-06$	$4.36E-08$	$3.65E-08$	—
^3P-$^5D^o$	5	3	1026.47*	—	$6.64E-08$	$1.60E-06$	$3.87E-08$	$7.14E-09$	—
^3P-$^5D^o$	3	5	1028.12*	—	$1.32E-08$	$1.76E-08$	$3.66E-08$	$1.10E-07$	—
^3P-$^5D^o$	3	3	1028.12*	—	$4.48E-08$	$1.33E-06$	—	$1.07E-07$	—
^3P-$^5D^o$	1	3	1028.83*	—	$3.85E-08$	$5.62E-07$	—	$1.44E-07$	—

*Wavelengths are from FFT.

Table 10: The same as Table 7 for 4 multiplets 2p^3(^{4}S) 3p P-2p^3(^{4}S) 3d D°

Terms	g_i	g_K	λ (Å)	gf(NIST)	gf(CW)	gf(SS)	gf(AS)	gf(FFT)	gf(HBGV)
^{5}P-^{3}D°	5	7	9204.93	1.48E − 04	1.05E − 05	1.15E − 05	1.83E − 05	8.44E − 05	1.50E − 04
^{5}P-^{5}D°	3	1	9260.81	5.74E − 01	6.73E − 01	5.42E − 01	1.08E − 01	5.90E − 01	6.75E − 01
^{5}P-^{5}D°	3	3	9260.85	1.29E + 00	1.51E + 00	1.22E + 00	2.43E − 01	1.33E + 00	1.52E + 00
^{5}P-^{5}D°	3	5	9260.94	1.00E + 00	1.18E + 00	9.49E − 01	1.89E − 01	1.03E + 00	1.18E + 00
^{5}P-^{5}D°	5	3	9262.58	4.29E − 01	5.05E − 01	4.07E − 01	8.09E − 02	4.43E − 01	4.32E − 01
^{5}P-^{5}D°	5	5	9262.67	1.67E + 00	1.96E + 00	1.58E + 00	3.15E − 01	1.72E + 00	1.68E + 00
^{5}P-^{5}D°	5	7	9262.78	2.67E + 00	3.13E + 00	2.53E + 00	5.04E − 01	2.75E + 00	2.69E + 00
^{5}P-^{5}D°	7	5	9265.83	1.91E − 01	2.24E − 01	1.81E − 01	3.60E − 02	1.97E − 01	1.92E − 01
^{5}P-^{5}D°	7	7	9265.93	1.33E + 00	1.57E + 00	1.27E + 00	2.52E − 01	1.38E + 00	1.34E + 00
^{5}P-^{5}D°	7	9	9266.01	5.15E + 00	6.05E + 00	4.88E + 00	9.73E − 01	5.31E + 00	5.19E + 00
^{3}P-^{3}D°	3	5	11286.32	2.22E + 00	1.97E + 00	9.64E − 01	3.07E − 01	2.02E + 00	2.20E + 00
^{3}P-^{3}D°	3	3	11286.41	7.40E − 01	6.58E − 01	3.22E − 01	1.02E − 01	6.74E − 01	7.32E − 01
^{3}P-^{3}D°	5	7	11286.91	4.13E + 00	3.68E + 00	1.80E + 00	5.71E − 01	3.77E + 00	4.10E + 00
^{3}P-^{3}D°	5	5	11287.03	7.40E − 01	6.56E − 01	3.22E − 01	1.02E − 01	6.74E − 01	7.32E − 01
^{3}P-^{3}D°	5	3	11287.12	4.93E − 02	4.38E − 02	2.14E − 02	6.80E − 03	4.49E − 02	4.88E − 02
^{3}P-^{3}D°	1	3	11287.32	9.86E − 01	8.77E − 01	4.28E − 01	1.37E − 01	8.98E − 01	9.76E − 01
^{3}P-^{5}D°	1	3	11373.92	2.32E − 05	1.11E − 06	8.55E − 07	3.56E − 07	1.43E − 05	2.00E − 05
^{3}P-^{5}D°	5	7	11374.01	1.37E − 04	1.32E − 06	6.93E − 07	1.44E − 05	5.16E − 04	1.50E − 04

CONCLUSIONS

The comparison between the energy levels calculated by the different atomic codes indicates that the agreement with NIST data is generally less than 20% except for the configuration 2p^3 3p where the SS gives 20% greater values than NIST and CW gives only 3% greater values than NIST.

We obtained, respectively, six and four energy levels corresponding to the configurations 2p^3 3p and 2p^3 3d which are not in the NIST atomic database.

TFF energy levels values are nearly the same compared to NIST ones with less than 1% difference but many energy levels are missed in these calculations.

When comparing oscillator strengths calculated by AS code, we obtain good agreement with NIST.

We can say that there is no best atomic code to use and we have to calculate atomic structure data with more than one code to compare results between them and with other theoretical and experimental sources.

ACKNOWLEDGMENTS

The project was supported by the Research Center, College of Science, King Saud University.

REFERENCES

1. A. K. Pradhan and H. E. Saraph, "Oscillator strengths for dipole transitions in neutral oxygen," Journal of Physics B: Atomic and Molecular Physics, vol. 10, no. 17, pp. 3365–3376, 1977.
2. W. Eissner, M. Jones, and H. Nussbaumer, "Techniques for the calculation of atomic structures and radiative data including relativistic corrections," Computer Physics Communications, vol. 8, no. 4, pp. 270–306, 1974.
3. S. S. Tayal and R. J. W. Henry, "Oscillator strengths and electron collisional excitation cross sections for atomic oxygen," Physical Review A, vol. 39, no. 9, pp. 4531–4536, 1989.
4. A. Hibbert, "CIV3—a general program to calculate configuration interaction wave functions and electric-dipole oscillator strengths," Computer Physics Communications, vol. 9, no. 3, pp. 141–172, 1975.
5. K. L. Bell and A. Hibbert, "Oscillator strengths for allowed transitions in atomic oxygen," Journal of Physics B, vol. 23, no. 16, pp. 2673–2685, 1990.
6. A. Hibbert, E. Biémont, M. Godefroid, and N. Vaeck, "E1 transitions of astrophysical interest in neutral oxygen," Journal of Physics B: Atomic, Molecular and Optical Physics, vol. 24, no. 18, pp. 3943–3958, 1991.
7. E. Biémont, A. Hibbert, M. Godefroid, N. Vaeck, and B. C. Fawcett, "Accurate oscillator strengths of astrophysical interest for neutral oxygen," The Astrophysical Journal, vol. 375, no. 2, pp. 818–822, 1991.
8. R. D. Cowan, Theory of Atomic Structure and Spectra, University of California Press, Berkeley, Calif, USA, 1981.
9. E. Biémont and C. J. Zeippen, "Electric dipole transitions in atomic oxygen and the lifetimes of the 2p(3) (4S0)3s5S0 and 3S0 states," Astronomy & Astrophysics, vol. 265, no. 2, pp. 850–856, 1992.
10. N. Zheng and T. Wang, "Radiative lifetimes and atomic transition probabilities for atomic carbon and oxygen," The Astrophysical Journal Supplement Series, vol. 143, no. 1, p. 231, 2002.
11. C. Froese Fischer, T. Brage, and P. Jönsson, Computational Atomic Structure: An MCHF Approach, Institute of Physics Publishing, Bristol, UK, 1997.

12. G. I. Tachiev and C. F. Fischer, "Breit-Pauli energy levels and transition rates for nitrogen-like and oxygen-like sequences," Astronomy and Astrophysics, vol. 385, no. 2, pp. 716–723, 2002.
13. C. Froese Fischer and G. Tachiev, "Breit-Pauli energy levels, lifetimes, and transition probabilities for the beryllium-like to neon-like sequences," Atomic Data and Nuclear Data Tables, vol. 87, no. 1, pp. 1–184, 2004.
14. J. Fan, N. W. Zheng, D. X. Ma, and T. Wang, "Calculation of the energy levels to high states in atomic oxygen," Physica Scripta, vol. 69, no. 5, pp. 398–402, 2004.
15. G. Çelik and Ş. Ateş, "The calculation of transition probabilities for atomic oxygen," European Physical Journal D, vol. 44, no. 3, pp. 433–437, 2007.
16. N. Ben Nessib, H. Elabidi, M. Cornille, and J. Dubau, "Radiative and collisional atomic data for neon-like silicon," Physica Scripta, vol. 72, no. 1, pp. 23–30, 2005.
17. H. Elabidi, S. Sahal-Bréchot, and N. B. Nessib, "Fine structure collision strengths for S VII lines," Physica Scripta, vol. 85, no. 6, Article ID 065302, 2012.
18. R. Hamdi and N. Ben Nessib, "Electric dipole transition probabilities in Al IV and Al V ions," Memorie della SAIt, vol. 7, pp. 228–229, 2005.
19. N. Ben Nessib, "Ab initio calculations of Stark broadening parameters," New Astronomy Reviews, vol. 53, no. 7-10, pp. 255–258, 2009.
20. R. Hamdi, N. Ben Nessib, M. S. Dimitrijević, and S. Sahal-Bréchot, "Stark broadening of Pb IV spectral lines," Monthly Notices of the Royal Astronomical Society, vol. 431, no. 2, pp. 1039–1047, 2013.
21. N. Ben Nessib, N. Alonizan, R. Qindeel, S. Sahal-Bréchot, and M. S. Dimitrijević, "The OIV 1407.3 Å /1401.1 Å emission-line ratio in a plasma," Advances in Space Research, vol. 54, no. 7, pp. 1190–1194, 2014.
22. A. Kramida, Yu. Ralchenko, J. Reader, and NIST ASD Team, NIST Atomic Spectra Database (ver. 5.3), National Institute of Standards and Technology, Gaithersburg, Md, USA, 2015, http://physics.nist.gov/asd.
23. http://cdsweb.u-strasbg.fr/topbase/topbase.html.
24. http://www.astronomy.ohio-state.edu/~nahar/nahar_radiativeatomicdata/index.html.
25. http://cdsweb.u-strasbg.fr/tipbase/home.html.
26. M. E. Galavis, C. Mendoza, and C. J. Zeippen, "Atomic data from the IRON Project-XXII. Radiative rates for forbidden transitions within the ground configuration of ions in the carbon and oxygen isoelectronic sequences," Astronomy and Astrophysics, vol. 123, no. 1, pp. 159–171, 1997.
27. http://www.chiantidatabase.org/chianti.html.
28. O. Zatsarinny and S. S. Tayal, "Electron collisional excitation rates for O I using the B-spline R-matrix approach," Astrophysical Journal, Supplement Series, vol. 148, no. 2, pp. 575–582, 2003.
29. A. K. Pradhan and S. N. Nahar, Atomic Astrophysics and Spectroscopy, Cambridge University Press, Cambridge, UK, 2011.
30. S. N. Nahar, W. Eissner, G.-X. Chen, and A. K. Pradhan, "Atomic data from the Iron Project LIII. Relativistic allowed and forbidden transition probabilities for Fe XVII," Astronomy and Astrophysics, vol. 408, no. 2, pp. 789–801, 2003.
31. H. Nussbaumer and P. J. Storey, "Bound-bound radiative transitions in C I," Astronomy & Astrophysics, vol. 140, no. 2, pp. 383–389, 1984.
32. N. R. Badnell, "On the effects of the two-body non-fine-structure operators of the Breit—Pauli Hamiltonian," Journal of Physics B: Atomic, Molecular and Optical Physics, vol. 30, no. 1, pp. 1–11, 1997.
33. N. R. Badnell, "A Breit-Pauli distorted wave implementation for autostructure," Computer Physics Communications, vol. 182, no. 7, pp. 1528–1535, 2011.
34. H. Elabidi and S. Sahal-Bréchot, "Excitation cross-sections by electron impact for OV and OVI levels,"Monthly Notices of the Royal Astronomical Society, vol. 436, no. 2, pp. 1452–1464, 2013.

CHAPTER 2

SELF-INTERACTION IN GREEN'S-FUNCTION THEORY OF THE HYDROGEN ATOM

W. Nelson[1], P. Bokes[2, 3] Patrick Rinke[3, 4] and R. W. Godby[1, 3]

[1]Department of Physics, University of York, Heslington, York YO10 5DD, United Kingdom

[2]Department of Physics, Faculty of Electrical Engineering and Information Technology, Slovak University of Technology, Ilkovičova 3, 841 04 Bratislava, Slovak Republic

[3]European Theoretical Spectroscopy Facility (ETSF)

[4]Fritz-Haber-Institut der Max-Planck-Gesellschaft, Faradayweg 4-6, 14195 Berlin, Germany

Atomic hydrogen provides a unique test case for computational electronic structure methods, since its electronic excitation energies are known analytically. With only one electron, hydrogen contains no electronic correlation and is therefore particularly susceptible to spurious self-interaction errors introduced by certain computational methods. In this paper we focus on many-body perturbation theory (MBPT) in Hedin's GW approximation. While the Hartree-Fock and the exact MBPT self-energy are free of self-interaction, the correlation part of the GW self-energy does not have this property. Here we use atomic hydrogen as a benchmark system for GW and show that the self-interaction part of the GW self-energy, while nonzero, is small. The effect of calculating the GW self-energy from exact wave functions and eigenvalues, as distinct from those from the local-density approximation, is also illuminating.

INTRODUCTION

Ab initio many-body quantum mechanical calculations are crucially important to our understanding of the behavior of atomic, molecular, and condensed matter systems. It is well-known that predicting the behavior of these systems requires the description of electronic correlation. While density-functional theory (DFT) in the local-density approximation (LDA) does this with startling success in many cases, it does so at the expense of a nonphysical electron self-interaction. For delocalized electron systems this self-interaction becomes negligible, but in atomic or strongly localized electronic systems it plays an important role. If one is interested in the calculation of quasiparticle excitation spectra, many-body perturbation theory (MBPT) is formally a correct way to proceed. For solids, MBPT in Hedin's GW approximation [1] has become the method of choice, but it is also increasingly being applied to molecular systems and clusters. The GW self-energy can be decomposed into correlation and exchange parts, where the latter is the same as the Fock operator encountered in Hartree-Fock theory and thus self-interaction free. While the exact self-energy must also be free of self-interaction, the correlation part of the GW self-energy does not have this property. To investigate the influence of self-interaction in the GW approach the hydrogen atom provides an ideal case because the exact solution is known analytically.

Hydrogen in its solid phase has previously been studied within the GW approximation by Li *et al.* [2], who analyzed the transition between the high-pressure solid phase and the low density, atomiclike limit. For individual atoms, GW electron removal and addition energies (we use the term "quasiparticle" energies by analogy with the solid-state situation) have been investigated by Shirley and Martin [3], Dahlen *et al.* [4,5],

Citation: W. Nelson, P. Bokes, Patrick Rinke and R. W. Godby, Self-interaction in Green's-function theory of the hydrogen atom, DOI: 10.1103/PhysRevA.75.032505

Stan *et al.* [6], and Delaney *et al.* [7], although hydrogen was not considered. These studies have shown that GW, in general, gives quasiparticle properties which are much improved over DFT and Hartree-Fock methods, even for atoms.

In this paper we use the hydrogen atom as a benchmark system to quantify the self-interaction error in the GW approach. Since the self-energy diagrams *beyond* GW, known as the vertex correction, must by definition correct this self-interaction error, our findings are relevant for research into vertex functions for the many-electron problem.

Attention has recently focused on the prospects for improving the usual non-self-consistent GW calculations by choosing an initial Green's function, G0, that is physically more reasonable than the LDA (e.g., [2, 8, 9]). We explore this here by determining the sensitivity of the self-interaction error to the use of the exact hydrogenic orbitals and energies in place of those from the local-density approximation (LDA). We also assess the error introduced into GW calculations by employing first-order perturbation theory in solving the quasiparticle equation (as opposed to the full numerical solution), and we analyze the quasiparticle wave functions that emerge from a full solution.

HARTREE-FOCK VERSUS DFT-LDA

In many-body perturbation theory the quasiparticle excitation energies $\epsilon_{i\sigma}$ and wave functions $\psi_{i\sigma}$ are the solutions of the quasiparticle equation

$$H_0(\mathbf{r})\psi_{i\sigma}(\mathbf{r}) + \sum_{\sigma'} \int d\mathbf{r}' M_{\sigma\sigma'}(\mathbf{r},\mathbf{r}';\epsilon_{i\sigma}^{\mathrm{qp}})\psi_{i\sigma'}(\mathbf{r}') = \epsilon_{i\sigma}^{\mathrm{qp}}\psi_{i\sigma}(\mathbf{r}), \tag{1}$$

where, in Hartree atomic units, $H_0(\mathbf{r})=-\frac{1}{2}\nabla^2+v_{ext}(\mathbf{r})$ and $v_{ext}(\mathbf{r})$ is the external potential. It is customary to divide the mass operator M into the local Hartree potential (v_H) and the nonlocal self-energy (Σ)

$$M_{\sigma\sigma'}(\mathbf{r},\mathbf{r}';\epsilon) = v_H(\mathbf{r})\delta(\mathbf{r}-\mathbf{r}')\delta_{\sigma\sigma'} + \Sigma_{\sigma\sigma'}(\mathbf{r},\mathbf{r}';\epsilon). \tag{2}$$

Omitting correlation contributions from Σ yields the exact-exchange or Hartree-Fock case, where the self-energy takes the form

$$\Sigma^x_{\sigma\sigma'}(\mathbf{r},\mathbf{r}') = -\sum_i^{occ} \frac{\psi_{i\sigma}(\mathbf{r})\psi^*_{i\sigma}(\mathbf{r}')}{|\mathbf{r}-\mathbf{r}'|}\delta_{\sigma\sigma'} \tag{3}$$

and the Hartree potential is given by

$$v_H(\mathbf{r}) = \sum_{i,\sigma}^{occ} \int d\mathbf{r}' \frac{\psi_{i\sigma}(\mathbf{r}')\psi^*_{i\sigma}(\mathbf{r}')}{|\mathbf{r}-\mathbf{r}'|}. \tag{4}$$

Since the sum runs over all occupied states the Hartree potential contains an artificial interaction of electrons with themselves. For the hydrogen atom this so-called self-interaction of the electron is the *only* content of the Hartree potential and may be calculated analytically using the exact ground-state wave function

$$\psi^{\uparrow}_{1s}(r) = \frac{1}{\sqrt{\pi}}e^{-r} \tag{5}$$

as

$$v_H(r) = \frac{1}{r}\{1-(1+r)e^{-2r}\}. \tag{6}$$

The self-interaction is a positive, decreasing function and hence tends to delocalize the wave function (i.e., incorrectly pushes its weight away from the nucleus).

In Hartree-Fock the self-interaction terms introduced in the Hartree potential are exactly canceled by the Fock operator Σ_x. This makes Hartree-Fock exact for one-electron systems such as the hydrogen atom. However, the lack of correlation renders Hartree-Fock unsuitable for many polyatomic systems of interest.

Of more practical use is Kohn-Sham [10] density-functional theory, which by virtue of the Hohenberg-Kohn theorem [11] establishes an exact and universal functional relationship between the ground-state density n(r) and the total energy E of a system. Mapping the system of interacting electrons onto a fictitious system of noninteracting electrons, that reproduces the exact density, yields the Kohn-Sham equations:

$$[H_0(\mathbf{r}) + v_H(\mathbf{r}) + v_{xc}(\mathbf{r})]\phi_{i\sigma}(\mathbf{r}) = \epsilon_{i\sigma}^{\mathrm{KS}}\phi_{i\sigma}(\mathbf{r}). \tag{7}$$

All electron-electron interactions beyond the Hartree mean field are encompassed by the exchange-correlation potential v_{xc}, which is formally given as the functional derivative of the exchange-correlation energy E_{xc}:

$$v_{xc}(\mathbf{r}) = \frac{\delta E_{xc}[n]}{\delta n(\mathbf{r})}. \tag{8}$$

In analogy to the quasiparticle equation (1) the Kohn-Sham eigenvalues $\epsilon_{i\sigma}^{\mathrm{KS}}$ are often interpreted as excitation energies, although this is not formally justified.

One of the most common approximations for E_{xc} is the local-density approximation (LDA) [10], in which the many-body exchange and correlation contributions to the total energy (E_{xc}) are included by comparison with the homogeneous electron gas (HEG):

$$E_{xc}^{\mathrm{LDA}}[n] = \int d\mathbf{r}n(\mathbf{r})\epsilon_{xc}^{\mathrm{HEG}}[n(\mathbf{r})]. \tag{9}$$

Here [12] we follow the parametrization of Perdew and Zunger [13] for the exchange-correlation energy density $\epsilon_{xc}^{\mathrm{HEG}}[n(\mathbf{r})]$ of the homogeneous electron gas based on the data of Ceperley and Alder [14].

The LDA has been remarkably successful at accounting for correlation even in systems that are highly inhomogeneous. However, it is well-known that the introduction of correlation in the LDA comes at the expense of the exact treatment of the self-interaction. Because the exchange functional is taken from the homogeneous electron gas it no longer cancels the spurious self-interaction present in the Hartree term. In most systems this is a minor effect and is more than compensated by the improved treatment of the electron correlation. The LDA can be improved by explicitly removing the self-interaction [13], however, this becomes increasingly difficult as the system's complexity is increased.

THE GW APPROXIMATION

In Hedin's GW approximation [1] the self-energy in Eq. (2) is given by

$$\Sigma_{\sigma\sigma}(\mathbf{r},\mathbf{r}';\epsilon) = \frac{i}{2\pi}\int_{-\infty}^{\infty} d\epsilon' e^{i\epsilon'\delta} G_{\sigma\sigma}(\mathbf{r},\mathbf{r}';\epsilon+\epsilon')W(\mathbf{r},\mathbf{r}';\epsilon), \tag{10}$$

where δ is an infinitesimal positive time. At the level of GW, spin flips are not accounted for [2], and the input Green's function is diagonal in its spin representation $G_{\sigma\sigma}=G_{\sigma\sigma'}\delta_{\sigma\sigma'}$.

In common with usual GW calculations, a Kohn-Sham Green's function G0 is used for G, given by [15]

$$G^0_{\sigma\sigma}(\mathbf{r},\mathbf{r}',\epsilon) = \sum_i \frac{\phi_{i\sigma}(\mathbf{r})\phi^*_{i\sigma}(\mathbf{r}')}{\epsilon - \epsilon^{KS}_{i\sigma} \mp i\delta}, \tag{11}$$

and makes the noninteracting polarizability

$$\chi^0_{\sigma\sigma}(\mathbf{r},\mathbf{r}',\epsilon) = -\frac{i}{\pi}\int_{-\infty}^{\infty} d\epsilon' G^0_{\sigma\sigma}(\mathbf{r},\mathbf{r}';\epsilon'-\epsilon)G^0_{\sigma\sigma}(\mathbf{r}',\mathbf{r};\epsilon') \tag{12}$$

and the dielectric function spin-dependent (though spin-diagonal). The inverse dielectric function in the random-phase approximation

$$\varepsilon^{-1}(\mathbf{r},\mathbf{r}',\epsilon) = \left[\delta(\mathbf{r}-\mathbf{r}') - \int d\mathbf{r}'' v(\mathbf{r},\mathbf{r}'')\sum_\sigma \chi^0_{\sigma\sigma}(\mathbf{r}'',\mathbf{r}',\epsilon)\right]^{-1} \tag{13}$$

and thus the screened Coulomb interaction

$$W_0(\mathbf{r},\mathbf{r}',\epsilon) = \int d\mathbf{r}''\varepsilon^{-1}(\mathbf{r},\mathbf{r}'',\epsilon)v(\mathbf{r}'',\mathbf{r}') \tag{14}$$

then emerge as spin-independent quantities, giving rise to the simple spin dependence in the GWself-energy [Eq. (10)].

For numerical convenience and physical insight we separate the GW self-energy (10) according to

$$\Sigma_{\sigma\sigma} = -i[G_{\sigma\sigma}v + G_{\sigma\sigma}(W-v)] = \Sigma^x_{\sigma\sigma} + \Sigma^c_{\sigma\sigma}. \tag{15}$$

The first term $(\Sigma^x_{\sigma\sigma})$ corresponds to the Fock operator in Eq. (3) and will exactly cancel the self-interaction introduced by the Hartree potential. It is therefore immediately clear that any deviation from the exact result for hydrogen can only come from the correlation part of the self-energy $(\Sigma^c_{\sigma\sigma})$.

The incorrect self-interaction affects the electron removal energies (here, the ionization potential). For electron addition energies such as the electron affinity, the entire Hartree potential has a physically reasonable interpretation, since it acts on the wave function of the originally unoccupied state which has not contributed to the electron density.

COMPUTATIONAL APPROACH

We solve the quasiparticle equation (1) with the G_0W_0 self-energy (10) for the quasiparticle energies and wave functions by fully diagonalizing the quasiparticle Hamiltonian in the basis of the single particle orbitals of the noninteracting system. Since the ground state of the hydrogen atom (5) is spherically symmetric, it is sufficient to describe all non-local operators in the GW formalism by two radial and one spin coordinates, r, r', and σ and one angular coordinate, θ, that denotes the angle between the vectors r and r'. The self-energy (10) then assumes the much simpler form

$$\Sigma_{\sigma\sigma}(r,r',\theta;\omega) = \sum_{l=0}^{\infty}[\Sigma_{l\sigma}(r,r';\omega)]P_l(\cos\theta)\delta_{\sigma,\sigma'}, \tag{16}$$

where $P_l(\cos\theta)$ is a Legendre polynomial of order l.

The Legendre expansion coefficients of the self-energy are calculated directly, thereby surpassing the need for a numerical treatment of the angular dependence. We use a real-space and imaginary time representation [16] to calculate the self-energy from the noninteracting Green's function G_0. The expression for the self-energy on the real frequency axis is obtained by analytic continuation [16]. The current implementation has been successfully applied to jellium clusters [17] and light atoms [7].

Our code allows us to solve the quasiparticle equation (1) for the GW self-energy with no further approximation. However, in order to separate the contribution that arises from the correlation part of the self-energy from that of the exchange part and the Hartree and exchange-correlation potential we *also* solve the quasiparticle equation with the frequently made approximation that the quasiparticle wave functions are given by the Kohn-Sham wave functions. The resulting equation for the quasiparticle energies is

$$\epsilon^{qp}_{i\sigma} = \epsilon^{KS}_{i\sigma} + \langle\Sigma^x_{\sigma\sigma}\rangle + \langle\Sigma^c_{\sigma\sigma}(\epsilon^{qp}_{i\sigma})\rangle - \langle v^{xc}_\sigma\rangle, \tag{17}$$

where the brackets ⟨⟩ denote matrix elements with respect to the Kohn-Sham wave function $\phi_{i\sigma}$.

In order to explore the role of the starting points for a GW calculation, two possible Kohn-Sham input Green's functions are chosen. First, the familiar LDA, and, second, the exact Kohn-Sham solution for the hydrogen atom which has the exact wave function of the hydrogen 1s state (5) and $v_{xc}(\mathbf{r})=-v_H(\mathbf{r})$. (This exact Kohn-Sham Green's function, incidentally, differs from the exact Green's function of the hydrogen atom because the exact Kohn-Sham unoccupied eigenvalues do not signify electron affinities. The exact Green's function cannot be constructed from any orthonormal set of one-particle wave functions).

RESULTS AND DISCUSSION

The calculated ionization potentials (from a full solution of the quasiparticle equation) are shown in Table I. The self-interaction errors in the two GW quasiparticle energies are seen to be fairly small: 0.95eV when the approximate LDA Kohn-Sham starting point is used, and the much smaller0.21eV when the exact Kohn-Sham starting point is used. Clearly the LDA is such a physically poor representation of the correct physics in this extreme system (owing to the large self-interaction present in the LDA calculation itself, as reflected in the large error in the LDA Kohn-Sham eigenvalue) that it forms a very unsuitable starting point for GW. However, a physically reasonable starting point reduces the GW self-interaction error to a small size.

Table 1. Quasiparticle energies (eV) for the 1s state of hydrogen (the ionization potential) obtained by diagonalizing the quasiparticle Hamiltonian (1). Two GW calculations are shown, starting from the LDA and from exact Kohn-Sham, respectively. For comparison, the Hartree-Fock (HF) and LDA eigenvalues are also shown

Exact	HF	LDA	LDA+GW	Exact+GW
−13.61	−13.61	−6.36	−12.66	−13.40

Since Σ_c gives a nonvanishing contribution to the hydrogen 1s state, even if the analytic solution is used as a starting point, the quasiparticle wave function will differ from the exact one. Figure 1 shows that the GW correlation gives rise to a slight delocalization of the quasiparticle wave function in this case. This relaxation, however, now makes the quasiparticle wave function an eigen function of the quasiparticle Hamiltonian. In the LDA the self-interaction error is much more pronounced and the wave function becomes significantly more delocalized. The GW self-energy corrects this to a small extent (as reflected in the quasiparticle wave function), but the remaining discrepancy reiterates the unsuitability of the LDA as a starting point for GW in this self-interaction-dominated atom.

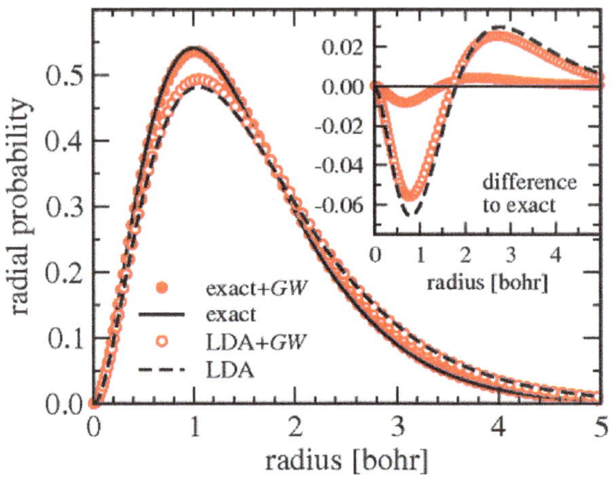

Figure. 1: (Color online) Radial probability distributions of the hydrogen 1s state: the quasiparticle wave function deviates only slightly from the exact wave function, when the latter is used as a starting point (exact+GW). The LDA wave function, on the other hand, is more delocalized as a result of the inherent self-interaction. Adding quasiparticle corrections (LDA+GW) brings the resulting quasiparticle wave function slightly closer to the exact one again. The inset shows the difference to the exact wave function.

For an analysis of the contributions to the self-energy we turn to the perturbative solution of the quasiparticle equation using Eq. (17), shown in Table II. When the exact Kohn-Sham wave function and eigenvalues are used, as in the Hartree-Fock case the exchange part of the self-energy is seen to cancel the self-interaction contribution from the Hartree potential exactly. The correlation part, on the other hand, is not zero, but amounts to a self-polarization of 0.25eV. When the LDA is used as the starting point the influence of the LDA wave function on the exchange operator becomes apparent and it reduces from −17.00eV in the exact case to −15.38eV. This corrects the highly overestimated LDA eigenvalue for the 1s state of −6.36eV (see Table 1) to −13.49eV. However, in this case the contribution from the correlation part of the GW self-energy is even larger than when starting from the exact case and increases the quasiparticle energy to −12.93eV.

Table 2. Quasiparticle energies (eV) for the 1s state of hydrogen obtained by solving Eq. (17). The contributions from the exchange $\langle\Sigma_x\rangle$ and correlation $\langle\Sigma_c\rangle$ part of the self-energy are compared to that of the exchange-correlation potential $\langle v_{xc}\rangle$ for the LDA and the exact case $(v_{xc}=-v_H)$ as a starting point. Exact value for ϵ_{1s} is −13.61eV.

Kohn-Sham G_0	ϵ_{1s}^{GW}	$\langle\Sigma_x\rangle$	$\langle v_{xc}\rangle$	$\langle\Sigma_c\rangle$
LDA	−12.93	−15.38	−8.25	0.56
Exact	−13.35	−17.00	−17.00	0.25

CONCLUSION

We have performed spin-resolved benchmark calculations for the GW formalism using the analytically known solutions of the hydrogen atom as a reference, making the self-interaction error introduced by the correlation part of the GW self-energy directly assessable. When the exact Kohn-Sham Green's function is used as the input to GW, the self-interaction error is small (0.21eV, 1⁄30the size of that in the LDA), but not negligible. If the LDA Kohn-Sham Green's function is used, as done in many GW calculations for more complex systems, a larger self-interaction error remains, inherited from the LDA starting point.

ACKNOWLEDGMENTS

This work was supported by the NATO Security Through Science Programme (EAP.RIG.981521), the Slovak grant agency VEGA (Project No. 1/2020/05), and the EU's 6th Framework Programme through the NANOQUANTA Network of Excellence (NMP4-CT-2004-500198).

REFERENCES

1. L. Hedin, Phys. Rev. 139, A796 (1965).
2. J.-L. Li, G.-M. Rignanese, E. K. Chang, X. Blase, and S. G. Louie, Phys. Rev. B 66, 035102 (2002).
3. E. L. Shirley and R. M. Martin, Phys. Rev. B 47, 15404 (1993).
4. N. E. Dahlen and U. von Barth, Phys. Rev. B 69, 195102 (2004).
5. N. E. Dahlen and R. van Leeuwen, J. Chem. Phys. 122, 164102 (2005).
6. A. Stan, N. E. Dahlen, and R. van Leeuwen, Europhys. Lett. 76, 298 (2006).
7. K. Delaney, P. García-González, A. Rubio, P. Rinke, and R. W. Godby, Phys. Rev. Lett. 93, 249701 (2004).
8. P. Rinke, A. Qteish, J. Neugebauer, C. Freysoldt, and M. Scheffler, New J. Phys. 7, 126 (2005).
9. J.-L. Li, G.-M. Rignanese, and S. G. Louie, Phys. Rev. B 71, 193102 (2005).
10. W. Kohn and K. J. Sham, Phys. Rev. 140, A1133 (1965).
11. P. Hohenberg and W. Kohn, Phys. Rev. 136, B864 (1964).
12. A local spin density calculation improves the eigenenergy of the hydrogen 1s only slightly [18].
13. J. P. Perdew and A. Zunger, Phys. Rev. B 23, 5048 (1981).
14. D. M. Ceperley and B. J. Alder, Phys. Rev. Lett. 45, 566 (1980).
15. The minus (plus) sign in the denominator applies to occupied (unoccupied) states.
16. H. N. Rojas, R. W. Godby, and R. J. Needs, Phys. Rev. Lett. 74, 1827 (1995).
17. P. Rinke, K. Delaney, P. García-González, and R. W. Godby, Phys. Rev. A 70, 063201 (2004).
18. S. Kotochigova, Z. H. Levine, E. L. Shirley, M. D. Stiles, and C. W. Clark, Phys. Rev. A 55, 191 (1997); 56, 5191 (1997); http://physics.nist.govPhysRefData/DFTdata/

CHAPTER 3

ADVANCED RELATIVISTIC MODEL POTENTIAL APPROACH TO CALCULATION OF RADIATION TRANSITION PARAMETERS IN SPECTRA OF MULTICHARGED IONS

A A Svinarenko, A V Ignatenko, V B Ternovsky, V V Nikola, S S Seredenko and T B Tkach

Odessa State Environmental University, L'vovskaya str.15, Odessa-16, 65016, Ukraine

ABSTRACT

The combined relativistic energy approach and relativistic many-body perturbation theory with the zeroth order optimized one-particle approximation are used for calculation of the Li-like ions (Z=11-42,69,70) energies and oscillator strengths of radiative transitions from the ground state to the low-excited and Rydberg states, in particular, $2s_{1/2}$ – $np_{1/2,3/2}$, $np_{1/2,3/2}$- $nd_{3/2,5/2}$(n=2-12). A comparison of the calculated oscillator strengths with available theoretical and experimental data is performed.

INTRODUCTION

The research on the spectroscopic and structural properties of highly ionized atoms has a fundamental importance in many fields of atomic physics (spectroscopy, spectral lines theory), astrophysics, plasma physics, laser physics and so on. It should be mentioned that the correct data about radiative decay widths, probabilities and oscillator strengths of atomic transitions are needed in astrophysics and laboratory, thermonuclear plasma diagnostics and in fusion research. In this light, a special interest includes studying the spectral characteristics of the He-, Li- etc. like ions. There have been sufficiently many reports of calculations and compilation of energies and oscillator strengths for the Li-like ions and other alkali-like ions (see, for example, [1–16]). Particularly, Martin and Wiese have undertaken a critical evaluation and compilation of the spectral parameters for Li-like ions (Z=3-28) [1,2]. The results of the high-precision non-relativistic calculations of the energies and oscillator strengths of 2s - 2p for Li-like systems up to Z = 50 are presented in Refs. [9-14]. The Hylleraas-type variational method and the 1/Z expansion method have been used. Chen Chao and Wang Zhi-Wen [14] have listed the nonrelativistic dipole-length, -velocity and -acceleration absorption oscillator strengths for the 1s22s–1s22p transitions of the Li I isoelectronic sequence on the basis of calculation within a full core plus correlation method with using multiconfiguration interaction wave functions. Fully variational nonrelativistic Hartree-Fock wave functions have been used by Bièmont in calculation of the 1s2n2L (n < Z, particularly for their high-excited (Rydberg) states, there is not enough precise information available in the literature. In our paper the combined relativistic energy approach [22-26] and relativistic manybody perturbation theory with the zeroth order optimized one-particle representation [26] are used for calculation the Li-like ions (Z=11-42,69,70) energies and oscillator strengths of radiative transitions from ground state to low-excited and Rydberg states. The comparison of the

Citation: A A Svinarenko, A V Ignatenko, V B Ternovsky, V V Nikola, S S Seredenko and T B Tkach, Advanced relativistic model potential approach to calculation of radiation transition parameters in spectra of multicharged ions, doi:10.1088/1742-6596/548/1/012047

calculated oscillator strengths with available theoretical and experimental data is performed.

THE THEORETICAL METHOD

In the relativistic energy approach [4, 5, 22-25] the imaginary part of electron energy shift of an atom is connected with the radiation decay possibility (transition probability). An approach, based on the Gell-Mann and Low formula with the QED scattering matrix, is used in treating the relativistic atom. The total energy shift of the state is usually presented in the form:

$$\Delta E = \mathrm{Re}\Delta E + i\,\Gamma/2 \tag{1}$$

where G is interpreted as the level width, and the decay probability P = G. For the a-n radiation transition the imaginary part of electron energy in the lowest order of perturbation theory is determined as [4]:

$$\mathrm{Im}\,\Delta E = -\frac{1}{4\pi}\sum_{\substack{\alpha>n>f\\ [\alpha<n\leq f]}} V_{\alpha n\alpha n}^{|\omega_{\alpha n}|}, \tag{2}$$

where $\omega_{\alpha n}$ is a frequency of the a-n radiation, (a>n>f) for electron and (a<n<f) for vacancy. The matrix element V is determined as follows:

$$V_{ijkl}^{|\omega|} = \iint dr_1 dr_2 \Psi_i^*(r_1)\Psi_j^*(r_2)\frac{\sin|\omega|r_{12}}{r_{12}}(1-\alpha_1\alpha_2)\Psi_k^*(r_2)\Psi_l^*(r_1) \tag{3}$$

The separated terms of the sum in (3) represent the contributions of different channels and a probability of the dipole transition is:

$$\Gamma_{\alpha_n} = \frac{1}{4\pi}\cdot V_{\alpha_n\alpha_n}^{|\omega_{\alpha_n}|} \tag{4}$$

The corresponding oscillator strength: $gf=\lambda^2\Gamma_{\alpha n}/6.67\cdot10^{15}$, where g is the degeneracy degree, l is a wavelength in angstrems (Å). When calculating the matrix elements in Equation (3) one should use the angle symmetry and write the expansion for potentials $\sin|\omega|r_{12}/r_{12}$ in spherical functions as follows [2]:

$$\frac{\sin|\omega|r_{12}}{r_{12}} = \frac{\pi}{2\sqrt{r_1 r_2}}\sum_{\lambda=0}^{\infty}(\lambda)J_{\lambda+\frac{1}{2}}(|\omega|r_1)J_{\lambda+\frac{1}{2}}(|\omega|r_2)P_\lambda(\cos\widehat{r_1 r_2}) \tag{5}$$

where J is the Bessel function of first kind and (l)= 2l + 1. This expansion is corresponding to usual multipole one for probability of radiative decay. Substitution of the expansion (5) to matrix element of interaction gives as follows [20]:

$$V_{1234}^{\omega} = [(j_1)(j_2)(j_3)(j_4)]^{1/2}\sum_{\lambda\mu}(-1)^{\mu}\begin{pmatrix} j_1 j_3 & \lambda \\ m_1 - m_3 & \mu \end{pmatrix}\times \mathrm{Im}Q_\lambda(1234),$$

$$Q_\lambda = Q_\lambda^{\mathrm{Qul}} + Q_\lambda^{\mathrm{Br}}, \tag{6}$$

where j_i is the total single electron momentums, m_i – the projections; Q^{Qul} is the Coulomb part of interaction, Q^{Br} - the Breit part. Their detailed definitions are presented in Refs. [4, 20]. The relativistic wave functions are calculated by solution of the Dirac equation with the potential, which includes the "outer electron- ionic core" potential and polarization potential [29]. In order to describe interaction of the outer electron with the He-like core the modified model potential [4] has been used. The calibration of the single model potential parameter has been performed on the basis of the special ab initio procedure within relativistic energy approach [24] (see also [5]). In Ref.[18] the lowest order multielectron effects, in particular, the gauge dependent radiative contribution Im dE_{ninv} for the certain class of the photon propagator calibration is treated. This value is considered to be representative for the correlation effects, whose minimization is a reasonable criterion in the searching for the optimal one-electron basis of the many-body perturbation theory. The minimization of the density functional Im dE_{ninv} leads to the integral-differential equation that can be solved using one of the standard numerical codes. It provides the construction of the optimized one-particle representation. All calculations are performed on the basis of the numeral code Superatom-ISAN (version 93).

RESULTS

We applied the above described approach to calculating the energies and oscillator strengths of transitions in spectra of the Li-like ions (Z=11-42, 69, 70). There are considered the radiative transitions from ground state to the low-excited and Rydberg states, particularly, $2s_{1/2} - np_{1/2,3/2}$, $np_{1/2,3/2}$-$nd_{3/2,5/2}$ (n=2-12). To test the obtained results, we compare our calculation results of the oscillator strengths values for some Li-like ions with the known theoretical and tabulated results [1, 2, 8]. As an example, in table 1 we present the oscillator strengths values for the 2s1/2 – 2p1/2,3/2 transitions in Li-like ions S^{13+}, Ca^{17+}, Fe^{23+}, Zn^{27+}, Zr^{37+}, Mo^{39+}, Sn^{47+}, Tm^{66+}, Yb^{67+}. The DF calculation data by Zilitis [6] and the "best" compillated (experimental) data [1, 2] for the low-Z Li-like ions are listed in table 1 for comparison too. Note that the experimental data on the oscillator strengths for many (especially, highZ) Li-like ions are missing. Overall, there is a physically reasonable agreement of the listed data. The important features of the approach used are using the optimized one-particle representation and account for polarization effects. It should be noted that an estimate of the gauge-non-invariant contributions (the difference between the oscillator strengths values calculated with using the transition operator in the form of "length" and "velocity") is about 0.3%, i.e., the results obtained with different photon propagator gauges (Coulomb, Babushkon, Landau) are practically equal.

Table 1. Oscillator strengths of the $2s_{1/2} - 2p_{1/2,\,3/2}$ transitions in Li-like ions

Method	DF [6]	DF [6]	[2]	[2]	Our	Our
Ion	$2s_{1/2}$–$2p_{1/2}$	$2s_{1/2}$–$2p_{3/2}$	$2s_{1/2}$–$2p_{1/2}$	$2s_{1/2}$–$2p_{3/2}$	$2s_{1/2}$–$2p_{1/2}$	$2s_{1/2}$–$2p_{3/2}$
S^{13+}	0.0299	0.0643	0.030	0.064	0.0301	0.0641
Ca^{17+}	0.0234	0.0542	0.024	0.054	0.0236	0.0541
Fe^{23+}	0.0177	0.0482	0.018	0.048	0.0179	0.0481
Zn^{27+}	0.0153	0.0477	–	–	0.0156	0.0475
Zr^{37+}	0.0114	0.0543	–	–	0.0118	0.0540
Mo^{39+}	–	–	0.011	0.056	0.0107	0.0556
Sn^{47+}	0.0092	0.0686	–	–	0.0095	0.0684
Tm^{66+}	–	–	–	–	0.0071	01140
Yb^{67+}	0.0067	0.1170	–	–	0.0069	0.1167

In Table 2 we present the oscillator strengths values for the $2s_{1/2}$ – npj (n=3-18, j=1/2) transitions in spectrum of the Li-like ion Zr^{37+}. The quantum defect approximation (QDA)

[6, 27], the DF oscillator strengths calculation results by Zilitis [6] and some tabulated (experimental) data by Martin-Weiss [1] are listed as well. It is self-understood that the QDA oscillator strengths data become more exact with the growth of the principal quantum number. At the same time the accuracy of the DF data may be decreased. The agreement between the Martin-Weiss data and our results is acceptable.

Table 2. Oscillator strengths of the $2s_{1/2} - np_{1/2}$ transitions in Zr^{37+}

Transition	QDA [6]	DF [6]	Our data
$2s_{1/2}$–$3p_{1/2}$	13.7	13.3	13.684
$-4p_{1/2}$	-	3.22	3.232
$-6p_{1/2}$	-	-	0.682
$-8p_{1/2}$	0.258	0.257	0.260
$-10p_{1/2}$	0.126	0.124	0.125
$-16p_{1/2}$	0.0291	0.0285	0.0287
$-18p_{1/2}$	-	-	0.0216

REFERENCES

1. Martin G A, Wiese W L 1976 *Phys.Rev.A*. 13 699; 1976 *J. Phys. Chem. Ref. Dat*. 5 538
2. Schweizer W, Faßbinder P, Gonzalez-Ferez R 1999 *Atom. Data. Nucl. Data. Tabl.* 72 33
3. Ivanova E P, Ivanov L N 1979 *Atom. Data. Nucl. Data. Tabl.* 24 95
4. Ivanov L N, Ivanova E P and Aglitsky E V 1988 *Phys. Rep.* 166 315
5. Glushkov A V, Ambrosov S V, Loboda A V, Gurnitskaya E P and Prepelitsa G P 2005 *Int. J. Quant.Chem*. 104 562
6. Zilitis V A 1983 *Opt. Spectr*. 55 215
7. Khetselius O Yu 2012 *J. of Phys.: Conf. Ser.* 397 012012
8. Glushkov A V 2012 *J. of Phys.: Conf. Ser*. 397 012011
9. Zong-Chao Yan and Drake G W F 1995 *Phys. Rev. A* 52 R4316
10. Lianhua Qu, Zhiwen Wang and Baiwen Li 1998 *J. Phys. B* 31 3601
11. Yan Z C, Tambasco M and Drake G W F 1998 *Phys. Rev. A* 57 1652
12. Xiaoxu Guan and Baiwen Li 2001 *Phys. Rev. A* 63 043413
13. Khetselius O Yu 2009 *Phys.Scr.* T.135 014023
14. Chen Chao and Wang Zhi-Wen 2005 *Commun. Theor. Phys.* 43 305
15. Glushkov A V, Malinovskaya S V, Prepelitsa G P and Ignatenko V M 2005 *J. of Phys.: Conf. Ser*. 11 199
16. Bièmont E 1977 *Astr. & Astrophys. Suppl.Ser*. 27 489
17. Malinovskaya S V, Glushkov A V, Khetselius O Yu, Loboda A V, Lopatkin Yu M, Svinarenko A A, Nikola L V and Perelygina T B 2011 *Int. J. Quant. Chem*. 111 288
18. Glushkov A V, Ivanov L N 1992 *Phys. Lett. A* 170 33
19. Yerokhin V A, Artemyev A N, Shabaev V M 2007 *Phys. Rev. A*. 75 062501
20. Glushkov A V, Malinovskaya S V, Sukharev D E, Khetselius O Yu, Loboda A V and Lovett L 2009 *Int. J. Quant. Chem*. 109 1717
21. Ivanova E P and Glushkov A V 1986 *J.Quant. Spectrosc. Rad.Transf.* 36 127
22. Glushkov A V, Khetselius O, Loboda A and Svinarenko A 2009 *Phys. Scr.* T153 014029
23. Ivanov L N, Ivanova E P and Knight L 1993 *Phys. Rev. A* 48, 4365
24. Glushkov A V, Khetselius O Yu and Malinovskaya S V 2008 *Europ. Phys. J.* ST 160 195
25. Khetselius O Yu 2009 *Int. J. Quant. Chem*. 109 3330
26. Glushkov A V, Ambrosov S, Ignatenko A and Korchevsky D 2004 *Int. J. Quant. Chem*. 99 936
27. Seaton M J 1983 *Rep. Prog. Phys*. 46 1983

CHAPTER 4

DETERMINATION OF TRACE AMOUNT OF CADMIUM BY ATOMIC ABSORPTION SPECTROMETRY IN TABLE SALT AFTER SOLID PHASE PRECONCENTRATION USING OCTADECYL SILICA MEMBRANE DISK MODIFIED BY A NEW DERIVATIVE OF PYRIDINE

Mahmood Payehghadr[1] Sousan Esmaeilpour[1, 2] Mohammad Kazem Rofouei[3] and Laleh Adlnasab[4]

[1]Department of Chemistry, Payame Noor University, Tehran 19395-4697, Iran

[2]Standard Research Institute, Karaj 31585-163, Iran

[3]Faculty of Chemistry, Tarbiat Moallem University, Tehran 15815-3587, Iran

[4]Department of Chemistry, Shahid Beheshti University, Evin, Tehran 1983963113, Iran

ABSTRACT

Silica-C18 bonded disk modified by a four-dentate Schiff base has been used for preconcentration of cadmium in table salt samples followed by flame atomic absorption spectrometry. The method is based on the adsorption of Cd on 1, 2-bis(pyridin-2-ylmethylene) hydrazine as Schiff base ligand on silica-C18 disk. The effects of several factors such as type and concentration of the eluent, pH of sample solution, amount of ligand, and breakthrough volume have been optimized based on one variable at a time. In optimum conditions (type of eluent, HNO_3 1.0 mol/L; volume of eluent, 5.0 mL; solution pH, 9.0; breakthrough volume, 1000.0 mL), preconcentration factor of the present method is about 200. The resultant limit of detection is 5.0 μg/L. Finally, the performance of the method has been evaluated for extraction and determination of Cd (II) in salt samples at milligram per liter concentration, and satisfactory results have been obtained (RSD $\leq$ 2.0%).

INTRODUCTION

The contamination of food with heavy metals could cause acute poisoning as well as long-term health problems. Table salt is an essential additive which is frequently added to certain foods for improving the taste, and it is used as a preservative as well. Therefore, due to the daily consumption of table salt, any contamination even in low level could put the consumer's health at risk. In fact, heavy metal ions are toxic, non-biodegradable, and tend to be accumulated in the human vital organs, where they can act progressively over a long period through food chains [1, 2], and the reported primary adverse health effects of them are lung cancer and kidney damage [3, 4]. Among these heavy metals, cadmium is one of the most toxic elements for human and animals, even at low concentrations, and the International Agency for Research on Cancer classified Cd as a human carcinogen [5]. The FAO/WHO (Food and Agriculture Organization/World Health Organization) and Joint Expert Committee on Food Activities recommended maximum tolerable daily intake of cadmium from all sources (food, air, and water) of 1.0–1.2 μg/g weight of body [6]. In Codex standard the permissible level of cadmium in table salt is 0.2 μg/g.

Citation: Mahmood Payehghadr, Sousan Esmaeilpour, Mohammad Kazem Rofouei, and Laleh Adlnasab, "Determination of Trace Amount of Cadmium by Atomic Absorption Spectrometry in Table Salt after Solid Phase Preconcentration Using Octadecyl Silica Membrane Disk Modified by a New Derivative of Pyridine," Journal of Chemistry, vol. 2013, Article ID 417085, 6 pages, 2013. doi:10.1155/2013/417085

Because of the complexity of sample matrix and low concentrations of cadmium in the environmental samples, preconcentration and matrix-removal steps are required to guarantee the accuracy and precision of the analytical results [1, 5, 7]. Many preconcentration methods followed by different quantitative instrumental techniques have been investigated in trace heavy metal in the environmental samples [8].

Although atomic absorption spectroscopy (AAS) and inductively coupled plasma atomic emission spectroscopy (ICP-AES) are the most widely used methods, their sensitivity is usually insufficient for monitoring the low level concentration of these ions in the environmental samples [9–12]. Therefore, element with lower concentrations than the detection limits of the instrumental techniques are a serious problem in the trace heavy metal determination in the environmental samples [13]. The most common techniques reported for preconcentration of Cd (II) are liquid-liquid extraction (LLE) [14, 15], solid-phase extraction (SPE) [16, 17], ion exchange [18], cloud point extraction [19, 20], and electroanalytical techniques [21].

Solid-phase extraction (SPE) has advantages over other preconcentration methods in terms of simplicity, economy, rapidity, reusability of the adsorbent, ease of automation, higher preconcentration factor, lower consumption of reagents, and, more importantly, its environmental friendliness (use of less organic solvents). The most extensively used SPE sorbents are modified C-18 silica [22], activated carbon [23], alumina [24], and Amberlite XAD resins [25, 26]. For example, silica and silica-bonded adsorbents offer good advantages in terms of thermal, mechanical, and chemical stability. In addition, they act selectively towards a particular metal ion. In fact, the modified silica gel surfaces are known to act as a weak cation exchanger via its weak silanol groups, through immobilization of organic complexing agents, either chemically or physically. These modified surfaces greatly enhance metal exchange capacity and improve selectivity of these phases to metal ion removal, separation, or preconcentration prior to their determination by X-ray fluorescence or AAS analysis [11, 27–30]. Chelating agents can be easily loaded on silica-bonded adsorbents with good stability and Schiff bases often used as chelating ligands and their metal complexes have been of great interest for many years. Chelating ligands containing N, S, and O donor atoms are of special interest because of the various ways in which they are bonded to the metal ions. The variety of possible Schiff base metal complexes with wide choices of ligands and coordination environment has encouraged us to choose a tetradentate Schiff base for modifying the adsorbent disk. The mobilization and immobilization of cadmium have been shown to depend significantly on the complexation of the metal center by chelating nitrogen donor ligands [8, 31–33].

In the present method, octadecyl silica disk modified with a new derivative of pyridine ligand (tetra-dentate Schiff base), (Figure 1) which is followed by flame atomic absorption spectrometry, has been applied for preconcentration and determination of cadmium in salt samples because of the toxic environmental impact of Cd. The effects of different variables such as type and concentration of eluent, sample solution pH, concentration of ligand, and effect of breakthrough volume have been investigated and optimized based on the method of one variable at a time. In the end, the optimized method has been applied for determination of Cd ions in real samples.

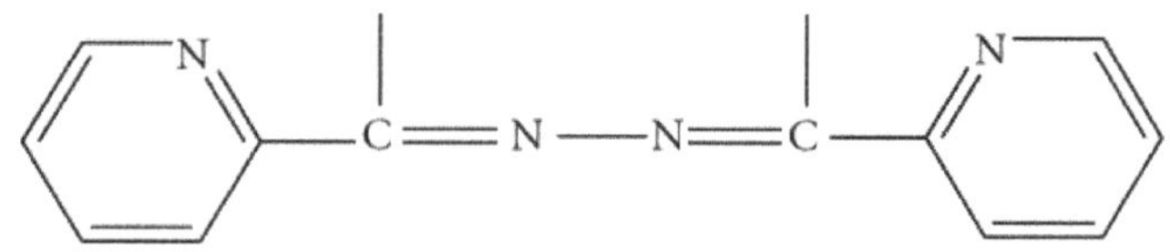

Figure 1: Structure of (1, 2-bis(pyridin-2-ylmethylene) hydrazine) ligand.

EXPERIMENTAL

Chemicals and Reagents

Acetonitrile (ACN), methanol, nitric acid, ethylene diamine tetraacetic acid (EDTA), acetic acid, sulfuric acid, hydrochloric acid, phosphoric acid, sodium dihydrogen phosphate, ammonium acetate, ammonia solution, and ammonium chloride salt are of analytical grade from Merck (Darmstadt, Germany). Double distillated water has been prepared by Mili-Q system from Millipore (Bedford, MA, USA).

Reference standard solution of cadmium having the metal concentration of 1000.0 ± 2.0 mg/L in 2% (v/v) of HNO_3 have been purchased from assurance company. The working standard solutions have been prepared by appropriate dilution of the stock solution with double distillated water, all solutions have been stored in ambient temperature. The new synthesized Schiff base (Figure 1) with highest purity has been used as the chelating ligand.

Reference standard solutions of As, Hg, Pb, Na, K, Ca, and Mg (each of them 1000.0 ± 2.0 mg/L in 2% (v/v) of HNO_3) have been bought from Merck company.

For preparing pH 2.0 and 3.0 buffer solutions, appropriate amount of phosphoric acid has been added to sodium dihydrogen phosphate solution (0.1 mol/L). Moreover, appropriate amount of acetic acid has been added to ammonium acetate solution (0.1 mol/L) to give buffer solutions of pH 4.0–6.0. Titrisol buffer solution from Merck ($C(KH_2PO_4)$ = 0.026 mol/L and $C(Na_2HPO_4)$ = 0.041 mol/L) has been used as pH 7.0, and an ammonium ammonia buffer solution has been prepared by adding an appropriate amount of ammonia to ammonium chloride solution (0.1 mol/L) to give a solution of pH 8.0–10.0.

Instrumentation

A Varian model AA-1275 flame atomic absorption spectrometer with a deuterium background corrector was used for determination of Cd ions, and a cadmium hollow cathode lamp was used as light source operated at 3.5 mA. The wavelength was set at 228.8 nm resonance line and the spectral band pass at 0.5 nm, and the measurements were carried out in an air/acetylene flame. A digital pH meter, Mettler Toledo model MP225, equipped with a combined glass calomel electrode was used for the pH adjustments.

Extraction was performed with 47 mm diameter ×0.5 mm thickness Empore membrane disks containing octadecyl-bonded silica (8 μm particle, 60-Å pore size, 3 M Co., St. Paul, MN) with a standard Millipore 47 mm filtration apparatus. The twin cylinder vacuum pump model of TC-2000 V (Taiwan) was used to provide the required pressure for extraction.

Extraction Procedure

The membrane disk was conditioned with 20.0 mL of methanol and 20.0 mL of water, respectively, to remove all contaminants arising from the manufacturing process and the environment. The disk was dried by passing the air through it for several minutes, and then 8 mg of ligand was dissolved in 3.0 mL ACN and passed through sorbent several times under gravity until the ligand penetrated the membrane completely. Then modified disk was applied for extraction step, 100.0 mL of standard solution containing 10.0 μg of Cd was prepared, and solution pH was adjusted with buffer solution at 9.0. Afterwards, this solution was loaded at a flow rate of about 4.0 mL/min with the aid of a vacuum pump, and the retained Cd ions were eluted with 5.0 mL of HNO_3 (1.0 mol/L) under gravity (about 2.0 mL/min) followed by FAAS for determination of Cd content. Five measurements were made for each sample and the results were subsequently averaged. After each extraction, the disk eluted with two times 5.0 mL of HNO_3 (1.0 mol/L) and with 5.0 mL of water, respectively.

Sample Preparation

One salt sample from mine (Rahrahak mine, Iran) and a commercial sample from Iranian market were collected as the real samples. 2.0 gr table salt was dissolved in 60.0 mL double distilled water, and solution pH was adjusted to 9.0 with 5.0 mL of ammonium chloride/ammonia buffer solution and diluted to 100.0 mL with water. This solution was applied for the preconcentration step.

RESULTS AND DISCUSSION

In the proposed method, solid phase extraction with silica-C18 disk, modified by a four-dentate Schiff base, which is followed by FAAS determination was applied for preconcentration of Cd ions in salt samples. An azine compound (1, 2-bis(pyridin-2-ylmethylene)hydrazine), (Figure 1), was used as a selective chelating ligand for cadmium complexation. The effect of different parameters on the extraction and elution of Cd ions was optimized by using the method of one variable at a time.

Effect of Solution pH

Among chemical variables, pH is seemed to be the most important parameter for adsorption of the metal ions on ligand. As Schiff bases are conjugate bases of weak acid groups, they have very strong affinity for hydrogen ions. The pH of solution will determine the values of conditional stability constants of metal complexes on the surface of the sorbent. In order to assess the effect of solution pH on extraction efficiency, the pH of 100.0 mL model solutions containing 0.1 mg/L of Cd ions was adjusted in the range of 3.0–10.0 with different buffer solutions. The retained Cd ions were eluted with 5.0 mL of HNO_3 (1.0 mol/L) followed by FAAS for determination of Cd content. Figure 2 reveals that extraction recovery of the analyte grows significantly by increasing the pH of sample solution, but it is important to know that the pH higher than 10.0 may cause the destruction of the disk. Therefore, pH=9.0 was selected as the best pH in the subsequent experiments.

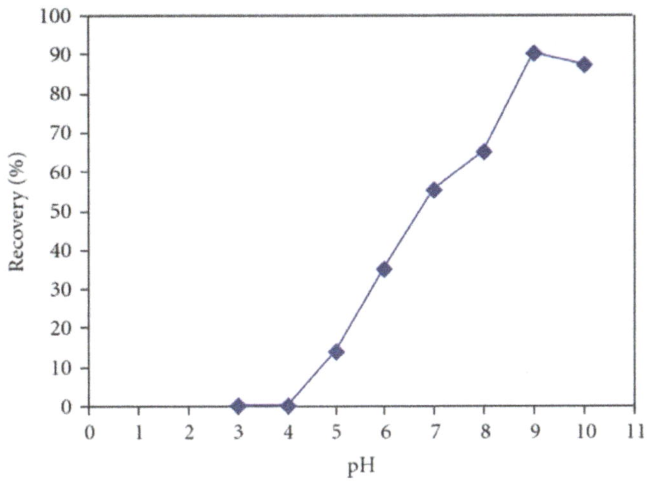

Figure 2: Effect of solution pH on recovery (%) (extraction conditions: sample volume, 100.0 mL; eluent solvent, 5.0 mL of HNO_3 (1.0 mol/L); amount of ligand, 8 mg; concentration of Cd ion, 0.10 mg/L).

Effect of Type and Concentration of Eluent

Complete stripping of Cd ions from ligand was examined with several eluents such as H_2SO_4, HNO_3, HCl, and EDTA. In fact, the selection of the eluent was really difficult due to the limitation of FAAS to tolerate organic solvents, and the eluent should not destroy the solid phase. In lower pH, the ligand–Cd complex dissociates to release Cd ions. Therefore, acidic eluents are the best choice to obtain efficient extraction. According to the results in Figure 3, HNO_3 was selected as the best eluent. HNO_3 at various concentrations was investigated for increasing recovery. The results show that HNO_3 with 1.0 mol/L concentration is the best solvent for elution of Cd ions.

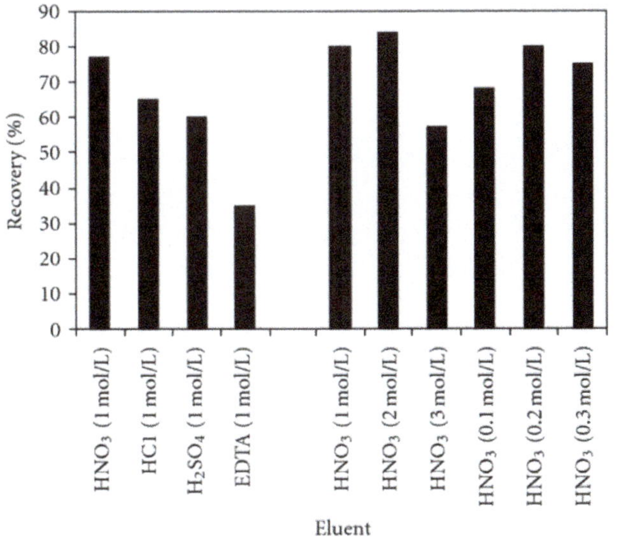

Figure 3: Effect of type and concentration of eluent on recovery (%), (Extraction conditions: sample volume, 100.0 mL; amount of ligand, 8 mg; sample solution pH, 9.0; concentration of Cd ion, 0.10 mg L^{-1}).

Effect of Ligand Concentration

Prior to adsorption of trace heavy metal ions on a solid phase for preconcentration, metal ions were converted to a suitable form including metal chelates or metal inorganic complexes. Because of this, different concentrations of ligand were investigated for increasing extraction efficiency. Different amounts (3, 5, 8, and 10 mg) of ligand in 3.0 mL of ACN were added to the disk, and quantitative recoveries were obtained for cadmium ions with 8 mg of ligand. In all further works, 8 mg of ligand in 3.0 mL ACN solution was applied for modification of the disk.

Stability of Solid Phase

The reuse of the modified disk was tested for its ability to perform the SPE for more than one sample. It was found that the use of the same disk modified with Schiff base for at least twenty times did not change the recovery of Cd ion. In addition, there was no decrease in recovery for interday and intraday extractions. However, for the SPE of trace amounts of Cd from real samples, it is recommended to use a new modified disk for each sample to protect them from possible contaminations.

Effect of the Breakthrough Volume

The breakthrough volume of sample solution was tested by treating standard solutions containing 10 µg Cd ions with different sample volumes (25.0, 50.0, 100.0, 250.0, 500.0, and 1000.0 mL). The effect of breakthrough volume on recovery has been showed in Figure 4. In all cases, the extraction by membrane disk was found to be quantitative. Thus, 1000.0 mL was the maximum sample volume in which quantitative extraction of Cd ions is possible. Sample volume more than 1000.0 mL was not tested due to increase of extraction time. To reduce the extraction time, 100.0 mL sample solution was selected.

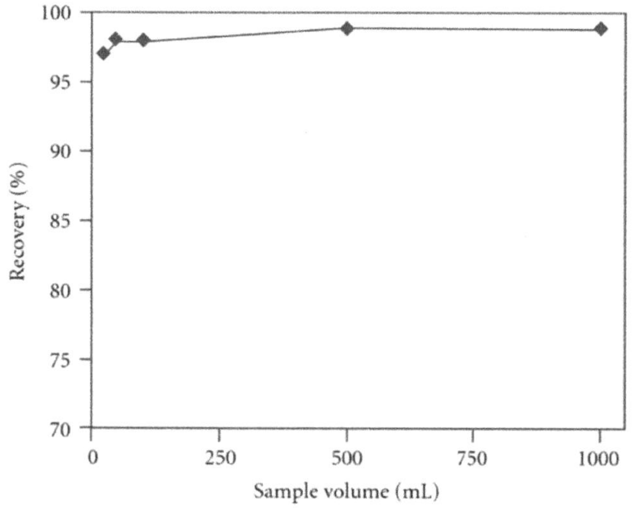

Figure 4: Effect of breakthrough volume on recovery (%) (extraction conditions: eluent solvent, 5.0 mL of HNO_3 (1.0 mol/L); amount of ligand, 8 mg; sample solution pH, 9.0; concentration of Cd ion, 0.10 mg L^{-1}).

Sorption Capacity

The maximum capacity of the membrane disk modified by a new Schiff base ligand was determined by passing 50.0 mL of an aqueous solution containing 500.0 µg Cd ions through the disk. The amount of the retained Cd ions determined using furnace atomic absorption spectrometry. The maximum capacity of the disk was found to be 325 µg Cd ion per 8 mg of ligand.

Influence of Interference Ions

The impact of a variety of cations found in salt samples on the determination of Cd has been studied. The limit of heavy metals (Pb, Cd, As, Hg, Cu, Fe) and other cations (K^+, Mg^{2+}, Ca^{2+}) is determined in Codex standard for table salt. Therefore, the above-mentioned cations have been chosen for the investigation of interference ions. For this purpose, Na^+, K^+, Mg^{2+}, Ca^{2+}, As^{2+}, Pb^{2+}, and Hg^{2+} with limited concentrations in accordance with Codex standard mentioned in Table 1 were added individually to 100.0 mL aliquots of solutions containing 10 µg of Cd. The same procedure for extraction on solid phase by Schiff base was followed. The effect of Cu and Fe ions was not investigated because of precipitate formation in the working pH.

Table 1: Extraction of Cd ion from binary mixtures

Interference ion	Concentration (mg/L)	Recovery of Cd (II) (%)
Mg^{+2}	300	96.0 ± 0.5
Ca^{+2}	1500	95.0 ± 1.0
K^+	100	94.0 ± 1.0
Hg^{+2}	100	94.0 ± 1.0
As^{+3}	100	95.0 ± 0.5
Pb^{+2}	100	89.0 ± 1.0

Analytical Performance

Regression equation, correlation coefficient (R^2), limit of detection (LOD), and preconcentrations factor (PF) were calculated in optimal conditions and summarized in Table 2.

Table 2: The limits of detection, regression equations, correlation coefficient, preconcentration factor, recovery, and relative standard deviation for extraction method

Analyte	Regression equation	R^2	LOD (µg/L)	PF	Recovery (%)	RSD (%)
Cd ion	$Y = 0.368X + 0.126$	0.9905	5.0	200	98.0 ± 0.5	≤2

The LOD was calculated as the analyte concentration equal to three times the standard deviation of the blank signal divided by the slope of the calibration curve. The LOD was obtained in the range of 5.0 µg/L.

The PF was calculated as the ratio of sample solution volume to eluent solution volume. The PF obtained for solution containing 100.0 µg/L of Cd (II) was 200.

The real samples were extracted by C18-bonded silica disk modified with the tetradentate Schiff base followed by FAAS. Analysis of samples showed that they contained trace amount of Cd (II) which were lower than permissible limit in Codex standard. Therefore, all the real salt samples were spiked with reference standard solution at 100.0 µg/L concentration to assess the matrix effects.

Table 3 shows that the results of the four replicate analyses of each sample obtained by the SPE/FAAS method are satisfactorily in agreement with the amounts of Cd (II) added. It is noticeable that the amount of Cd ion was not detectable in all samples without spiking.

Table 3: Determination of Cd ions in the salt samples

Sample	C_{added} ($\mu g\,L^{-1}$)	C_{found} ($\mu g\,L^{-1}$)	Recovery (%)	RSD (%)
Rahrahak sample	100.0	93.7	93.7	1.7
Commercial sample	100.0	95.4	95.4	1.4

CONCLUSION

This paper describes a preconcentration/separation procedure based on solid phase extraction of cadmium in salt samples using a octadecyl silica disk modified by a new pyridine derivative ligand followed by flame atomic absorption spectrometric determination. The proposed method has advantages such as high sample cleanup, efficiency, selectivity, and preconcentration factor. In conclusion, the performance of this procedure has been excellent in extraction of trace amounts of Cd (II) from different salt samples. Furthermore, this method is very environmental friendly, and it can be used in complex matrices with high selectivity.

REFERENCES

1. F. Xie, X. Lin, X. Wu, and Z. Xie, "Solid phase extraction of lead (II), copper (II), cadmium (II) and nickel (II) using gallic acid-modified silica gel prior to determination by flame atomic absorption spectrometry," Talanta, vol. 74, no. 4, pp. 836–843, 2008.
2. M. Tuzen and M. Soylak, "Multi-element coprecipitation for separation and enrichment of heavy metal ions for their flame atomic absorption spectrometric determinations," Journal of Hazardous Materials, vol. 162, no. 2-3, pp. 724–729, 2009.
3. A. M. Cheraghali, F. Kobarfard, and N. Faeizy, "Heavy metals contamination of table salt consumed in iran," Iranian Journal of Pharmaceutical Research, vol. 9, no. 2, pp. 129–132, 2010.
4. C. Duran, A. Gundogdu, V. N. Bulut et al., "Solid-phase extraction of Mn(II), Co(II), Ni(II), Cu(II), Cd(II) and Pb(II) ions from environmental samples by flame atomic absorption spectrometry (FAAS),"Journal of Hazardous Materials, vol. 146, no. 1-2, pp. 347–355, 2007.
5. Y. Zhai, Y. Liu, X. Chang, S. Chen, and X. Huang, "Selective solid-phase extraction of trace cadmium(II) with an ionic imprinted polymer prepared from a dual-ligand monomer," Analytica Chimica Acta, vol. 593, no. 1, pp. 123–128, 2007.
6. H. Ciftci, "Separation and solid phase extraction method for the determination of cadmium in environmental samples," Desalination, vol. 263, no. 1–3, pp. 18–22, 2010.
7. G. Yang, W. Fen, C. Lei, W. Xiao, and H. Sun, "Study on solid phase extraction and graphite furnace atomic absorption spectrometry for the determination of nickel, silver, cobalt, copper, cadmium and lead with MCI GEL CHP 20Y as sorbent," Journal of Hazardous Materials, vol. 162, no. 1, pp. 44–49, 2009.
8. B. Dede, F. Karipcin, and M. Cengiz, "Novel homo- and hetero-nuclear copper(II) complexes of tetradentate Schiff bases: synthesis, characterization, solvent-extraction and catalase-like activity studies," Journal of Hazardous Materials, vol. 163, no. 2-3, pp. 1148–1156, 2009.
9. A. Malekpour, S. Hajialigol, and M. A. Taher, "Study on solid-phase extraction and flame atomic absorption spectrometry for the selective determination of cadmium in water and plant samples with modified clinoptilolite," Journal of Hazardous Materials, vol. 172, no. 1, pp. 229–233, 2009.
10. k. J. Brodie and J. P. Matousek, "Determination of cadmium in air by non-flame atomic absorption spectrometry," Analytica Chimica Acta, vol. 69, no. 1, pp. 200–202, 1974.
11. E. Melek, M. Tuzen, and M. Soylak, "Flame atomic absorption spectrometric determination of cadmium(II) and lead(II) after their solid phase extraction as dibenzyldithiocarbamate chelates on Dowex Optipore V-493," Analytica Chimica Acta, vol. 578, no. 2, pp. 213–219, 2006.
12. S. P. Dolan and S. G. Capar, "Multi-element analysis of food by microwave digestion and inductively coupled plasma-atomic emission spectrometry," Journal of Food Composition and Analysis, vol. 15, no. 5, pp. 593–615, 2002.
13. M. Soylak, A. U. Karatepe, L. Elci, and M. Dogan, "Column preconcentration/separation and atomic absorption spectrometric determinations of some heavy metals in table salt samples using amberlite XAD-1180," Turkish Journal of Chemistry, vol. 27, pp. 235–242, 2003.
14. A. N. Anthemidis, G. A. Zachariadis, C. G. Farastelis, and J. A. Stratis, "On-line liquid-liquid extraction system using a new phase separator for flame atomic absorption spectrometric determination of ultra-trace cadmium in natural waters," Talanta, vol. 62, no. 3, pp. 437–443, 2004.
15. I. Komjarova and R. Blust, "Comparison of liquid-liquid extraction, solid-phase extraction and co-precipitation preconcentration methods for the determination of cadmium, copper, nickel, lead and zinc in seawater," Analytica Chimica Acta, vol. 576, no. 2, pp. 221–228, 2006.
16. V. Camel, "Solid phase extraction of trace elements," Spectrochimica Acta B, vol. 58, no. 7, pp. 1177–1233,

2003.

17. N. Pourreza and H. Zavvar Mousavi, "Determination of cadmium by flame atomic absorption spectrometry after preconcentration on naphthalene-methyltrioctylammonium chloride adsorbent as tetraiodocadmate (II) ions," Analytica Chimica Acta, vol. 503, no. 2, pp. 279–282, 2004.
18. T. Kumamaru, H. Matsuo, Y. Okamoto, and M. Iked, "Sensitivity enhancement for inductively-coupled plasma atomic emission spectrometry of cadmium by suction-flow on-line ion-exchange preconcentration," Analytica Chimica Acta, vol. 181, pp. 271–275, 1986.
19. M. Ghaedi, A. Shokrollahi, K. Niknam, E. Niknam, A. Najibi, and M. Soylak, "Cloud point extraction and flame atomic absorption spectrometric determination of cadmium(II), lead(II), palladium(II) and silver(I) in environmental samples," Journal of Hazardous Materials, vol. 168, no. 2-3, pp. 1022–1027, 2009.
20. J. L. Manzoori and G. Karim-Nezhad, "Development of a cloud point extraction and preconcentration method for Cd and Ni prior to flame atomic absorption spectrometric determination," Analytica Chimica Acta, vol. 521, no. 2, pp. 173–177, 2004.
21. P. Kuban, R. Guchardi, and P. C. Hauser, Trace-metal analysis with separation methods, Trends Anal Chem., vol. 24, pp. 192–198, 2005.
22. J. Otero-Romaní, A. Moreda-Piñeiro, A. Bermejo-Barrera, and P. Bermejo-Barrera, "Evaluation of commercial C18 cartridges for trace elements solid phase extraction from seawater followed by inductively coupled plasma-optical emission spectrometry determination," Analytica Chimica Acta, vol. 536, no. 1-2, pp. 213–218, 2005.
23. S. L. C. Ferreira, H. M. C. Andrade, and H. C. dos Santos, "Characterization and determination of the thermodynamic and kinetic properties of the adsorption of the molybdenum(VI)-calmagite complex onto active carbon," Journal of Colloid and Interface Science, vol. 270, no. 2, pp. 276–280, 2004.
24. G. Absalan and M. A. Mehrdjardi, "Separation and preconcentration of silver ion using 2-mercaptobenzothiazole immobilized on surfactant-coated alumina," Separation and Purification Technology, vol. 33, no. 1, pp. 95–101, 2003.
25. L. Elci, A. A. Kartal, and M. Soylak, "Solid phase extraction method for the determination of iron, lead and chromium by atomic absorption spectrometry using Amberite XAD-2000 column in various water samples," Journal of Hazardous Materials, vol. 153, no. 1-2, pp. 454–461, 2008.
26. V. N. Bulut, A. Gundogdu, C. Duran et al., "A multi-element solid-phase extraction method for trace metals determination in environmental samples on Amberlite XAD-2000," Journal of Hazardous Materials, vol. 146, no. 1-2, pp. 155–163, 2007.
27. E. M. Soliman, M. E. Mahmoud, and S. A. Ahmed, "Synthesis, characterization and structure effects on selectivity properties of silica gel covalently bonded diethylenetriamine mono- and bis-salicyaldehyde and naphthaldehyde schiff's bases towards some heavy metal ions," Talanta, vol. 54, no. 2, pp. 243–253, 2001.
28. N. Soltanzadeh and A. Morsali, "Metal-organic supramolecular assemblies generated from bismuth(III) bromide and polyimine ligands," Polyhedron, vol. 28, no. 4, pp. 703–710, 2009.
29. G. Mahmoudi and A. Morsali, "Crystal-to-crystal transformation from a weak hydrogen-bonded two-dimensional network structure to a two-dimensional coordination polymer on heating," Crystal Growth and Design, vol. 8, no. 2, pp. 391–394, 2008.
30. G. Mahmoudi and A. Morsali, "Mercury(II) iodide coordination polymers generated from polyimine ligands," Polyhedron, vol. 27, no. 3, pp. 1070–1078, 2008.
31. E. Melek, M. Tuzen, and M. Soylak, "Flame atomic absorption spectrometric determinationof cadmium(II) and lead(II) after their solid phase extraction as dibenzyldithiocarbamate chelates on Dowex Optipore V-493," Analytica Chimica Acta, vol. 578, no. 2, pp. 213–219, 2006.
32. L. A. Saghatforoush, A. Aminkhani, S. Ershad, G. Karimnezhad, S. Ghammamy, and R. Kabiri, "Preparation of zinc (II) and cadmium (II) complexes of the tetradentate Schiff base ligand 2-((E)-(2-(2-(pyridine-2-yl)-ethylthio)ethylimino)methyl)-4- bromophenol (PytBrsalH)," Molecules, vol. 13, no. 4, pp. 804–811, 2008.
33. D. Kara, A. Fisher, and S. J. Hill, "Determination of trace heavy metals in soil and sediments by atomic spectrometry following preconcentration with Schiff bases on Amberlite XAD-4," Journal of Hazardous Materials, vol. 165, no. 1–3, pp. 1165–1169, 2009.

CHAPTER 5

FLAME ATOMIC ABSORPTION SPECTROMETRIC DETERMINATION OF TRACE AMOUNTS OF SILVER AFTER SOLID-PHASE EXTRACTION WITH 2-MERCAPTOBENZOTHIAZOLE IMMOBILIZED ON MICROCRYSTALLINE NAPHTHALENE

Farid Shakerian, Ali Mohammad Haji Shabani, Shayessteh Dadfarnia, and Mahdieh Shabani

Department of Chemistry, Faculty of Science, Yazd University, Yazd 89195-741, Iran

ABSTRACT

A simple and sensitive solid-phase extraction (SPE) procedure combined with flame atomic absorption spectrometry (FAAS) was designed for the extraction and determination of trace amounts of silver. A column of immobilized 2-mercaptobenzothiazole (MBT) on microcrystalline naphthalene was used as the sorbent. Silver was quantitatively retained on the column in the pH range of 0.5–6.0. After extraction, the solid mass consisting of silver complex and naphthalene was dissolved out of the column with 5.0 mL of dimethylformamide, and the analyte was determined by flame atomic absorption spectrometry (FAAS). Under the optimum experimental conditions, the adsorption capacity was found to be 1.18 mg of silver per gram of the sorbent. A sample volume of 800 mL resulted in a preconcentration factor of 160. The relative standard deviation obtained for ten replicate determinations at a concentration of 0.8 μg L^{-1} was 1.4%, and the limit of detection was 0.02 μg L^{-1}. The method was successfully applied to the determination of silver in radiology film, waste water, and natural water samples. The accuracy was examined by recovery experiments, independent analysis by electrothermal atomic absorption spectrometry, and analysis of two certified reference materials.

INTRODUCTION

Silver and its compounds have an essential role in electronic and electrical applications, photographic film production, dental and pharmaceutical preparations, and manufacturing of fungicides [1, 2]. Silver is also used for the preparation of high-strength and corrosion-resistance alloys and jewelry. These widespread applications have resulted in an increased silver content of the environmental samples. Silver can also enter the environment via industrial waters because it is often an impurity in zinc, copper, antimony, and arsenic ores [3]. On the other hand, silver has been recognized as a toxic element to many aquatic organisms even at low concentrations [4]. Moreover, Ag accumulation in the human body may lead to a permanent blue-gray skin discoloration (known as Argyria), and the threshold values regulating the maximum amount of silver

Citation: Farid Shakerian, Ali Mohammad Haji Shabani, Shayessteh Dadfarnia, and Mahdieh Shabani, "Flame Atomic Absorption Spectrometric Determination of Trace Amounts of Silver after Solid-Phase Extraction with 2-Mercaptobenzothiazole Immobilized on Microcrystalline Naphthalene," Journal of Chemistry, vol. 2013, Article ID 465825, 6 pages, 2013. doi:10.1155/2013/465825

permitted in different kinds of samples have been established in many countries [5]. Therefore, its trace determination in water and environmental samples is important.

Flame atomic absorption spectrometry is often accepted as a suitable instrumental technique for the measurement of silver because of its speed and ease of operation. However, it has the major drawback of low sensitivity for direct determination of silver in complex matrices [6]. Therefore, a separation/preconcentration step is often required prior to the ultra-trace determination of silver. The most widely used methods for the separation and preconcentration of the trace amounts of silver ions are solid-phase extraction (SPE) [7, 8], solvent extraction [9], cloud-point extraction (CPE) [4], and dispersive liquid-liquid microextraction [10]. Among these methods, solid-phase extraction has gained rapid acceptance due to its simplicity, flexibility, high enrichment factor, and short analysis time. Appropriate organic ligands can be chemically or physically supported on various substrates, thus providing complexing or chelating sorbents for the separation and preconcentration of metal ions. Naphthalene is known as an excellent extractant at high temperature, but the procedure is time consuming and cannot be applied to thermally unstable complexes [11]. The extraction of metal chelates with microcrystalline naphthalene by adding an acetone solution of naphthalene is more rapid [12]. In recent years, column extraction of metal ions with modified microcrystalline naphthalene has attracted many researchers [13, 14].

2-Mercaptobenzothiazole (MBT) with sulfur and nitrogen donor atoms is known to form stable complexes with various metal ions. MBT is insoluble in water but soluble in many organic solvents. Thus, it can be applied to the separation and preconcentration of trace amounts of metals present in aqueous solutions. MBT has been used as the chelating agent for the preconcentration and spectrometric determination of Ag(I) [15–17], Cu(II) [18], Hg(II) [19, 20], Au(III), Pt(IV), and Pd(II) [21]. It has also been used in the construction of ion-selective and modified electrodes [22, 23].

In this study, a simple and efficient method for the selective separation and preconcentration of silver from a large volume of aqueous solution using a column of immobilized 2-mercaptobenzothiazole on microcrystalline naphthalene is described. The solid mass consisting of the silver complex and microcrystalline naphthalene is easily dissolved out of the column with a small volume of dimethylformamide (DMF), and the analyte is determined by flame atomic absorption spectrometry.

EXPERIMENTAL

Reagents

All of the chemicals were of the highest purity available from the Merck Company (Darmstadt, Germany) and were used without any further purification. Double distilled water was used throughout the experiments. A stock solution of 1000 mg L^{-1} of silver ion was prepared by dissolving an appropriate amount of $AgNO_3$ in water. Working solutions were prepared daily from the stock solution by appropriate dilution with distilled water.

Apparatus

An Analytikjena novAA 300 (model 330, Germany) atomic absorption spectrometer furnished with a silver hollow-cathode lamp and air-acetylene flame was used for all of the absorption measurements. The hollow-cathode lamp current was 4.0 mA, the wavelength was set at 328.1 nm, and the slit width was 1.2 nm. The pH measurements were carried out by a Metrohm pH meter (model 691, Switzerland) using a combined glass-calomel electrode.

Preparation of the Sorbent

MBT (0.4 g) and naphthalene (20 g) were dissolved in a hundred milliliters of acetone and then mixed with a magnetic stirrer at 35°C for 5 minutes. The mixture was then slowly added to 1000 mL of double distilled water at room temperature. The mixture was stirred for about 1 hour and was left aside for 120 minutes. Then, it was filtered through a sintered-glass funnel using the vacuum pump, and the residue was washed with distilled water for several times. Finally, the prepared sorbent was air dried and kept in a closed brown bottle for the subsequent use. The color of the prepared sorbent was yellow and it was stable at least for two months.

Procedure

The pH of an aliquot of the sample or standard solution containing 0.2–20.0 µg of silver was adjusted at ~1 with an appropriate amount of nitric acid. The solution was passed through a glass column packed with the MBT immobilized on the microcrystalline naphthalene (20 mm × 10 mm i.d.) at a flow rate of 10.0 mL $min^{-1}$with the aid of a suction pump. The column was washed with a small amount of water, and the sorbent was pushed down with a flat glass rod to remove the excess of water attached to naphthalene. Finally, the solid mass consisting of the metal complex and naphthalene was dissolved with 5.0 mL of DMF, and the silver concentration in the resulting solution was determined by flame atomic absorption spectrometer.

Preparation of Water Samples

Water samples were filtered through 0.45 µm membrane Millipore filter. The pH was adjusted to ~1.0 with nitric-acid solution, and the analyte was determined according to the given procedure.

Preparation of Radiology Film

The proper amount of radiology film was washed with distilled water and dry ashed in a muffle furnace at 550°C for 60 min. The residue was treated with 10 mL of nitric-acid solution (6 mol L^{-1}) and evaporated to dryness at a low heating rate. The residue was dissolved in 80 mL of distilled water, filtered, and adjusted to pH 1.0 using nitric-acid solution. The clear solution was transferred into a 100 mL volumetric flask and diluted to mark with distilled water.

Certified Reference Materials

To proper amount of CPB-1 (Composition: Pb = 64.74 ± 0.12%, S = 17.8 ± 0.2%, Fe = 8.48 ± 0.06%, Zn = 4.42 ± 0.04%, SiO_2 = 0.74 ± 0.04%, Sb = 0.36 ± 0.03%, Al_2O_3 = 0.28 ± 0.02%,

Cu=0.254 ± 0.004%, As=0.056 ± 0.004%, Mn=0.039 ± 0.002%, Bi=0.023 ± 0.002%, Sn=0.019 ± 0.005%, Cd=0.0143 ± 0.002%, Ag=626 ± 6 µg g^{-1}, Se=30 ± 3 µg g^{-1}, and Hg=5.5 ± 0.5 µg g^{-1}) or BCR No. 288 (Composition: Ag=30.5 ± 0.5 µg g^{-1}, As=55.7 ± 1.6 µg g^{-1}, Bi=215.8 ± 2.4 µg g^{-1}, Cd=33.3 ± 0.9 µg g^{-1}, Cu=19.3 ± 0.4 µg g^{-1}, Ni=4.57 ± 0.11 µg g^{-1}, Sb=32.5 ± 0.9 µg g^{-1}, Se<0.2 µg g^{-1}, Sn=30.6 ± 1.5 µg g^{-1}, Te=32.8 ± 1.3 µg g^{-1}, Tl=2.3 ± 0.1 µg g^{-1}, and Zn=8.2 ± 0.4 µg g^{-1}), 5 mL of concentrated nitric acid was added and the solution was heated. Then, 3 mL of hydrogen peroxide was added, and the mixture was heated near to dryness. The solution was diluted with distilled water and was filtered. The pH was adjusted to ~1, and the solution was diluted to 100 mL in a conical flask.

RESULTS AND DISCUSSION

The aim of this study was to develop a sensitive method for the separation and preconcentration of trace amounts of silver ions from an aqueous solution based on the selective adsorption of silver on a column packed with MBT immobilized on the microcrystalline naphthalene. MBT forms a strong water-insoluble complex with silver ions and permits the separation of silver at low pH. In order to obtain the best conditions for the extraction of silver, the procedure was optimized by the univariable method.

Influence of pH

The effect of the sample pH on the retention of silver was studied by varying the pH within the range of 0.5–9.0. The results showed (Figure 1) that the recovery of silver was maximized in the pH range of 0.5–6.0. The decrease in the extraction efficiency at pH>6.0 is probably due to the precipitation of silver as its hydroxide. Thus, in order to achieve the maximum efficiency and selectivity of the silver extraction, a pH of ~1.0 was selected for the subsequent studies.

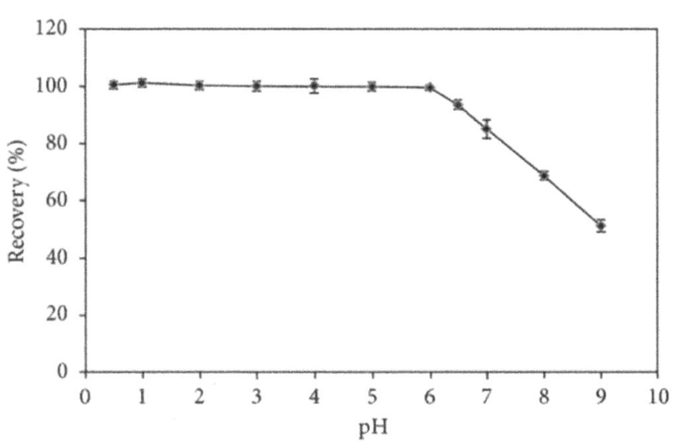

Figure 1: Effect of pH on the recovery of silver. Conditions: amount of silver 10 µg; sample volume 50 mL; sample flow rate 5.0 mL min^{-1}; solvent for dissolving adsorbent DMF (5.0 mL).

Choice of Solvent

The choice of an appropriate solvent for dissolving the Ag(I)-MBT along with naphthalene is an important factor. The solvent should completely dissolve the content of the column and should not interfere with the detection system; thus, with FAAS determination it must efficiently burn during the analysis of the sample. Various solvents were tested for dissolving the complex of Ag(I)-MBT immobilized on naphthalene. The solid material was insoluble in toluene, n-hexane, methyl isobutyl ketone, dioxane, and chloroform; however, it was readily dissolved in acetone, acetonitrile, and dimethylformamide (DMF). DMF was chosen because of its high capability in dissolving the sorbent, high stability, and compatibility with the FAAS. Furthermore, 5.0 mL of this solvent was found to be sufficient for complete dissolution of the solid mass.

Effect of Sample Flow Rate

Another important factor influencing the extraction efficiency and the speed of the analysis is the sample flow rate. In order to have a good precision, as well as sensitivity and speed, it is necessary to select a flow rate that guarantees the achievement of the equilibrium between the sample and the sorbent. The effect of the flow rate on the extraction recovery of silver was examined by varying the flow rate from 0.5 to 25.0 mL min^{-1} under constant experimental conditions. The outcome showed that (Figure 2) the extraction was relatively fast, and up to the sample flow rate of 12.0 mL min^{-1} the uptake of silver was constant and independent of the flow rate. Therefore, a sample flow rate of 10.0 mL min^{-1} was selected in further studies.

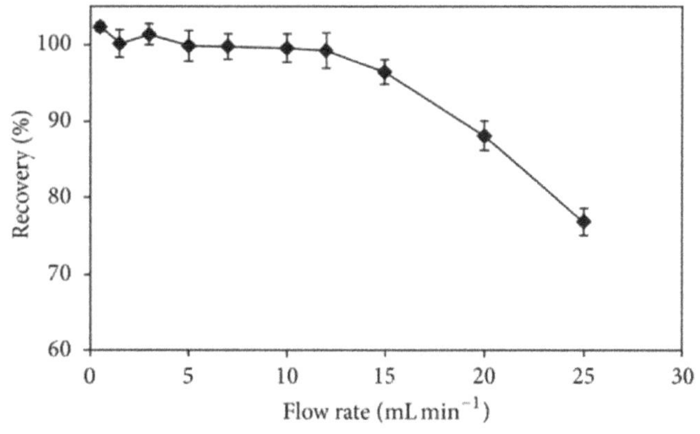

Figure 2: Effect of sample flow rate on the recovery of silver. Conditions: amount of silver 10 µg; sample volume 50 mL; pH ~ 1.0; solvent for dissolving adsorbent DMF (5 mL).

Effect of Sample Volume

In order to explore the possibility of enriching the trace amount of silver from the large sample volume, different sample volume (50–1000 mL) containing 10 µg of silver was passed through the column. The retained analyte was then eluted with 5.0 mL of DMF, and the concentration of silver was determined. The results (Figure 3) showed that, up to an aqueous phase of 800 mL, the recovery was quantitative (≥95%). Thus, the method has the capability of achieving a high preconcentration factor for silver ions.

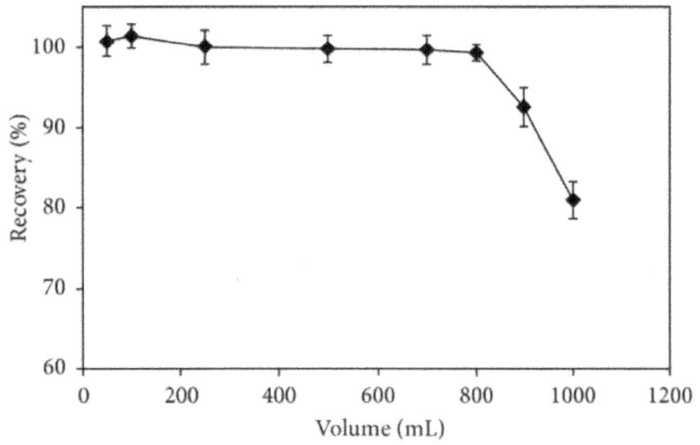

Figure 3: The influence of sample volume on the recovery of silver. Conditions: amount of silver 10 µg; sample flow rate 10.0 mL min^{-1}; pH ~ 1.0; solvent for dissolving adsorbent DMF (5.0 mL).

Interferences Study

A possible concern was that whether the method can be employed for the determination of silver in real samples where other cations or anions might compete with the analyte and lower the extraction efficiency. For this reason, the effect of various ions on the recovery of 5 µg of silver from 100 mL of aqueous sample solution at an initial mole ratio of 1000 (ion/silver) was studied. When the interference was observed, the concentration of the interfering ion was lowered. A relative error of less than 5% was considered to be within the range of experimental error. The results of these studies (Table 1) indicated that the presence of high concentrations of the possible interfering ions in the sample had no significant effect on the recovery of silver at trace levels. Furthermore, the effect of common ions present in the matrix of investigated samples such as Na^+, K^+, Ca^{2+}, Mg^{2+}, and Pb^{2+}at higher mole ratio (10000) was also considered, and no interference was observed. Thus, the method offers a high selectivity for silver ions.

Table 1: Effect of foreign ions on the recovery of 5 µg of silver ions from 100 mL of aqueous solution

Foreign ion	Mole ratio (M^{n+}/Ag^+)	Recovery (%)
K^+	1000	102.1 ± 3.1
Na^+	1000	99.9 ± 1.5
Mg^{2+}	1000	100.3 ± 1.4
Ca^{2+}	1000	99.7 ± 3.5
Ba^{2+}	1000	99.2 ± 1.9
Zn^{2+}	1000	99.4 ± 2.6
Co^{2+}	1000	102.8 ± 1.4
Ni^{2+}	1000	99.5 ± 1.0
Pb^{2+}	1000	98.1 ± 3.3
Cu^{2+}	1000	97.7 ± 2.1
NO_3^-	1000	97.6 ± 2.8
CH_3COO^-	1000	99.6 ± 3.0
SO_4^{2-}	500	99.1 ± 2.1
Cr^{3+}	500	99.0 ± 2.0
Cd^{2+}	500	99.3 ± 2.9
Fe^{3+}	500	104.3 ± 1.8
Cl^-	50	99.2 ± 1.0
Hg^{2+}	30	97.3 ± 1.0

Sorbent Capacity

The capacity of the sorbent for the retention of silver was determined. For this purpose, under the optimum pH, 1.0 g of the sorbent was added to the 100 mL of the solution containing 1500 µg of silver, and it was mixed for 30 minutes. The sorbent was then separated and the concentration of silver remaining in solution was determined by FAAS. The capacity of the sorbent for silver was determined from the differences in the amount of the analyte in the initial and final solutions. The capacity of the sorbent for silver was found to be 1.18 mg g^{-1} of the sorbent.

Analytical Performance

Different concentrations of silver solution (800 mL) were processed according to the procedure, and it was found that the calibration graph exhibited linearity over the range of 0.15–25 µg L^{-1} of silver with a correlation coefficient of 0.9995. The equation of the calibration curve was A=0.0205C+0.0009 (where A is the absorbance and C is the concentration of silver in µg L^{-1}). The preconcentration factor defined as the ratio of the sample volume to the eluent volume was 160. The relative standard deviation (RSD) for ten replicate measurements of 0.8 µg L^{-1} of silver was 1.4%. The limit of detection defined as $3s_b/m$ (where s_b is the standard deviation of the blank and m is the slope of the calibration curve) was 0.02 µg L^{-1}.

Application

The procedure was applied to the determination of silver ions in rain water, the Damavand spring water, well water, river water (taken from Karaj rood, Karaj, Iran), waste water, and the radiology film sample. The reliability was checked by the recovery experiments, and the comparison of the results with the data was obtained by means of electrothermal atomic absorption spectrometry. The results of this investigation are indicated in Table 2. It can be seen that the recovery of the spiked sample is good, and at 95% confidence level there is no significant difference between the results of this study and the data obtained by electrothermal atomic absorption spectrometry. Furthermore, the proposed procedure was applied to the determination of silver in two certified reference materials, CPB-1 and BCR No. 288 with the silver concentration of 626.0 ± 6.0 µg g^{-1} and 30.5 ± 0.5 µg g^{-1}, respectively. The concentrations of silver in CPB-1 and BCR No. 288 were found to be 618.0 ± 2.5 and 30.2 ± 0.8 µg g^{-1}, respectively, which are in good agreement with the accepted values. Thus, the method is suitable for the determination of silver in a wide range of samples.

Table 2: Determination of silver in water samples and radiology film

Samples	Added/µg L^{-1}	Found[a]/µg L^{-1}	Recovery (%)	GFAAS[a]/µg L^{-1}
Well water	0	1.71 ± 0.11	—	1.68 ± 0.06
	10	11.50 ± 0.36	97.9	—
Rain water	0	1.86 ± 0.15	—	1.91 ± 0.08
	10	11.78 ± 0.23	99.2	—
River water	0	4.21 ± 0.12	—	4.16 ± 0.19
	10	14.15 ± 0.18	99.4	—
Spring water	0	4.28 ± 0.10	—	4.47 ± 0.16
	10	14.23 ± 0.27	99.5	—
Waste water	0	10.25 ± 0.35	—	10.16 ± 0.45
	10	20.06 ± 0.54	98.1	—
Radiology film[b]	0	2.11 ± 0.16	—	2.18 ± 0.23
	1	3.07 ± 0.35	96.0	—

[a]Mean and standard deviation of three independent determinations; [b]mg g^{-1}.

Comparison of the Method with Other SPE Methods

The figures of merit of the proposed method and some other SPE methods combined with flame atomic absorption spectrometric for the determination of silver are summarized in Table 3. The proposed method in comparison with other methods showed some advantages such as a higher preconcentration factor and a lower detection limit.

Table 3: A comparison between the figures of merit of the proposed method with similar SPE methods for Ag determination by FAAS

Sorbent material	Reagent	pH	PF	DL/μg L^{-1}	Reference
Polystyrene-divinylbenzene	6-Mercaptopurine	5.5–6.0	—	29	[24]
Silica gel	2,4,6-Trimorpholino-1,2,3-triazin	3.0–6.0	130	—	[25]
Silica gel	MBT	AM	—	0.66	[15]
Alumina	MBT	1.0–6.0	100	—	[16]
Alumina	Dithizone	1.0–9.0	100	—	[26]
Alumina	DDTC	1.0–7.0	125	0.7	[7]
Multiwalled carbon nanotubes	—	7.0–9.0	50	0.6	[27]
Moringa oleifera seeds	—	6.0–8.0	35	0.22	[28]
Naphthalene	MBT	0.5–6.0	160	0.02	This work

PF: preconcentration factor; DL: detection limit; Pb-DDTC: lead diethyl dithiocarbamate; MBT: 2-mercaptobenzothiazole; DDTC: diethyl dithiocarbamate; AM: acidic medium (0.05–6 mol L^{-1} nitric acid).

CONCLUSION

The immobilized MBT on microcrystalline naphthalene is an effective sorbent for the separation and preconcentration of the trace amounts of silver ions from the aqueous solutions. The sorbent can selectively adsorb Ag(I) from the solution at low pH. So, the other heavy metals do not significantly interfere. The main advantages of the proposed method are the following: the ease of preparation of the sorbent, the high preconcentration factor (160), and the low detection limit (0.02 µg L^{-1}) which is suitable for the determination of silver in different real samples.

REFERENCES

1. M. Grayson, Kirk-Othmer Encyclopedia of Chemical Technology, vol. 21, John Wiley & Sons, New York, NY, USA, 3rd edition, 1980.
2. I. C. Smith and B. L. Carson, Trace Metals in the Environment, vol. 2, Ann Arbor Science Publishers, Ann Arbor, Mich, USA, 1977.
3. R. Soager, Metallic Raw Materials Dictionary, Bank Tobel, Zurich, Switzerland, 1984.
4. J. L. Manzoori, H. Abdolmohammad-Zadeh, and M. Amjadi, "Ultra-trace determination of silver in water samples by electrothermal atomic absorption spectrometry after preconcentration with a ligand-less cloud point extraction methodology," Journal of Hazardous Materials, vol. 144, no. 1-2, pp. 458–463, 2007.
5. M. Resano, M. Aramendía, E. García-Ruiz, C. Crespo, and M. A. Belarra, "Solid sampling-graphite furnace atomic absorption spectrometry for the direct determination of silver at trace and ultratrace levels," Analytica Chimica Acta, vol. 571, no. 1, pp. 142–149, 2006.
6. C. K. Christou and A. N. Anthemidis, "Flow injection on-line displacement/solid phase extraction system coupled with flame atomic absorption spectrometry for selective trace silver determination in water samples," Talanta, vol. 78, no. 1, pp. 144–149, 2009.
7. S. Dadfarnia, A. M. Haji Shabani, and M. Gohari, "Trace enrichment and determination of silver by immobilized DDTC microcolumn and flow injection atomic absorption spectrometry," Talanta, vol. 64, no. 3, pp. 682–687, 2004.
8. A. A. Hill, R. J. Lipert, and M. D. Porter, "Determination of colloidal and dissolved silver in water samples using colorimetric solid-phase extraction,"Talanta, vol. 80, no. 5, pp. 1606–1610, 2010.
9. T. Koh and T. Sugimoto, "Extractive spectrophotometric determination of silver(I) at the 10^{-7} M level using 1,10-phenanthroline and tetrabromophenolphthalein ethyl ester," Analytica Chimica Acta, vol. 333, no. 1-2, pp. 167–173, 1996.
10. S. Jafarvand, A. Bidari, P. Hemmatkhah, M. R. M. Hosseini, and Y. Assadi, "Dispersive liquid-liquid microextraction of silver prior to determination by microsample introduction-flame atomic absorption spectrometry," Analytical Letters, vol. 42, no. 14, pp. 2214–2231, 2009.
11. A. Wasey, R. K. Bansal, B. K. Puri, and A. L. J. Rao, "Spectrophotometric determination of osmium with 1-phenyl-4,4,6-trimethyl-(1H,4H)-2-pyrimidinethiol and extraction into molten naphthalene," Talanta, vol. 31, no. 3, pp. 205–206, 1984.
12. R. K. Dubey, A. Bhalotra, M. K. Gupta, and B. K. Puri, "Differential pulse polarographic determination of rhodium(III) and ruthenium(III) in synthetic samples after preconcentration of their quinolin-8-olate complexes onto microcrystalline naphthalene," Microchemical Journal, vol. 58, no. 1, pp. 117–126, 1998.
13. A. M. H. Shabani, S. Dadfarnia, and N. Nasirizadeh, "Speciation analysis of mercury in water samples by cold vapor atomic absorption spectrometry after preconcentration with dithizone immobilized on microcrystalline naphthalene," Analytical and Bioanalytical Chemistry, vol. 378, no. 5, pp. 1388–1391, 2004.
14. B. Rezaei, S. Meghdadi, and N. Majidi, "Preconcentration of thallium(III) with 2,6-bis(N-phenyl carbamoyl)

pyridine on microcrystalline naphthalene prior to its trace determination in human serum spectrophotometri cally,"Spectrochimica Acta A, vol. 67, no. 1, pp. 92–97, 2007.

15. Q. Pu and Q. Sun, "Application of 2-mercaptobenzothiazole-modified silica gel to on-line preconcentration and separation of silver for its atomic absorption spectrometric determination," Analyst, vol. 123, no. 2, pp. 239–243, 1998.
16. G. Absalan and M. A. Mehrdjardi, "Separation and preconcentration of silver ion using 2-mercaptobenzothiazole immobilized on surfactant-coated alumina," Separation and Purification Technology, vol. 33, no. 1, pp. 95–101, 2003.
17. A. Safavi, N. Iranpoor, and N. Saghir, "Directly silica bonded analytical reagents: synthesis of 2-mercaptobenzothiazole-silica gel and its application as a new sorbent for preconcentration and determination of silver ion using solid-phase extraction method," Separation and Purification Technology, vol. 40, no. 3, pp. 303–308, 2004.
18. Y. S. Choi and H. S. Choi, "Sensitized spectrophotometric determination of trace Cu(II) in cationic surfactant media," Bulletin of the Korean Chemical Society, vol. 24, no. 2, pp. 222–224, 2003.
19. J. Chwastowska, A. Rogowska, E. Sterlińska, and J. Dudek, "Chelating 2-mercaptobenzothiazole loaded resin. Application to the separation of inorganic and alkylmercury species for their atomic absorption spectrometry determination in natural waters," Talanta, vol. 49, no. 4, pp. 837–842, 1999.
20. D. Pérez-Quintanilla, I. del Hierro, M. Fajardo, and I. Sierra, "Preparation of 2-mercaptobenzothiazole-derivatized mesoporous silica and removal of Hg(II) from aqueous solution," Journal of Environmental Monitoring, vol. 8, no. 1, pp. 214–222, 2006.
21. Q. Pu, Z. Su, Z. Hu, X. Chang, and M. Yang, "2-Mercaptobenzothiazole-bonded silica gel as selective adsorbent for preconcentration of gold, platinum and palladium prior to their simultaneous inductively coupled plasma optical emission spectrometric determination," Journal of Analytical Atomic Spectrometry, vol. 13, no. 4, pp. 249–253, 1998.
22. S. M. Golabi and J. Mohammadi, "Wire-coated silver(I) ion-selective electrode based on 2-mercaptobenzothiazole (MBT) ionophore: application to the determination of silver in real samples," Analytical Sciences, vol. 19, no. 6, pp. 877–881, 2003.
23. R. K. Shervedani and M. K. Babadi, "Application of 2-mercaptobenzothiazole self-assembled monolayer on polycrystalline gold electrode as a nanosensor for determination of Ag(I)," Talanta, vol. 69, no. 3, pp. 741–746, 2006.
24. B. C. Mondal, D. Das, and A. K. Das, "Application of a new resin functionalised with 6-mercaptopurine for mercury and silver determination in environmental samples by atomic absorption spectrometry," Analytica Chimica Acta, vol. 450, no. 1-2, pp. 223–230, 2001.
25. T. Madrakian, A. Afkhami, M. A. Zolfigol, and M. Solgi, "Separation, preconcentration and determination of silver ion from water samples using silica gel modified with 2,4,6-trimorpholino-1,3,5-triazin," Journal of Hazardous Materials, vol. 128, no. 1, pp. 67–72, 2006.
26. G. Absalan and A. Aghaei Goudi, "Optimizing the immobilized dithizone on surfactant-coated alumina as a new sorbent for determination of silver,"Separation and Purification Technology, vol. 38, no. 3, pp. 209–214, 2004.
27. Q. Ding, P. Liang, F. Song, and A. Xiang, "Separation and preconcentration of silver ion using multiwalled carbon nanotubes as solid phase extraction sorbent," Separation Science and Technology, vol. 41, no. 12, pp. 2723–2732, 2006.
28. C. S. T. Araújo, V. N. Alves, H. C. Rezende, and N. M. M. Coelho, "Development of a flow system for the determination of low concentrations of silver using Moringa oleifera seeds as biosorbent and flame atomic absorption spectrometry," Microchemical Journal, vol. 96, no. 1, pp. 82–85, 2010.

CHAPTER 6

TOPOLOGICAL INDICES STUDY OF MOLECULAR STRUCTURE IN ANTICANCER DRUGS

Wei Gao[1] Weifan Wang[2] and Mohammad Reza Farahani[3]

[1]School of Information Science and Technology, YunnanNormalUniversity, Kunming 650500, China

[2]Department of Mathematics, Zhejiang Normal University, Jinhua 321004, China

[3]Department of Applied Mathematics, Iran University of Science and Technology, Narmak, Tehran 16844, Iran

ABSTRACT

Numerous studies indicate that there is strong inherent relationship between the chemical characteristics of chemical compounds and drugs (e.g., boiling point and melting point) and their molecular structures. Topological indices defined on these chemical molecular structures can help researchers better understand the physical features, chemical reactivity, and biological activity. Thus, the study of the topological indices on chemical structure of chemical materials and drugs can make up for lack of chemical experiments and can provide a theoretical basis for the manufacturing of drugs and chemical materials. In this paper, we focus on the family of smart polymer which is widely used in anticancer drugs manufacturing. Several topological indices are determined in view of edge dividing methods, and these results remedy the lack of chemical and medicine experiments thus providing the theoretical basis for pharmaceutical engineering.

INTRODUCTION

In this era of rapid technological development, chemical and pharmaceutical techniques in recent years have been rapidly evolved, and thus a large number of new nanomaterials, crystalline materials, and drugs emerge every year. To determine the chemical properties of such a large number of new compounds and new drugs requires a large amount of chemical experiments, thereby greatly increasing the workload of the chemical and pharmaceutical researchers. Fortunately, the chemical based experiments found that there was strong connection between topology molecular structures and their physical behaviors, chemical characteristics, and biological features, such as melting point, boiling point, and toxicity of drugs (see Wiener [1] and Katritzky et al. [2] as examples).

The topological index of a molecule structure can be considered as a nonempirical numerical quantity which quantitates the molecular structure and its branching pattern. In this point of view, topological index can be regarded as a score function which maps each molecular structure to a real number

Citation: Wei Gao, Weifan Wang, and Mohammad Reza Farahani, "Topological Indices Study of Molecular Structure in Anticancer Drugs," Journal of Chemistry, vol. 2016, Article ID 3216327, 8 pages, 2016. doi:10.1155/2016/3216327

and used as a descriptor of the molecule under testing. There are several famous indices applied in chemical engineering (e.g., QSPR/QSAR study) for grasping the relationships between the molecular structure and the potential physicochemical characteristics, such as PI index, Zagreb index, harmonic index, Wiener index, and connectivity index (see Yan et al. [3], Gao and Shi [4], and Gao and Wang [5, 6] for more details).

In theoretical chemistry setting, chemical compounds, materials, and drugs are expressed as (molecular) graphs in which each vertex represents an atom of molecule structure and each edge implies covalent bounds between two atoms. Let $G = ((G), E(G))$ be a (molecular) graph with vertex set $V(G)$ and edge set $E(G)$, respectively. We assume that all the graphs considered in this paper are simple graphs, that is, no loop and multiple edge. The notations and terminologies used but not clearly undefined in this paper can be found in [7].

There are several degree based indices introduced to test the properties of compounds and drugs, which have been widely used in chemical and pharmacy engineering. Bollobas and Erdos [8] introduced the general Randic index stated as

$$R_k(G) = \sum_{e=uv} (d(u)d(v))^k, \tag{1}$$

where k is a real number and d(u) is the degree of vertex u. Li and Liu [9] studied the first three minimum general Randic indices of tree structures, and the corresponding extremal trees are described. Liu and Gutman [10] estimated the general Randic index and its special situations, the ordinary index and modified Zagreb index. In what follows, we always assume that k is a real number.

By setting k=1 and k = -1, formula (1) becomes the second Zagreb index ($M_2(G)$) and the modified second Zagreb index ($M_2^*(G)$) which are stated as

$$M_2(G) = \sum_{e=uv} d(u)d(v), \tag{2}$$

$$M_2^*(G) = \sum_{e=uv} \frac{1}{d(u)d(v)}, \tag{3}$$

respectively.

The sum connectivity index ($\chi(G)$) of molecular graph was defined by

$$\chi(G) = \sum_{e=uv} (d(u)+d(v))^{-1/2}. \tag{4}$$

Few years ago, Zhou and Trinajstic [11] extended this concept and introduced the general sum connectivity as follows:

$$\chi_k(G) = \sum_{e=uv} (d(u)+d(v))^k. \tag{5}$$

Note that Shirdel et al. [12] introduced a new version of Zagreb indices named Hyper-Zagreb index as

$$HM(G) = \sum_{e=uv} (d(u)+d(v))^2. \tag{6}$$

Obviously, Hyper-Zagreb index is just a special case of general sum connectivity when k=2.

The harmonic index for a molecular graph G is defined as

$$H(G) = \sum_{e=uv} \frac{2}{d(u)+d(v)}. \tag{7}$$

Favaron et al. [13] researched the relationship between the eigenvalues of molecular graphs and harmonic index. Zhong [14] obtained the minimum and maximum values of the harmonic index for connected molecular structures and trees, and the corresponding extremal molecular graphs are described. Wu et al. [15] yielded the minimum harmonic index of molecular graphs with $\delta(G) \geq 2$. Liu [16] gave several relations between the harmonic index and diameter of molecular graphs.

Very recently, in order to extend harmonic index for more chemical engineering applications, Yan et al. [3] introduced the general version of harmonic index which was formulated by

$$H_k(G) = \sum_{e=uv} \left(\frac{2}{d(u)+d(v)}\right)^k. \tag{8}$$

Eliasi and Iranmanesh [17] proposed the ordinary geometric-arithmetic index as the extension of geometric-arithmetic index which was stated as

$$OGA_k(G) = \sum_{e=uv} \left(\frac{2\sqrt{d(u)d(v)}}{d(u)+d(v)}\right)^k. \tag{9}$$

Azari and Iranmanesh [18] raised the generalized version of Zagreb index as

$$M_{t_1,t_2}(G) = \sum_{e=uv} \left(d(u)^{t_1} d(v)^{t_2} + d(u)^{t_2} d(v)^{t_1}\right), \tag{10}$$

where parameters t_1 and t_2 are arbitrary nonnegative integers.

Several polynomials closely related to degree based indices are also introduced. For example, the first and the second Zagreb polynomials corresponding to first and the second Zagreb indices are expressed as

$$M_1(G,x) = \sum_{e=uv} x^{d(u)+d(v)}, \tag{11}$$

$$M_2(G,x) = \sum_{e=uv} x^{d(u)d(v)}, \tag{12}$$

respectively.

Additionally, the third Zagreb index and third Zagreb polynomial were defined as

$$M_3(G) = \sum_{e=uv} |d(u)-d(v)|,$$

$$M_3(G,x) = \sum_{e=uv} x^{|d(u)-d(v)|}. \tag{13}$$

As degree based topological indices, the multiplicative version of these Zagreb indices (the first multiplicative Zagreb index $PM_1(G)$ and the second multiplicative Zagreb index $PM_2(G)$) of a (molecular) graph were introduced by Gutman [19] and Ghorbani and Azimi [20] as

$$PM_1(G) = \prod_{e=uv\in E(G)} (d(u) + d(v)),$$
$$PM_2(G) = \prod_{e=uv\in E(G)} (d(u)d(v)). \quad (14)$$

Several advances on $PM_1(G)$ and $PM_2(G)$ can be referred to Eliasi et al. [21], Xu and Das [22], and Farahani [23].

Furthermore, the redefined version of Zagreb indices of a (molecular) graph G was introduced by Ranjini et al. [24] as follows:

$$ReZG_1(G) = \sum_{e=uv\in E(G)} \frac{d(u) + d(v)}{d(u)d(v)},$$
$$ReZG_2(G) = \sum_{e=uv\in E(G)} \frac{d(u)d(v)}{d(u) + d(v)},$$
$$ReZG_3(G) = \sum_{e=uv\in E(G)} (d(u)d(v))(d(u) + d(v)). \quad (15)$$

Here, $ZG_1(G)$, $ZG_2(G)$,, and $ZG_3(G)$ are first, second, and third redefined Zagreb indices.

In the following contents, we first introduce the smart polymers of Dox-loaded micelle comprising PEG-PAsp block copolymer with chemically conjugated Dox and explain the importance of this molecular structure. This is the reason why we focus on the topological indices computation on this family of molecular graphs. Then, in Section 3, we present our main results and detailed proofs.

MOTIVATION

As a special class of macromolecules, smart polymers manifest impressive response to physiochemical change if their circumstance has slight interference, for example, changes on PH value, ionic disturbance, magnetic field, light, and temperature (see Hai and Broekmann [25], do Nascimento Marques et al. [26], and Kroning et al. [27]). Thus, smart polymers are also denoted by environmentally responsive systems or stimuli responsive ones. As good delivery systems, these structures have wide applications in biomedical field, for instance, smart polymers with nucleic acid or protein delivery to intracellular targets just like nucleus or ribosome in tissue engineering (see Chonkar et al. [28], Hrubý et al. [29], Duro-Castano et al. [30], and Khandare and Calderón [31]). As a special class of smart polymer, polymeric micelles (e.g., Dox-conjugated PEG-b-poly(aspartate) (PEG-PAsp) block copolymers; see Shanthi et al. [32] and Osada et al. [33]) are widely applied in delivering anticancer drug.

The Dox-loaded micelle containing PEG-PAsp block copolymer with chemically conjugated Dox (see Figure 1 for its detailed structure) is a famous family of smart polymer which is used as anthracycline anticancer antibiotic and applied in treating various kinds of cancers. It is employed as excipients and drug delivery carriers for strengthening the stability and the times of drug retention. Thus, it has powerful anticancer activity and is widely considered in the pharmacy field (see Nishiyama and Kataoka [34] and Butt et al. [35] for more details).

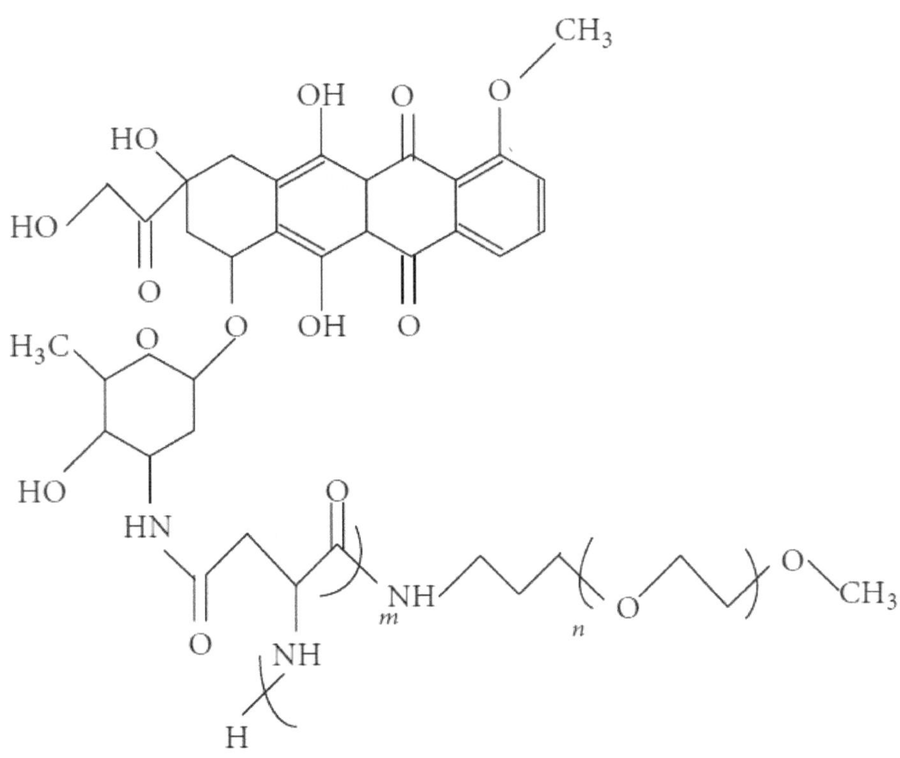

Figure 1: Dox-loaded micelle comprising PEG-PAsp block copolymer with chemically conjugated Dox $S[n]$.

As can be seen in Figure 1, the integer number n is step of growth in this kind of polymers. We depict the structure of Dox-loaded micelle containing PEG-PAsp block copolymer with chemically conjugated Dox when n = 1, 2, and 3 (see Figures 2, 3, and 4, resp.).

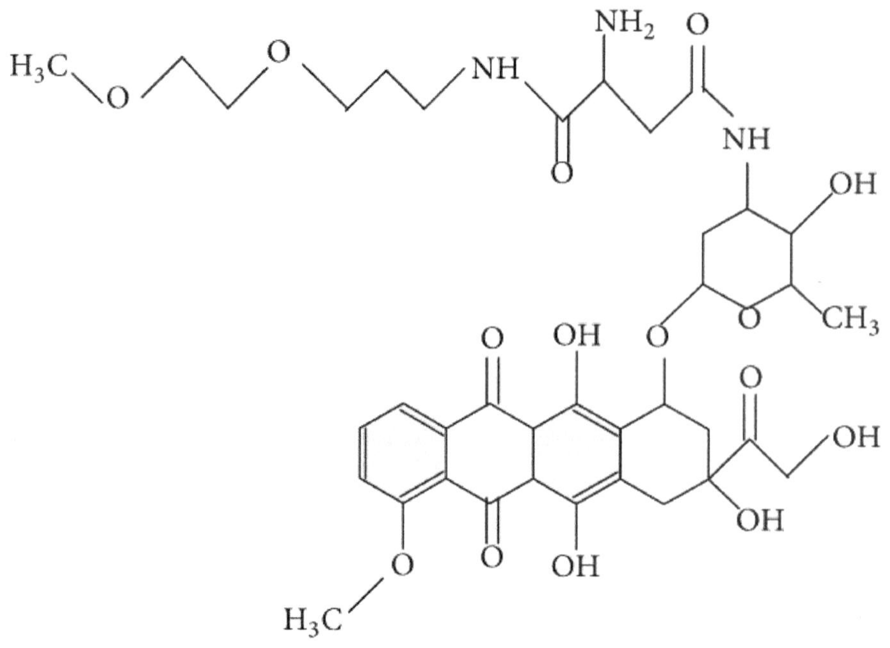

Figure 2: The molecular structure of *S*[1].

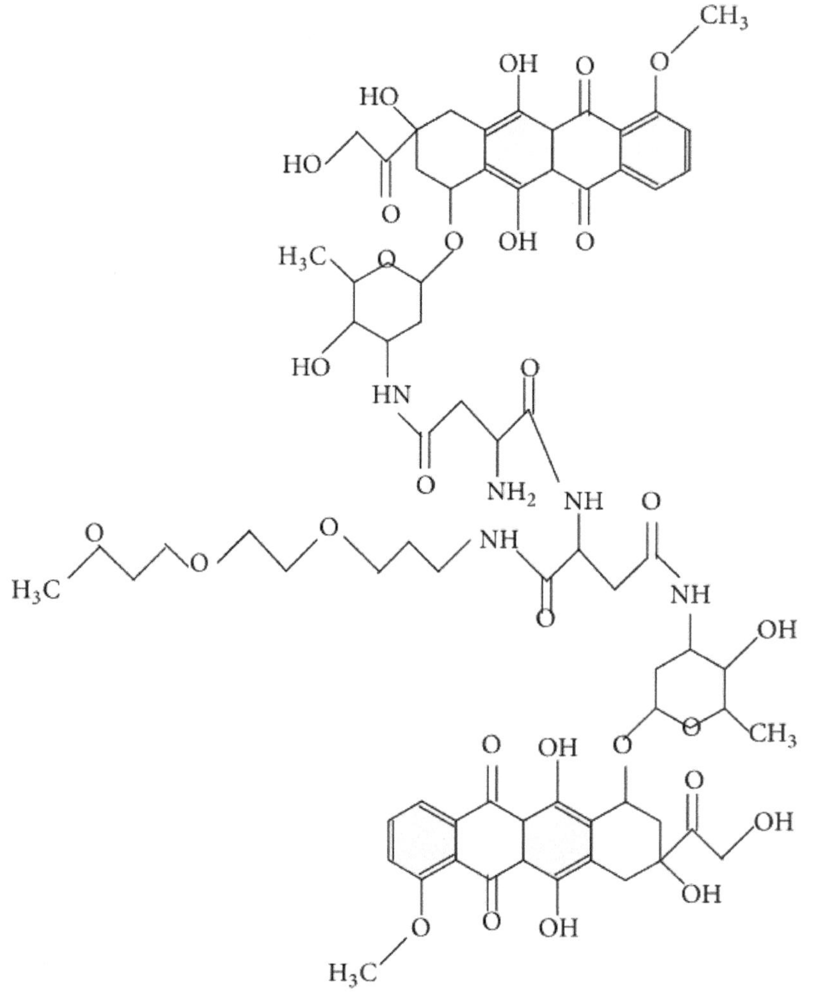

Figure 3: The molecular structure of *S*[2].

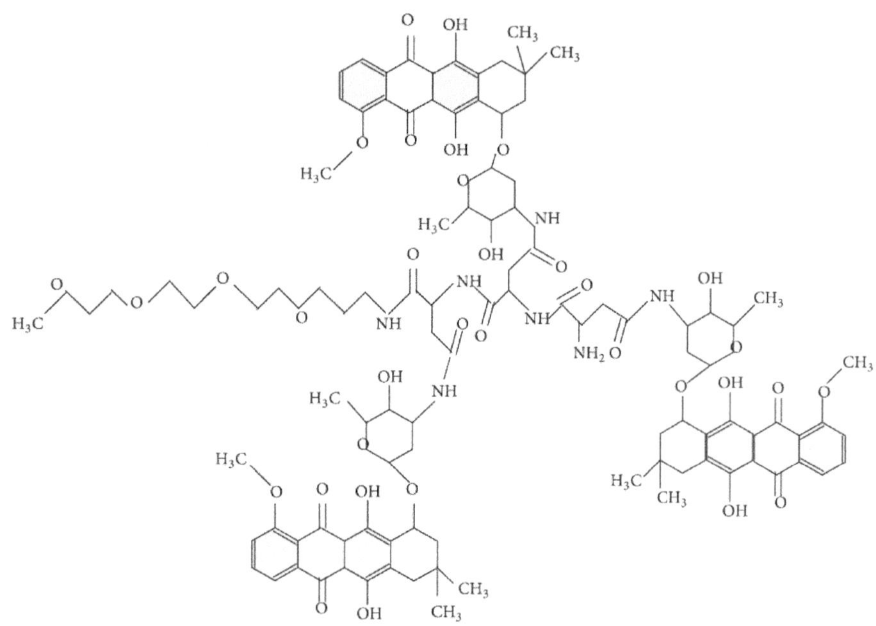

Figure 4: The molecular structure of $S[3]$.

Although several advances have been made in PI index, Zagreb index, Wiener index, hyper-Wiener index, and sum connectivity index of different kinds of molecular graphs, the study of topological indices for Dox-loaded micelle comprising PEG-PAsp block copolymer with chemically conjugated Dox has been largely limited. In addition, this kind of smart polymer structures is widely used in medical science and pharmaceutical field. For example, it plays a key role in delivery system and in the development of anticancer drugs. Based on these reasons, industrial interest and tremendous academic interest have been attracted to research the topological indices of this molecular structure from a mathematical point of view.

MAIN RESULTS AND PROOFS

In this section, we present the main results of this paper. The techniques we used here to get the main conclusion are followed by the tricks of edge dividing.

Theorem 1. Let $S[n]$ be the Dox-loaded micelle comprising PEG-PAsp block copolymer with chemically conjugated Dox. One has

$$\chi_k(SP[n]) = (2n+1)\,3^k + (14n+2)\,4^k + (19n-1)\,5^k + 18n\cdot 6^k + n\cdot 7^k,$$

$$R_k(SP[n]) = (2n+1)\,2^k + (9n+1)\,3^k + (6n+4)\,4^k + (18n-1)\,6^k + 2n\cdot 8^k + 16n\cdot 9^k + n\cdot 12^k. \quad (16)$$

Proof. Let δ and Δ be the minimum degree and maximum degree of $S[n]$, respectively. In what follows, we assume that the edge set $(SP[n])$ can be divided into several partitions:

(i) for any $i, 2\delta(G) \le i \le 2\Delta(G)$, let $E_i = \{e = uv \in E(G) \mid d(u) + d(v) = i\}$;

(ii)for any $j, (\delta)^2 \le j \le (\Delta)^2$, let $E_j^* = \{e = uv \in E(G) \mid d(u)d(v) = j\}$.

Specifically, by observing and computing, the edge set of $S[n]$ can be divided into the following edge subsets:

(i) E_3 (or E_2^*): $d(u) = 1$ and $d(v) = 2$;

(ii) E_3^*: $d(u) = 1$ and $d(v) = 3$;

(iii) $E_5 \cap E_4^*$: $d(u) = 1$ and $d(v) = 4$;

(iv) $E_4 \cap E_4^*$: $d(u) = 2$ and $d(v) = 2$;

(v) E_6^*: $d(u) = 2$ and $d(v) = 3$;

(vi) E_8^*: $d(u) = 2$ and $d(v) = 4$;

(vii) E_9^*: $d(u) = d(v) = 3$;

(viii) E_7 (or E_{12}^*): $d(u) = 3$ and $d(v) = 4$.

In terms of further calculating, we check that $|V(SP[n])| = 49n + 6$ and $|E(SP[n])| = 54n + 5$. More specifically, we deduce $|E_3| = |E_2^*| = 2n + 1$, $|E_3^*| = 9n + 1$, $|E_5 \cap E_4^*| = |E_7| = |E_{12}^*| = n$, $|E_4 \cap E_4^*| = 5n+4$, $|E_6^*| = 18n-1$, $|E_8^*| = 2n$, and $|E_9^*| = 16n$.

Therefore, by the definition of general sum connectivity and general Randic index, we get

$$\begin{aligned}\chi_k(SP[n]) &= \sum_{e=uv}(d(u)+d(v))^k \\ &= \sum_{e=uv\in E_3}(d(u)+d(v))^k + \sum_{e=uv\in E_3^*}(d(u)+d(v))^k + \sum_{e=uv\in E_5\cap E_4^*}(d(u)+d(v))^k \\ &\quad + \sum_{e=uv\in E_4\cap E_4^*}(d(u)+d(v))^k + \sum_{e=uv\in E_6^*}(d(u)+d(v))^k + \sum_{e=uv\in E_8^*}(d(u)+d(v))^k \\ &\quad + \sum_{e=uv\in E_9^*}(d(u)+d(v))^k + \sum_{e=uv\in E_7}(d(u)+d(v))^k \\ &= (2n+1)3^k + (9n+1)4^k + n\cdot 5^k + (5n+4)4^k + (18n-1)5^k + 2n\cdot 6^k + 16n\cdot 6^k \\ &\quad + n\cdot 7^k = (2n+1)3^k + (14n+5)4^k + (19n-1)5^k + 18n\cdot 6^k + n\cdot 7^k, \\ R_k(SP[n]) &= \sum_{e=uv}(d(u)d(v))^k \\ &= \sum_{e=uv\in E_3}(d(u)d(v))^k + \sum_{e=uv\in E_3^*}(d(u)d(v))^k + \sum_{e=uv\in E_5\cap E_4^*}(d(u)d(v))^k \\ &\quad + \sum_{e=uv\in E_4\cap E_4^*}(d(u)d(v))^k + \sum_{e=uv\in E_6^*}(d(u)d(v))^k + \sum_{e=uv\in E_8^*}(d(u)d(v))^k \\ &\quad + \sum_{e=uv\in E_9^*}(d(u)d(v))^k + \sum_{e=uv\in E_7}(d(u)d(v))^k \\ &= (2n+1)2^k + (9n+1)3^k + n\cdot 4^k + (5n+4)4^k + (18n-1)6^k + 2n\cdot 8^k + 16n\cdot 9^k \\ &\quad + n\cdot 12^k \\ &= (2n+1)2^k + (9n+1)3^k + (6n+4)4^k + (18n-1)6^k + 2n\cdot 8^k + 16n\cdot 9^k + n\cdot 12^k.\end{aligned} \tag{17}$$

Hence, we get the desired conclusion.

In what follows, $S[n]$ is denoted as the Dox-loaded micelle comprising PEG-PAsp block copolymer with chemically conjugated Dox, and we will not explain this notation again.

Theorem 2. The general harmonic index of $S[n]$ is

$$\begin{aligned}H_k(SP[n]) &= (2n+1)\left(\frac{2}{3}\right)^k + (14n+5)\left(\frac{1}{2}\right)^k \\ &\quad + (19n-1)\left(\frac{2}{5}\right)^k + 18n\left(\frac{1}{3}\right)^k \\ &\quad + n\left(\frac{2}{7}\right)^k.\end{aligned} \tag{18}$$

Proof. By the definition of general harmonic index and what we have discussed in Theorem 1, we infer

$$\begin{aligned}H_k(SP[n]) &= \sum_{e=uv}\left(\frac{2}{d(u)+d(v)}\right)^k \\ &= \sum_{e=uv\in E_3}\left(\frac{2}{d(u)+d(v)}\right)^k + \sum_{e=uv\in E_3^*}\left(\frac{2}{d(u)+d(v)}\right)^k + \sum_{e=uv\in E_5\cap E_4^*}\left(\frac{2}{d(u)+d(v)}\right)^k \\ &\quad + \sum_{e=uv\in E_4\cap E_4^*}\left(\frac{2}{d(u)+d(v)}\right)^k + \sum_{e=uv\in E_6^*}\left(\frac{2}{d(u)+d(v)}\right)^k \\ &\quad + \sum_{e=uv\in E_8^*}\left(\frac{2}{d(u)+d(v)}\right)^k + \sum_{e=uv\in E_9^*}\left(\frac{2}{d(u)+d(v)}\right)^k + \sum_{e=uv\in E_7}\left(\frac{2}{d(u)+d(v)}\right)^k \\ &= (2n+1)\left(\frac{2}{3}\right)^k + (9n+1)\left(\frac{1}{2}\right)^k + n\left(\frac{2}{5}\right)^k + (5n+4)\left(\frac{1}{2}\right)^k + (18n-1)\left(\frac{2}{5}\right)^k \\ &\quad + 2n\left(\frac{1}{3}\right)^k + 16n\cdot 6^k + n\left(\frac{2}{7}\right)^k \\ &= (2n+1)\left(\frac{2}{3}\right)^k + (14n+5)\left(\frac{1}{2}\right)^k + (19n-1)\left(\frac{2}{5}\right)^k + 18n\left(\frac{1}{3}\right)^k + n\left(\frac{2}{7}\right)^k.\end{aligned} \tag{19}$$

We complete the proof.

By taking k=1 in (18), we yield the harmonic index of $S[n]$.

Corollary 3. Consider the following:

$$H(SP[n]) = \frac{2333}{105}n + \frac{83}{30}. \tag{20}$$

Using similar fashion, we get the following conclusions on the exact expression of several important indices and polynomials of $S[n]$.

Theorem 4. The ordinary geometric-arithmetic index of $S[n]$ is

$$\begin{aligned}\mathrm{OGA}_k(SP[n]) &= (4n+1)\left(\frac{2\sqrt{2}}{3}\right)^k \\ &\quad + (9n+1)\left(\frac{2\sqrt{3}}{4}\right)^k + n\left(\frac{4}{5}\right)^k \\ &\quad + (21n+4) + (18n-1)\left(\frac{2\sqrt{6}}{5}\right)^k \\ &\quad + n\left(\frac{4\sqrt{3}}{7}\right)^k.\end{aligned} \tag{21}$$

Theorem 5. The generalized version of Zagreb index of $S[n]$ is

$$\begin{aligned}M_{t_1,t_2}(SP[n]) &= (2n+1)\left(2^{t_1}+2^{t_2}\right) \\ &\quad + (9n+1)\left(3^{t_1}+3^{t_2}\right) + n\left(4^{t_1}+4^{t_2}\right) \\ &\quad + (5n+4)\,2^{t_1+t_2+1} \\ &\quad + (18n-1)\left(2^{t_1}3^{t_2}+2^{t_2}3^{t_1}\right) \\ &\quad + 2n\left(2^{t_1+2t_2}+2^{t_2+2t_1}\right) + 16n \\ &\quad \cdot 3^{t_1+t_2+1} + n\left(3^{t_1}4^{t_2}+3^{t_2}4^{t_1}\right).\end{aligned} \tag{22}$$

Theorem 6. The first and second Zagreb polynomials of $S[n]$ are

$$\begin{aligned}M_1(SP[n],x) &= (2n+1)x^3 + (14n+5)x^4 \\ &\quad + (19n-1)x^5 + 18nx^6 + nx^7, \\ M_2(SP[n],x) &= (2n+1)x^2 + (9n+1)x^3 \\ &\quad + (6n+4)x^4 + (18n-1)x^6 \\ &\quad + 2nx^8 + 16nx^9 + nx^{12}.\end{aligned} \tag{23}$$

Theorem 7. The third Zagreb index and polynomial of $S[n]$ are

$$\begin{aligned}M_3(SP[n]) &= 46n+2, \\ M_3(SP[n],x) &= 21nx + (11n+1)x^2 + nx^3 \\ &\quad + (21n+4).\end{aligned} \tag{24}$$

Theorem 8. The multiplicative Zagreb indices of $S[n]$ are

$$\begin{aligned}PM_1(SP[n]) &= 3^{2n+1}4^{14n+2}5^{19n-1}6^{18n}7^n, \\ PM_2(SP[n]) &= 2^{2n+1}3^{9n+1}4^{6n+4}6^{18n-1}8^{2n}9^{16n}12^n.\end{aligned} \tag{25}$$

Theorem 9. The redefined Zagreb indices of $S[n]$ are

$$\begin{aligned}\mathrm{ReZG}_1(SP[n]) &= 49n+6, \\ \mathrm{ReZG}_2(SP[n]) &= \frac{8941}{140}n + \frac{253}{60}, \\ \mathrm{ReZG}_3(SP[n]) &= 1804n+52.\end{aligned} \tag{26}$$

CONCLUSION

In this paper, inspired by widely used Dox-loaded micelle comprising PEG-PAsp block copolymer with chemically conjugated Dox for anticancer drug study in pharmaceutical engineering applications, we focus on the theoretical analysis of topological indices expression for this molecular structure. By means of edge dividing approaches, we present the exact expression of several important indices, including general sum connectivity, general Randic index, general harmonic index, ordinary geometric-arithmetic index, generalized version of Zagreb index, Zagreb polynomials, multiplicative Zagreb indices, and redefined Zagreb indices. The results obtained in our paper illustrate the promising application prospects in chemical and pharmacy engineering.

CONFLICT OF INTERESTS

The authors declare that there is no conflict of interests regarding the publication of this paper.

ACKNOWLEDGMENT

The research is partially supported by NSFC (nos. 11401519, 11371328, and 11471293).

REFERENCES

1. H. J. Wiener, "Structural determination of paraffin boiling points," Journal of the American Chemical Society, vol. 69, no. 1, pp. 17–20, 1947.
2. A. R. Katritzky, R. Jain, A. Lomaka, R. Petrukhin, U. Maran, and M. Karelson, "Perspective on the relationship between melting points and chemical structure," Crystal Growth & Design, vol. 1, no. 4, pp. 261–265, 2001.
3. L. Yan, W. Gao, and J. S. Li, "General harmonic index and general sum connectivity index of polyomino chains and nanotubes," Journal of Computational and Theoretical Nanoscience, vol. 12, no. 10, pp. 3940–3944, 2015.
4. W. Gao and L. Shi, "Wiener index of gear fan graph and gear wheel graph,"Asian Journal of Chemistry, vol. 26, no. 11, pp. 3397–3400, 2014.
5. W. Gao and W. Wang, "Second atom-bond connectivity index of special chemical molecular structures," Journal of Chemistry, vol. 2014, Article ID 906254, 8 pages, 2014.
6. W. Gao and W. F. Wang, "The vertex version of weighted wiener number for bicyclic molecular structures," Computational and Mathematical Methods in Medicine, vol. 2015, Article ID 418106, 10 pages, 2015.
7. J. A. Bondy and U. S. R. Murty, Graph Theory, Spring, Berlin, Germany, 2008.
8. B. Bollobas and P. Erdos, "Graphs of extremal weights," Ars Combinatoria, vol. 50, pp. 225–233, 1998.
9. B. Li and W. Liu, "The smallest Randic index for trees," Proceedings—Mathematical Sciences, vol. 123, no. 2, pp. 167–175, 2013.
10. B. L. Liu and I. Gutman, "Estimating the Zagreb and the general randic indices," MATCH Communications in Mathematical and in Computer Chemistry, vol. 57, no. 3, pp. 617–632, 2007.
11. B. Zhou and N. Trinajstic, "On general sum-connectivity index," Journal of Mathematical Chemistry, vol. 47, no. 1, pp. 210–218, 2010.
12. G. H. Shirdel, H. RezaPour, and A. M. Sayadi, "The hyper-Zagreb index of graph operations," Iranian Journal of Mathematical Chemistry, vol. 4, no. 2, pp. 213–220, 2013.
13. O. Favaron, M. Maho, and J.-F. Sacle, "Some eigenvalue properties in graphs (conjectures of Graffiti- II)," Discrete Mathematics, vol. 111, no. 1–3, pp. 197–220, 1993.
14. L. Zhong, "The harmonic index for graphs," Applied Mathematics Letters, vol. 25, no. 3, pp. 561–566, 2012.
15. R. Wu, Z. Tang, and H. Deng, "A lower bound for the harmonic index of a graph with minimum degree at least two," Filomat, vol. 27, no. 1, pp. 51–55, 2013.
16. J. X. Liu, "On harmonic index and diameter of graphs," Journal of Applied Mathematics and Physics, vol. 1, no. 3, pp. 5–6, 2013.
17. M. Eliasi and A. Iranmanesh, "On ordinary generalized geometric-arithmetic index," Applied Mathematics Letters, vol. 24, no. 4, pp. 582–587, 2011.
18. M. Azari and A. Iranmanesh, "Generalized Zagreb index of graphs," Studia Universitatis Babe□-Bolyai, vol. 56, no. 3, pp. 59–70, 2011.
19. I. Gutman, "Multiplicative Zagreb indices of trees," Bulletin of the Veterinary Institute in Pulawy, vol. 1, pp. 13–19, 2011.
20. M. Ghorbani and N. Azimi, "Note on multiple Zagreb indices," Iranian Journal of Mathematical Chemistry, vol. 3, pp. 137–143, 2012.
21. M. Eliasi, A. Iranmanesh, and I. Gutman, "Multiplicative versions of first Zagreb index," MATCH Communications in Mathematical and in Computer Chemistry, vol. 68, no. 1, pp. 217–230, 2012.
22. K. Xu and K. C. Das, "Trees, unicyclic, and bicyclic graphs extremal with respect to multiplicative sum Zagreb index," MATCH Communications in Mathematical and in Computer Chemistry, vol. 68, no. 1, pp. 257–272, 2012.
23. M. R. Farahani, "Multiplicative versions of Zagreb indices of TUSC4C8(S),"Journal of Chemistry and Materials Research, vol. 2, pp. 67–70, 2015.
24. P. S. Ranjini, V. Lokesha, and A. Usha, "Relation between phenylene and hexagonal squeeze using harmonic index," International Journal of Graph Theory, vol. 1, pp. 116–121, 2013.
25. N. T. M. Hai and P. Broekmann, "Smart hybrid polymers for advanced damascene electroplating: combination of superfill and leveling properties,"ChemElectroChem, vol. 2, no. 8, pp. 1096–1099, 2015.
26. N. do Nascimento Marques, A. M. da Silva Maia, and R. de Carvalho Balaban, "Development of dual-sensitive smart polymers by grafting chitosan with poly (N-isopropylacrylamide): an overview," Polímeros, vol. 25, no. 3, pp. 237–246, 2015.
27. A. Kroning, A. Furchner, D. Aulich et al., "In situ infrared ellipsometry for protein adsorption studies on ultrathin smart polymer brushes in aqueous environment," ACS Applied Materials & Interfaces, vol. 7, no. 23, pp. 12430–12439, 2015.

28. A. Chonkar, U. Nayak, and N. Udupa, "Smart polymers in nasal drug delivery," Indian Journal of Pharmaceutical Sciences, vol. 77, no. 4, pp. 367–375, 2015.

29. M. Hrubý, S. K. Filippov, and P. Štěpánek, "Smart polymers in drug delivery systems on crossroads: which way deserves following," European Polymer Journal, vol. 65, pp. 82–97, 2015.

30. A. Duro-Castano, J. Movellan, and M. J. Vicent, "Smart branched polymer drug conjugates as nano-sized drug delivery systems," Biomaterials Science, vol. 3, no. 10, pp. 1321–1334, 2015.

31. J. Khandare and M. Calderón, "Dendritic polymers for smart drug delivery applications," Nanoscale, vol. 7, no. 9, pp. 3806–3807, 2015.

32. K. Shanthi, K. Vimala, D. Gopi, and S. Kannan, "Fabrication of a pH responsive DOX conjugated PEGylated palladium nanoparticle mediated drug delivery system: an in vitro and in vivo evaluation," RSC Advances, vol. 5, no. 56, pp. 44998–45014, 2015.

33. K. Osada, R. J. Christie, and K. Kataoka, "Polymeric micelles from poly(ethylene glycol)-poly(amino acid) block copolymer for drug and gene delivery," Journal of the Royal Society Interface, vol. 6, no. 3, pp. S325–S339, 2009.

34. N. Nishiyama and K. Kataoka, "Polymeric micelle drug carrier systems: PEG-PAsp(Dox) and second generation of micellar drugs," in Polymer Drugs in the Clinical Stage, vol. 519 of Advances in Experimental Medicine and Biology, pp. 155–177, Springer, 2003.

35. A. M. Butt, M. C. Iqbal, M. Amin, and H. Katas, "Synergistic effect of pH-responsive folate-functionalized poloxamer 407-TPGS-mixed micelles on targeted delivery of anticancer drugs," International Journal of Nanomedicine, vol. 10, pp. 1321–1334, 2015.

CHAPTER 7

DENSITY FUNCTIONAL THEORY STUDY ON THE MOLECULAR STRUCTURE OF LOGANIN

Anoop Kumar Pandey[1] Shamoon Ahmad Siddiqui[2] Apoorva Dwivedi[1] Kanwal Raj[3] and Neeraj Misra[1, 4]

[1]Department of Physics, University of Lucknow, Lucknow, India

[2]Department of Physics, Centre for Advanced Materials and Nanoengineering, Najran University, Najran, Kingdom of Saudi Arabia

[3]Medicinal and Process Chemistry Division, CDRI, Lucknow, India

[4]B-981, Sector-A, Mahanagar, Lucknow 226006, India

ABSTRACT

The computational Quantum Chemistry (QC) has been used for different types of problems, for example: structural biology, surface phenomena and liquid phase. In this paper we have employed the density functional method for the study of molecular structure of loganin. The equilibrium geometry, harmonic vibrational frequencies and infrared intensities were calculated by B3LYP/6-311G (d, p) method and basis set combinations. It was found that the optimized parameters obtained by the DFT/B3LYP method are very near to the experimental ones. A detailed conformational analysis was carried out. A detailed interpretation of the infrared spectra of loganin is also reported in the present work. The FT-IR spectra of loganin were recorded in solid phase. The thermodynamic calculations related to the title compound were also performed at B3LYP/6-311G (d, p) level of theory.

INTRODUCTION

The traditional medicine used for the management of liver disorders has helped in the identification of the natural product responsible for the hepatoprotection [1,4,8]. One of the most important class of natural products found useful for the management of hepatic disorder is iridoid glycosides [30]. The monoterpene glucoside, loganin is a key intermediate in the biosynthesis of corynanthe, Aspidosperma, Iboga, Cinchona, Ipecacuanha and pyrolochinolin alkaloids [9] and of simple monoterpene glucoside [13]. It was first isolated from strychnos nux vomica [10]. Its structure [2] has been confirmed by the total synthesis [5]. Earlier we reported the immunostimulant and hepatoprotective activity of loganin and its modified analogoues [36, 42, 43]. Here, in the present communication, the experimental FT-IR frequencies of the title molecule are compared with the theoretical frequencies obtained by Density Functional method. As a general approach to the vibrational problems of organic molecules, we have calculated harmonic frequencies of loganin and compared the results with the observed fundamental vibrational frequencies. This method is useful for obtaining a reliable assignment of the vibrational spectra. The purpose of this study assumes importance because of the further work on loganin, in which presumably, some

Citation: Anoop Kumar Pandey, Shamoon Ahmad Siddiqui, Apoorva Dwivedi, Kanwal Raj, and Neeraj Misra, "Density functional theory study on the molecular structure of loganin," Spectroscopy, vol. 25, no. 6, pp. 287-302, 2011. doi:10.3233/SPE-2011-0517

modification by chemical reaction such as oxidation/reduction on polyfunctional bioactive natural product will generate new reactive sites in the molecule, which can be exploited for elaborating and appending pharmacophores to generate patentable compounds of biological interest. The aim of this study is to study and interpret the simulation of IR spectra of the title compound with the use of standard 6-311G (d, p) basis set. To the best of our knowledge quantum chemical calculations for loganin have not been reported so far in the literature.

EXPERIMENTAL SECTION

Fresh fruit of plant Strychnos nux-vomica L (strychanaceae) (10 kg, collected from NBRI, Lucknow) were extracted with 95% ethanol (4×10 l) and was concentrated in vacuo at temperature below 50°C. It was diluted with water (0.5 l) and fractionated with n-hexane (3×300 ml), chloroform (3×300 ml) and n-butanol (3 × 200 ml), successively. On concentration, butanol fraction in vaccuo, the crude loganin was precipitated out, which was filtered. This was crystalized with ethanol to give pure loganin (50 g), m.p. 223°C. The mother liquors were mixed and additional quantity of loganin (10 g) was isolated by column chromatography and it was identified from its spectral data (yield: 0.6%; m.p. 223°C); IR (KBr): 1714 cm^{-1} (>C=O); ^{1}H-NMR (400 MHz, D2O): δ 7.42 (s, 1H, H-3), 5.41 (d, 1H, J = 4, H-1), 4.81 (d, 1H, J = 8, H-1′), 4.14 (bs, 1H, H-7), 3.92 (d, 1H, J = 12, H-6′), 3.31 (d, 1H, J = 12, H-6′), 3.73 (m, 1H, H-5′), 3.72 (s, 3H, OCH_3), 3.49 (t, 1H, J = 8, H-3′), 3.39 (t, 1H, J = 8, H-4′), 3.27 (t, 1H, J = 8, H-2′), 3.07 (m, 1H, H-5), 2.15 (m, 1H, H-6α), 2.15 (me, 1H, H-9), 1.92 (m, 1H, H-8), 1.75 (m, 1H, H-6β), 1.06 (d, 3H, J = 6.6, H-10); MS (FAB): m/z 391 (M^+ + 1), 413 (M^+ + Na); analysis calculated for $C_{17}H_{26}O_{16}$: C, 52.30; H, 6.66. Found: C, 52.26: H, 6.49%; $[\alpha]_D^{25} = -87.81$ (C = 0.28 in methanol) [35]. The above data is also supported by another work on loganin, in which anti-amnesic (cognitive enhancing) activity has been reported [26]. The FT-IR spectra of the sample of loganin have been recorded in CsI on a Perkin Elmer 1800 spectrophotometer. Spectroscopic preparation of the sample was carried out under an atmosphere of prepurified nitrogen. The FT-IR spectra of title molecule is recorded in the region 4000–500 cm^{-1} and is shown in Fig. 1.

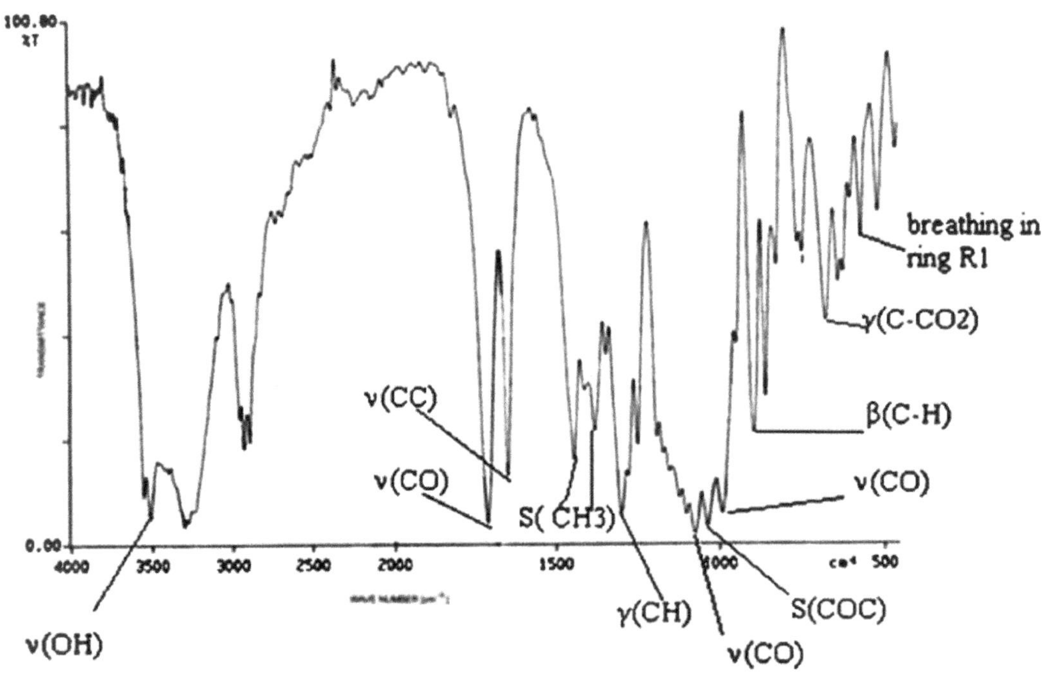

Figure. 1. FT-IR spectrum of loganin.

COMPUTATIONAL DETAILS

The DFT calculations were mainly carried out in the framework of the Becke–Lee–Yang–Parr (B3LYP) functionals, in which the exchange functional is a local spin density exchange with Becke gradient correction and the correlation functional is that of Lee, Yang and Parr with both local and nonlocal terms employing 6-31G (d, p) [3,16,17,25,31,44] as the basis set, using Gaussian 03 suite of programs [12], involving gradient optimized geometry [38]. The basis set 6-311G (d, p) augmented by d polarization functions on heavy atoms and p polarization functions on hydrogen atoms as well as diffuse functions for both hydrogen and heavy atoms were used. The Gauss View 3.0 [11] program has been employed to get a visual animation and also for the study of the normal modes description. It is well known that vibrational frequencies obtained by quantum chemistry calculation are typically larger than that of their experimental counterpart and thus experimental scaling factors are usually employed to have a better agreement with the experimental vibrational frequencies [14]. The scaling factor depends on both basis set and the method used in the calculation and is determined from the mean deviation between calculated and experimental value of the frequencies [37, 39]. On the other hand B exchange functional have an advantage of standard frequency very close to unity so B based procedures can be often used without scaling [18, 29, 32]. The absolute infrared absorption intensities were calculated in the harmonic approximation, at the same level of theory as used for the optimized geometries, from the derivatives of the dipole moment and polarizability of each normal mode, respectively. The optimized structural parameters of loganin calculated by B3LYP method with the 6-311G (d, p) basis set are listed in Tables 1 and 2, in accordance with the atom numbering scheme as given in Fig. 2.

Table 1 Optimized bond lengths of loganin using B3LYP/6-311G (d, p) method

S. No.	Loganin parameter bond lengths	X-ray (Å)	Calculated B3LYP/6-311G (d, p) (Å)
1	C1–H2	–	1.101
2	C1–C3	1.511	1.536
3	C1–O14	1.346	1.402
4	C3–H4	–	1.100
5	C3–C5	1.515	1.525
6	C3–O15	1.419	1.414
7	C5–H6	–	1.107
8	C5–C7	1.529	1.536
9	C5–O17	1.414	1.412
10	C7–H8	–	1.095
11	C7–C9	1.530	1.532
12	C7–O19	1.415	1.429
13	C9–H10	–	1.105
14	C9–C11	1.507	1.524
15	C9–O23	1.442	1.434
16	C11–H12	–	1.101
17	C11–H13	–	1.092
18	C11–O21	1.418	1.414
19	O14–C24	–	1.415
20	O15–H16	–	0.968
21	O17–O18	–	0.969
22	O19–H20	–	0.969
23	O21–H22	–	0.967
24	C24–H25	–	1.095
25	C24–C26	1.432	1.437
25	C24–C52	1.509	1.517
26	O26–C27	1.357	1.358
27	C27–H28	–	1.083
28	C27–C29	1.333	1.346
29	C29–C30	1.482	1.475
30	C29–C37	1.510	1.513
31	C30–O31	1.341	1.358
32	C30–C36	1.119	1.218
33	O31–C32	1.449	1.435
34	C32–H33	–	1.089
35	C32–H34	–	1.093
36	C32–C35	–	1.092
37	C37–H38	–	1.093
38	C37–C39	1.553	1.567
39	C37–C52	1.549	1.559
40	C39–H40	–	1.094
41	C39–H41	–	1.093
42	C39–C42	1.522	1.536
43	C42–H43	–	1.100

44	C42–O44	1.439	1.430
45	C42–C46	1.522	1.537
46	O44–H45	–	0.965
47	C46–H47	–	1.099
48	C46–C48	1.528	1.529
49	C46–C52	1.535	1.547
50	C48–H49	–	1.095
51	C48–H50	–	1.092
52	C48–H51	–	1.095
53	C52–H53	–	1.094

RESULTS AND DISCUSSION

Geometry Optimization

The conformer study was done by Monte Carlo Multiple Minimum method. Twenty three conformers were generated. Out of the 23 conformers, the minimum energy calculated by using HF/3-21G method and global minimum energy was found for the conformer number 19. In this conformer both hydrogen atoms attached with oxygen in R1 rotate by 180° to minimize the repulsion and other hydrogen attached to oxygen rotate nearly perpendicular to the plane of ring. All the 24 figures of the conformers and their corresponding energies (Table 6) have been given as the supplementary material. After finding out the most stable geometry the calculations were carried out by DFT method with a higher basis set. The calculations converged to optimized geometries, which correspond to the true energy minima, as also revealed by the lack of imaginary frequencies in the vibrational mode calculation. Subsequently, the local minimum energy obtained for loganin with 6-311G (d, p) basis set was found approximately to be −1415.690 A.U. In case of loganin molecule there is C_1 molecular symmetry. The dihydropyran ring is in sofa conformation with C_1 out of plane with the other atom. The torsion angles in the ring show a mean absolute deviation of 4° than the ideal sofa. The torsion angles in C(1)–O(2)–C(3)–C(4) is negative and glucosyloxy group is in the quasiaxial position [6]. Experimental data is taken from the literature [15]. Nearly all the calculated bond lengths match well with the experimental values. For example, the optimized bond length of C–C in ring R3, falls in the range from 1.56 to 1.54 Å, which is in good agreement with those of the experimental bond length (1.550–1.520 Å). In the ring, carbon–carbon bond length is not the same (same type of hybridization) and this is due to the antibonding repulsion occurring between carbon and oxygen atom present in the ring and also the group attached to the ring. To achieve molecular stability, the strain was minimized and consequently, some bond length got enlarged and some bond lengths were contracted. The optimized bond length of C–O in ring R2 and R3 falls in the range from 1.358 to 1.437 Å, which is also in good agreement with the experimental bond lengths (1.357–1.442 Å). The other calculated bond length also show an excellent agreement with the experimental value. However, by allowing the relaxation of all parameters, some intermolecular interaction, and (hydrogen bonding) affects the bond angle and bond length so a comparison of calculated and experimental result gave important information about the intermolecular forces.

Table 2 Optimized bond angles of loganin using B3LYP/6-311G (d, p) method

S. No.	Loganin parameter bond angle	X-ray (in degree)	Calculated B3LYP/6-311G (d, p) (in degree)
1	H2–C1–C3	–	109
2	H2–C1–O14	–	110
3	H2–C1–O23	–	110
4	C3–C1–O14	109	107
5	C3–C1–O23	110	111
6	O14–C1–O23	107	108
7	C1–C3–H4	–	107
8	C1–C3–C5	108	109
9	C1–C3–O15	109	109
10	H4–C3–O5	–	108
11	C5–C3–O15	108	109
12	C5–C3-150	108	109
13	H6–C5–C7	–	109
14	C3–C5–H6	–	107
15	C3–C5–C7	111	110
16	C3–C5–O17	109	108
17	H6–C5–O17	–	110
18	C7–C5–O17	111	110
19	C5–C7–H8	–	108
20	C5–C7–C9	110	111
21	C5–C7–O19	111	109
22	H8–C7–C9	–	108
23	H8–C7–O19	–	105

24	C9–C7–O19	112	112
25	C7–C9–H10	–	109
26	C7–C9–C11	112	113
27	C7–C9–O23	108	108
28	H4–C3–O15	–	108
29	H10–C9–C11	–	108
30	H10–C9–O23	–	109
31	C11–C9–O23	107	106
32	C9–C11–H12	–	108
33	C9–C11–H13	–	109
34	C9–C11–O21	112	111
35	H12–C11–H13	–	107
36	H12–C11–O21	–	111
37	H13–C11–O21	–	107
38	C1–O14–C24	114	115
39	C3–O15–H16	–	106
40	C5–O17–H18	–	105
41	C7–O19–H20	–	107
42	C11–O21–H22	–	105
43	C1–O23–C9	111	112
44	O14–C24–H25	–	110
45	O14–C24–O26	109	109
46	O14–C24–O52	106	107
47	H25–C24–O26	–	103
48	H25–C24–C52	–	112
49	O26–C24–H52	113	112
50	C24–O26–C27	115	115
51	O26–C27–H28	–	112
52	O26–C27–C29	124	124
53	H28–C27–C29	–	123
54	C27–C29–C30	120	120
55	C27–C29–C37	123	122
56	C30–C29–C37	116	117
57	C29–C30–O31	113	113
58	C29–C30–O36	123	123
59	31O–30C–36O	123	122
60	30C–31O–32C	115	115
61	31O–32C–33H	–	105
62	31O–32H–34H	–	110
63	31O–32C–35H	–	110
64	33H–32C–34H	–	110
65	33H–32C–35H	–	110
66	34H–32C–35H	–	108
67	29C–37C–38H	–	108
68	29C–37C–39C	112	111
69	29C–37C–52C	110	111
70	38H–37C–39C	–	109
71	38H–37C–52C	–	110
72	39C–37C–52C	104	104
73	37C–39C–40H	–	110
74	37C–39C–41H	–	111
75	37C–39C–42H	–	109
76	41H–39C–42C	–	112
77	39C–42C–43H	–	112
78	39C–42C–44O	111	111
79	39C–42C–46C	103	103
80	43H–42C–44O	–	110
81	43H–42C–46C	–	111
82	44O–42C–46C	106	106
83	42C–44O–45H	–	107
84	42C–46C–47H	–	106
85	24C–52C–53H	–	106
86	37C–52C–46C	103	105
87	37C–52C–53H	–	109
88	46C–52C–53H	–	107

89	42C–46C–48C	114	115
90	42C–46C–52C	101	104
91	47H–46C–48C	–	109
92	47H–46C–52C	–	108
93	48C–46C–52C	115	115
94	46C–48C–49H	–	111
95	4C–48C–50H	–	110
96	46C–48C–51H	–	110
97	49H–48C–50H	–	108
98	49H–48C–51H	–	107
99	50H–48C–51H	–	107
100	24C–52C–37C	113	113
101	24C–52C–46C	114	114

Note: Experimental data is taken from [15].

In case of loganin, there are four intermolecular hydrogen bonding interactions between 21O ... 15O– 16H, 15O ... 19O–20H, 44O ... 22H–21O, 23O ... 45H–44O which have been taken from the crystallographic data [15]. In general, bond between carbon and oxygen are overestimated but at the position where hydrogen bonding occurs, the bond between oxygen and carbon is underestimated, e.g., bond length where there is no hydrogen bonding are C24–O26, O26–C27, C30–O31, C30–O36 and is found to be 1.432, 1.357, 1.341 and 1.197 Å, respectively, and in case of experimental data corresponding calculated bond lengths are 1.437, 1.358, 1.358 and 1.218 Å (Table 5 of Supplementary material; see http://www.mediafire.com/?mdls5a5nueyi74i). Whereas, in places where hydrogen bonding occurs, the bond length between 23O–9C, 23O–1C, 15O–3C, 44O–42C, 21O–11C was found to be 1.442, 1.433, 1.419, 1.439, 1.418 Å and their corresponding calculated values are 1.434, 1.421, 1.414, 1.430 and 1.414 Å using B3LYP/DFT method. In experimental data refinement proceed to R = 0.11 (isotropic) and R = 0.08 (anisotropic), at which stage all H atom were revealed by different synthesis [15]. In final stages C–H distance were fixed at 0.96 Å and H–C–H angle at 109.5◦ [15]. As revealed by the Gauss view, no ring lies in plane and also, a single hydrogen atom of each methyl unit lies in the plane of adjacent ring while the other two are symmetrically positioned above and below the plane of adjacent ring. The methyl group is attached with the R3 ring and the C–H bond length which is in the plane and is smaller than the other which is nonplanar, e.g., bond length in between carbon and hydrogen which are nonplanar are 48C–49H, 48C–51H and it is 1.095 and 1.095 Å, respectively. While bond length between carbon and hydrogen (in case of CH3 which is attached to ring R3), in plane is 1.092 Å. Since, there is a large deviation from the experimental bond length X–H, i.e, bond length arising from low scattering factor of hydrogen atom in X-ray diffraction experiment hence, we have not discussed the C–H bond length. Several authors have reported that B3LYP method predicts bond length, which is systematically too large, particularly C–H bond lengths [24, 27, 28].

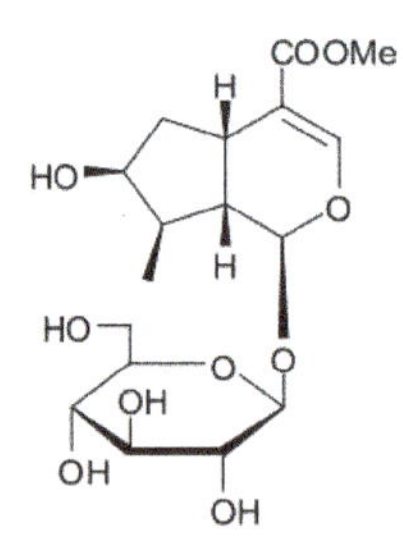

Loganin

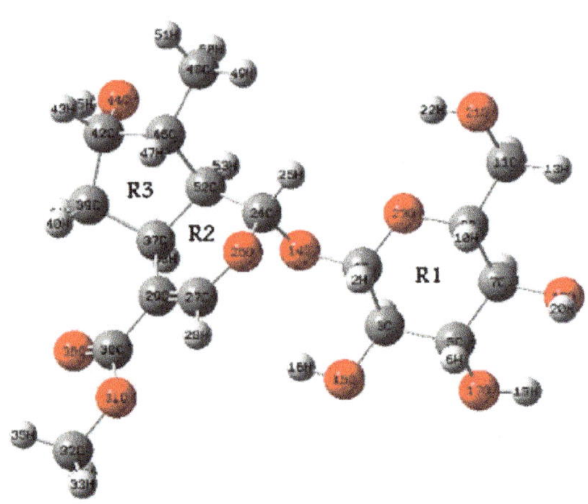

Figure. 2. Model molecular structure of loganin (conformer 19). (Colors are visible in the online version of the article; http://dx.doi. org/10.3233/SPE-2011-0517.)

The angular changes in benzene ring geometry have also proved to be a sensitive indicator of the interaction between the substituent and the benzene ring [19]. The structural changes in the carbon skeleton involve both, the bond distances and bond angles. They are most pronounced at the place of substitution and depend on the electro negativity as well as on the σ/π donor/acceptor character of the substituent. The bond distances have already been discussed above and now the bond angles needs to be discussed.

The calculated substituent impact on the C–C–C bond angles of the benzene ring is summarized in Table 2. The data reflects the well-known trends observed for various substituents: groups with π-acceptor substituent tend to increase in α_{ipso}, decrease α_{ortho} and this results in a small increase in α_{para}, e.g., in the title molecule the bond angle between C37–C29–C27; C29–C37–C52; and C37–C52–C24 are 122.21°; 111.62° and 113.0° and their corresponding experimental data are at 123.2°; 112.3° and 113.1°, respectively, which are found to be in good agreement. Another feature concerning the bond angle between carbon and hydrogen in methyl group is that the bond angle which lies in a plane is different from the one which lies out of plane so methyl group gets distorted from the regular tetrahedral shape, e.g., bond angle between in plane carbon and hydrogen (34H–32C–35H, 50H–48C–49H) are 108.6°, 108.7° and the corresponding out of plane, 33H–32C–35H, 33H–32C–34H and 50H–48C–51H, 51H–48C–49H are 107.8°, 107.15° and 110.7°, 110.7° respectively calculated by B3LYP/6-311G (d, p).

Vibrational Assignments

The molecule under study has 53 atoms and 153 normal modes of fundamental vibration. Out of 153 normal modes, N − 1 modes correspond to the stretching modes and remaining are bending modes. All the modes are divided into two parts: one corresponding to below 1000 cm^{-1} called the finger print region and the other corresponding to the above 1000 cm^{-1} called the functional group region. The assignments of the experimental wavenumbers are based on an animated view of normal mode description and a comparison with the theoretically scaled ones as listed in the Table 3. It needs to be emphasized that the calculated frequencies represent vibrational signatures of the molecules in its gas phase. Hence, the experimentally observed spectra of the solid/liquid samples may differ to some extent from the calculated spectrum. Moreover, the calculated harmonic force constants and frequencies are usually higher than the corresponding experimental quantities; due to combination of electron correlation effect [33] and basis set deficiencies. This is the reason for using scaling factors for theoretical calculations. Nevertheless, after applying the uniform scaling factor the theoretical calculation reproduce the experimental data well. The observed slight disagreement between the theory and the experiment could be a consequence of the anharmonicity [34] and of the general tendency of the quantum chemical methods to overestimate the force constants at the exact equilibrium geometry. The experimental FT-IR spectra are shown in Fig. 1. Vibrational frequencies at B3LYP/6-311G (d, p) level were scaled by 0.9682 above 800 cm−1 and 1.0119 below 800 cm−1 [29].

A good agreement between the theoretical and experimental wavenumbers for the majority of bands is evident. Some important modes of vibrations are discussed which are as follows.

Carbonyl Absorption

Carbonyl absorptions are sensitive and both the carbon and oxygen atoms of the carbonyl group move during the vibration having nearly equal amplitude. In the present study a highly intense polarized IR peak appears due to C=O stretching in the plane of the ring R2 and is observed at 1726 cm^{-1}, which is in good agreement with the experimental frequency obtained at 1710 cm^{-1}.

Table 3 Vibrational wavenumbers obtained for loganin at B3LYP/6-311G (d, p) in cm−1, experimental frequencies from FT-IR spectra in cm^{-1}, IR intensities (K_m mol^{-1}) and assignment

S. No.	Wavenumber		Experimental frequencies IR	IR intensity	Assignment
	Unscal.	Scal.			
1	15	15		0.014	Twisting in Ring R1 and R2 with respect to 14O
2	20	20		0.031	Twisting in Ring R1 and R2 with respect to 14O
3	34	34		0.063	Twist ring R3 with respect R2
4	40	40		0.296	All ring bend
5	59	59		1.826	Twist(O–CH_3)adjR2 + τ(CC)R1, R2&R3
6	63	63		0.216	Twist(O–CH_3)adjR2
7	73	73		0.288	γ(CO)adjR2R1 + τ(CC)R1, R2&R3
8	96	96		4.877	Twist(O–CH_3)adjR2 + twist(CH_3OH)adjR1
9	106	107		1.697	R(C–O–C)R3 + R(CH_2)R3 + γ(O–CH_3)adjR2 + γ(C–O)adjR1+ γ(C–COH3)adjR1
10	109	110		0.5748	Twist(O–CH_3)adjR2 + γ(CCC)R1 + γ(CO)adjR1
11	118	119		3.131	Twist(O–CH_3)adjR2 + (C–CH_2OH)adiR1
12	125	126		1.264	τ(CC)R1, R2&R3 + twist(O–CH_3)adjR2
13	133	134		1.066	Twist(O–CH_3)adjR2
14	151	152		0.179	R(CH_2)R3 + γ(CCC)R3 + γ(COC)R1 + γ(C–CH_3)adjR3
15	173	175		1.035	Ring R1 breathing + Ring R2&R3 bending with respect to 14O
16	183	185		2.065	γ(C–O–CH_3)adjR2 + τ(C–COH_3)adjR1
17	193	195		0.296	τ(CO)adjR3&R2 + τ(CC)R2, R3&adjR2&R3 + γ(OH)R1
18	205	207		2.071	Twist(C–CH_3)R3 + γ(O–H)adjR1
19	216	218		3.055	Twist(C–CH_3)R3
20	254	257		8.740	S(C–C–O)R1 + γ(CCC)R1
21	261	264		2.005	β(C–CH_3)adjR3 + β(O–H)adjR3
22	265	268		17.59	γ(O–H)R3 + S(CCC)R1
23	272	275		7.306	γ(C–O–C)R2 + γ(O–H)R3
24	289	292		86.22	γ(O–H)R3
25	294	297		24.715	γ(C–CH_3)adjR3 + γ(C–O–C)R1 + γ(OH)R1 adjR3 + γ(CH)R1
26	304	307		3.518	β(C–CH_3)adjR3 + γ(COC)R1 + β(OH)adjR1&R3 + β(CH)adjR1

27	321	324		9.321	Ring R3 breathing + γ(O–CH_3)R3
28	333	336		25.76	β(C–CH_3)adjR2 + β(C–C–O)R2
29	345	349		13.835	Ring breathing R1 + γ(O–H)R1
30	361	365		4.800	γ(O–H)adjR1
31	383	387		34.437	γ(O–H)adjR1 + R(CH_2)R3 + γ(CCC)R2
32	384	388		32.19	R(CH_2)R3 + γ(OH)R1 adjR1 + γ(CH)R2
33	400	404		83.931	γ(C–H)R1
34	418	422		59.959	Ring R1 breathing + β(O–H)adjR1 + γ(CCC)R2 + R(CH_2)R3
35	420	424		35.999	γ(O–H)adjR1 + R(CH_2)R3 + β(CCC)R1 + γ(CCC)R2
36	427	432		61.548	γ(C–H)R1 + ring R1 breathing
37	459	464		7.953	Ring R1, R2&R3 bend + β(O–H)adjR1 + R(CH_2)R3
38	465	470		3.867	oopring R2 bend + ring R1 breathing + γ(OH)adjR1 + S(CCO)R2
39	473	478		72.049	Ring bend R1 + β(O–H)R1
40	482	487		106.199	γ(O–H)R1 + γ(C–H)R1
41	521	527		14.315	Ring R1 breathing + ring R2 bend + β(C–CH_3)adjR3
42	536	542		3.393	Ring R2, R1&R3breathing + β(OH)adjR1 + β(C–H)adjR1
43	546	552	550	0.450	γ(CCC)R2 + ω(CH_2)adjR3
44	571	577	580	25.365	Ring breathing + ω(CH_3)adjR2, R3
45	604	611	615	15.112	Ring R1 breathing + β(O–H)adjR1 + β(C–H)adjR1
46	619	626		4.494	Ring R2 breathing + γ(CCC)R2 + R(CH_2)adjR1+ β(C–H)adjR1&R2
47	634	641	630	7.140	Bending in R3 + R(CH_2)R3
48	648	655	645	24.447	Breathing in R1 + β(O–H)adjR1 + β(C–H)adjR1
49	672	679	665	7.813	Ring R2&R3 breathing + γ(CCC)R3 + β(O–H)adjR1+ β(C–H)adjR1
50	701	709	720	0.317	β(CCC)R2R3 + R(CH_2)R3 + β(CO)adjR2
51	755	763	745	14.198	γ(C–CO_2)adjR2 + Ring bend R2, R3
52	766	775		9.512	Ring bend R2 + γ(C–CO_2)adjR2
53	793	802		5.857	Ring R1, R2 bendig + S(CO_2)adjR3
54	836	809	800	14.715	Bending R2, R3 + τ(CH_2)R3
55	858	830		6.775	Ring R2, R3 bending + R(CH_2)R3 + ω(CH_3)R3
56	878	850	852	70.412	β(C–H)R2&R3 + ν(CO)R2 + ν(CC)R2R3 + ω(CH_3)R3+
57	897	868		13.258	β(C–H)R1 + R(CH_2)adjR1 + γ(C–H–O)R1
58	904	875		15.131	Ring R3 bend + R(CH_3)adjR1 + τ(CH_2)R3
59	912	882	890	85.186	γ(CCC)R3 + γ(C–H)R3 + R(CH_3)R3 + R(CH_2)R3
60	939	909		12.226	γ(CH)R2
61	973	942	943	51.100	ν(CH_3–O)adjR2 + Ring breding R2, R3
62	987	955		23.572	ω(CH_3)R3 + β(CCC)R3 + β(CH)R3
63	996	964		120.111	Ring breathing R1
64	1005	973		20.279	ν(O–CH_3)adjR1 + ring bend + R(CH_3)R3
65	1017	984		41.175	β(CO)adjR1 + bending in R1
66	1031	998	990	98.574	ν(CO)R1 + ν(CC)R1 + β(O–H)adjR1 + β(C–H)adjR1+ γ(CCC)R2&R3 + R(CH_2)R3
67	1040	1006		27.608	γ(CO)R1 + ν(CC)R3 + γ(CH)R1 + γ(CC)R1
68	1056	1022		75.666	γ(CC)R3 R1 + bending in R2
69	1063	1029		220.491	Ring breathing R2, R3 + ν(O–CH_2)adjR1 + bending R1
70	1069	1035		118.528	β(O–H)adjR1 + bending R1
71	1075	1040		77.439	β(O–H)adjR1 + ν(CO)R1 + bending R1
72	1085	1050	1040	247.071	S(COC)adjR1 + R(CH_2)adjR1 + γ(CH)R3
73	1098	1063		104.792	γ(O–C–C)R1 + τ(CH_2)R3 + β(OH)R2 R3
74	1101	1065		197.857	Ring breathing + ν(C–CH_3)R3
75	1107	1071		9.0008	ν(C–CH_3)R3 + β(C–H)R2&R3 + ν(C–OH)adjR1
76	1118	1082	1083	41.897	ν(CO)R1 adjR1 + β(HO)R1 + ν(CC)R1
77	1122	1086		96.415	ν(CC)R1 + ν(C–CH_3)R3 + ν(O–CH_3)R1 + β(CH)R3 + γ(CCC)R3
78	1130	1094		80.320	ν(O–CH_3)adjR2 + ν(CC)R1 + ring bend R2
79	1135	1098		92.081	ν(CC)R1 + ring bending R1
80	1148	1111	1105	69.213	β(OH)adjR1 + ν(CC)R3 + β(C–CH_3)R3
81	1151	1114		62.674	β(CH)R3R2 + ring R1 bend + ν(C–CH_3)R2
82	1158	1121		11.173	β(CH)R3R2 + ω(CH_3)adjR3 + t(CH_2)R1
83	1172	1134	1122	67.610	Ring bend R1 + β(CH)adjR3R2 + γ(CH)R3

84	1178	1140		0.572	ω(CH$_3$)adjR2
85	1192	1154	1150	18.220	β(CH)adjR3R2 and R1R3 + β(OH)R1
86	1202	1163		10.924	β(CH)adjR3R2 and R1R3 + β(OH)R1
87	1207	1168	1170	111.894	ν(CO)R2 + ω(CH$_3$)adjR2 + β(CH)adjR3, R2 and R1
88	1213	1174		3.331	ω(CH$_3$)adj R2 + t(CH$_2$)R3
89	1216	1177		29.455	ω(CH$_3$)R2 + t(CH$_2$)R3 + β(CH)adjR3, R2
90	1232	1192		15.816	β(CH)R1 + ring R1 bending + β(OH)adjR1
91	1248	1208	1190	32.748	t(CH$_2$)adjR1 + β(OH)adjR1 + β(CH)R1
92	1277	1236		55.784	β(OH)adjR1 + β(CH)R3
93	1280	1239		13.299	β(OH)adjR1 + β(CH)R1 adjR1
94	1289	1248		1.455	β(CH)adjR2, R3 and R3
95	1302	1260	1250	85.053	β(OH)adjR1 + β(CH)R1
96	1310	1268		10.485	β(CH)adjR2, R3 + ω(CH$_2$)R2
97	1314	1272		6.435	β(OH)adjR1 + β(CH)R1
98	1320	1278		94.621	ν(CC)adjR2 + ring bend R2, R3 + γ(CH)R3
99	1324	1281	1270	289.885	β(CH)adjR3R2 + bending R2 + ν(CC)R2
100	1336	1293	1295	14.397	β(CH)R1
101	1345	1302		13.893	β(CH)R2 and R2, R3
102	1352	1309		6.439	β(OH)adjR + β(CH)R1, R3 and R2, R3, R1
103	1367	1323		18.785	β(CH)R1, R3 adjR1 + β(OH)adjR1+
104	1376	1332	1334	42.011	β(CH)R1, R2, R3 adjR1 + β(OH)adjR1
105	1385	1340		10.029	β(CH)R1, R3 + β(OH)adjR1
106	1390	1345		4.716	β(CH)R1 + β(OH)adjR1
107	1394	1349		21.453	β(CH)R1 adjR1 + β(OH)R1
108	1397	1352		14.565	β(CH)R1, R2 adjR1 + β(OH)adjR1
109	1399	1354		15.495	β(CH)R1, R2 adjR1 + β(OH)adjR1
110	1407	1362		22.477	γ(CH)R2 + β(CH)adjR1 + S(OCH)R2
111	1415	1370		3.630	γ(CH)R2, R1, R3 + S(CH$_3$)R3
112	1419	1373	1373	33.415	γ(CH)R1 + ν(C–CH$_3$)R3, R1
113	1427	1381		12.055	γ(CH)R2, R1 + ν(CC)R3, R1 + ν(C–CH$_3$)R3+
114	1429	1383		34.486	γ(CH)R1 + S(CH$_3$)adjR3
115	1430	1384		13.143	S(CH$_3$)adjR3 + γ(CH)R1 + β(OH)R1
116	1440	1394		33.696	ω(CH$_2$)adjR1 + β(OH)adjR1 + β(OH)R2
117	1446	1400		2.664	γ(CH)R1 + β(CH)R2
118	1472	1425	1410	8.021	β(CH) + β(CO)R1
119	1482	1434	1425	24.256	S(CH$_3$)adjR2
120	1491	1443		3.592	S(CH$_2$)R3
121	1494	1446	1440	5.678	S(CH$_3$)adjR2
122	1509	1461		4.953	β(CH$_3$)adjR3
123	1511	1462		7.015	S(CH$_2$)adjR2
124	1515	1466		0.791	S(CH$_3$)adjR1
125	1517	1468		4.561	S(CH$_3$)adjR3
126	1603	1644	1600	142.5	ν(CC)R2
127	1783	1726	1710	250.7	ν(CO)adjR2
128	2927	2833		30.57	ν(CH)R1 adjR1
129	2960	2865		30.318	ν(CH)R1 adjR1
130	2999	2903		31.336	ν(CH)R1 adjR1
131	3000	2904		32.867	ν(CH)R1
132	3002	2906	2900	61.338	ν(CH)R3
133	3020	2923	2912	43.307	ν(CH)R1
134	3030	2933	2940	19.196	ν(CH)R3
135	3048	2951	2958	26.034	ν(CH$_2$)adjR3
136	3062	2964		41.096	ν(CH$_3$)adjR2
137	3072	2974		29.165	ν(CH$_2$)R3
138	3080	2982		22.367	ν(CH)R1
139	3082	2983		4.570	ν(CH)R2
140	3096	2997		10.272	ν_{as}(CH$_3$)adjR3 + ν(CH)R2
141	3102	3003		28.561	ν_{as}(CH$_3$)adjR3 + ν(CH)R2
142	3109	3010		34.194	ν_{as}(CH$_2$)*R3* + ν(CH)R2
143	3117	3017		22.469	ν_{as}(CH$_2$)R3
144	3121	3021	3030	21.447	ν(CH)adjR1
145	3138	3038		20.707	ν(CH)adjR1
146	3139	3039		16.856	ν_{as} CH)R3
147	3171	3070		16.372	ν_{as}(CH$_3$)adjR2
148	3237	3134		02.022	ν(CH)R2
149	3773	3653	3520	39.715	ν(OH)R2
150	3776	3655		10.52	ν(OH)adjR1
151	3796	3675	3625	40.28	ν(OH)adjR1
152	3801	3680	3685	28.14	ν(OH)adjR1
153	3886	3699		18.89	ν(OH)R3

Notes: ν – stretching; ν_s – symmetric stretching; ν_{as} – asymmetric stretching; β – in-plane-bending; γ – out-of-plane bending; ω – wagging; R – rocking; t – twisting; τ – torsion, s – scissoring.

C–H Vibrations

The change in the frequency also reflects the change in the geometry. Significant changes in frequency and intensity take place in the H–X stretching vibration. As the magnitude of the shift in H–X stretching vibrational frequency is known to represent a measure of the intermolecular interaction, the difference in frequency is in good agreement with the interaction energy. The hetero aromatic structure shows the presence of C–H stretching vibrations in the region 2800–3100 cm^{-1}, which is the characteristic region for the ready identification of the C–H stretching vibration [41]. In the present study the C–H stretching vibration of the title compound is observed at 2923 and 2906 cm^{-1}, which are in good agreement with the calculated frequencies obtained at 2912 and 2900 cm^{-1}. Some other C–H stretching vibrations are also calculated at 2933, 2904, 2865 and 2833 cm^{-1}, which are also supported by the literature [22]. Some bending vibrations of C–H are also calculated and supported by literature. At lower frequency region some significant intensity peaks of polarized IR due to out of plane bending in C–H are also observed.

Methylene Group Vibrations

The asymmetric CH_2 stretching vibrations are generally observed in the region 3100–3000 cm^{-1}, while the symmetric stretching vibrations are generally observed between 3000–2900 cm−1 [22]. The CH_2 asymmetric stretching vibrations are calculated at 3010 and 3017 cm^{-1}, whereas CH_2 symmetric stretching vibrations are also calculated at 2951 and 2974 cm^{-1}. The bands corresponding to different bending vibrations of CH_2 group are summarized in Table 3 and are supported by the literature [22]. An intense IR band occurred at 1121, 1174, 1177 and 1208 cm^{-1}, due to the twisting of CH_2 in R3. Generally, twist occurs in the low frequency region but due to the hinging of ring R3 by R2 and also the spatial arrangement of molecule in plane which is disturbed by the twist in the planarity, it occurs in some higher region.

C–C Vibrations

The C–C aromatic stretch known as semi-circle stretching, is calculated at 1644, 1281, 1278, 1114, 1111, 1098 cm^{-1} etc., respectively, and is supported by the literature [23]. The theoretically calculated C–C–C bending modes and C–C torsional modes have also been found to be consistent with the recorded spectral values and the literature [23].

Methyl Group Vibrations

The asymmetric CH_3 stretching vibrations are calculated at 2997, 3003 and 3070 cm^{-1}, while symmetric CH_3 stretching vibrations are calculated at 2964 cm^{-1}. These assignments are also supported by the literature [40]. In the present study various bending vibrations of CH_3 group are also summarized in Table 3 and are supported by the literature [40].

C–O Vibrations

In this study the C–O stretching vibrations are observed at 990, 1083 and 1170 cm^{-1}, which are in good agreement with the calculated frequencies obtained at 998, 1082 and 1168 cm^{-1}, respectively. The various bending and torsional vibrations assigned in this study are also supported by the literature [7]. A lot of discrepancies observed between experiment and theory are due to the intermolecular hydrogen bonding. Some mixing of different modes of vibration along with C=O plane and out of plane bending is also observed in the lower side of spectra.

O–H Vibrations

In vibrational spectra, the strength of hydrogen bond determines the position of O–H band. Usually, the O–H stretching appears at high frequency ranging from 3600–3400 cm^{-1} [20] because of lower reduced mass. In this study, loganin showed very strong absorption peaks at 3699, 3680, 3675, 3655 and 3653 cm^{-1}, which is due to the O–H stretching vibration. The observed frequency range is different from experimental one and probably this is due to the low scattering factor for hydrogen and also the intermolecular interaction occurring in the molecule which overestimates the stretching modes. The various bending vibrations of the hydroxyl groups are also found to be in good agreement with the observed spectra and literature [20].

C-group Attached Out of Ring

Vibration belonging to the bond between ring and other groups are also important as mixing of vibration is possible due to the presence of heavy atom which shows lower absorption frequencies as compared to C–H, due to the increase reduced mass [21]. C–C stretching in ($C–CH_3$) is calculated at 1071 cm^{-1} and 1086 cm^{-1}, in ring R3 and O–CH stretching is observed at some lower modes than the previous ones due to the higher reduced mass.

CONCLUSION

Equilibrium geometries and harmonic frequencies of loganin were determined and analyzed at DFT/B3LYP level of theory using 6-311G (d, p) basis set. The vibrational frequency calculations proved that the structure is stable (no imaginary frequencies). We found the geometry obtained by the B3LYP method to be very accurate. The difference between the observed and scaled wavenumbers values of most of the fundamentals is very small. Any discrepancy noted between the observed and the calculated frequencies may be due to the fact that the calculations have been actually done on a single molecule in the gaseous state contrary to the experimental values recorded in the presence of intermolecular interactions. Some intensity of lower modes do not match well and presumably, this is due to the inherent impurity of the sample, packing fraction (molecular–molecular interaction), modes of vibrations are so near that it is difficult task to identify these modes as they look like a band and also the basis set deficiencies, anharmonicity can be held responsible for it.

ACKNOWLEDGEMENT

The corresponding author (NM) is grateful to the University Grants Commission, New Delhi for providing financial assistance.

REFERENCES

1. R.A. Ansari, B.S. Aswal, R. Chander, B.N. Dhawan, N.K. Garg, D.K. Kulshreshta, H.S. Medhi, B.N. Mehrotra, G.K. Patnaik and S.K. Sharma, Indian J. Med. Res. 87 (1988), 401.
2. A.R. Battersby, E.S. Hall and R. Southgate, J. Chem. Soc. C (1969), 721.
3. A.D. Becke, J. Chem. Phys. 98 (1993), 5648.
4. R. Bentley and H. Trimen, Medicinal Plants 3 (1983), 197.
5. G. Buchi, J.A. Carlson, J.E. Powell, Jr. and L.F.J. Tietze, Amer. Chem. Soc. 95 (1973), 540.
6. R. Bucourt and D. Hainaut, Bull. Soc. Chim. France (1965), 1366.
7. S.J. Bunce, H.G. Edwards, A.F. Johnson, I.R. Lewis and

P.H. Turner, Spectrochimica Acta A49 (1993), 775.

8. K.K. Chakraborti and S.S. Handa, Indian Drugs 27 (1989), 19.
9. G.A. Cordenn, Lloydia 37 (1974), 219.
10. W.R. Dustan and F.W. Short, Pharm. J. Trans. 14 (1884), 1025.
11. A. Frisch, A.B. Nelson and A.J. Holder, GAUSSVIEW Inc., Pittsburgh PA, 2000.
12. M.J. Frisch, G.W. Trucks, H.B. Schlegel, G.E. Scuseria, M.A. Robb, J.R. Cheeseman, J.A. Montgomery, Jr., T. Vreven, K.N. Kudin, J.C. Burant, J.M. Millam, S.S. Iyengar, J. Tomasi, V. Barone, B. Mennucci, M. Cossi, G. Scalmani, N. Rega, G.A. Petersson, H. Nakatsuji, M. Hada, M. Ehara, K. Toyota, R. Fukuda, J. Hasegawa, M. Ishida, T. Nakajima, Y. Honda, O. Kitao, H. Nakai, M. Klene, X. Li, J.E. Knox, H.P. Hratchian, J.B. Cross, C. Adamo, J. Jaramillo, R. Gomperts, R.E. Stratmann, O. Yazyev, A.J. Austin, R. Cammi, C. Pomelli, J.W. Ochterski, P.Y. Ayala, K. Morokuma
13. G.A. Voth, P. Salvador, J.J. Dannenberg, V.G. Zakrzewski, S. Dapprich, A.D. Daniels, M.C. Strain, O. Farkas, D.K. Malick, A.D. Rabuck, K. Raghavachari, J.B. Foresman, J.V. Ortiz, Q. Cui, A.G. Baboul, S. Clifford, J. Cioslowski, B.B. Stefanov, G. Liu, A. Liashenko, P. Piskorz, I. Komaromi, R.L. Martin, D.J. Fox, T. Keith, M.A. Al-Laham, C.Y. Peng, A. Nanayakkara, M. Challacombe, P.M.W. Gill, B. Johnson, W. Chen, M.W. Wong, C. Gonzalez and J.A. Pople, Gaussian 03 Revision C.02, Gaussian Inc., Pittsburgh, PA, 2003.
14. D. Gross, Fortschr. Chem. Org. Naturst. 28 (1970), 109.
15. F. Jensen, Introduction to Computational Chemistry, Wiley, New York, 1999, p. 162.
16. P.G. Jones, G.M. Sheldrick, K.-H. Glusenkamp and L.F. Tietze, Acta Cryst. B36 (1980), 481.
17. R.O. Jones and O. Gunnarson, Rev. Mol. Physics 61 (1989), 689.
18. W. Kohn and L.J. Sham, Phys. Rev. A 140 (1965), 1133.
19. T.M. Kolev and B.A. Stamboliyski, Spectrochim. Acta A58 (2002), 3127.
20. A. Kovacs and I. Hargittai, Struct. Chem. 11(2,3) (2000), 193.
21. V. Krishnakumar and R. Ramasamy, Spectrochim. Acta A61 (2005), 673.
22. V. Krishnakumar and R. Ramasamy, Spectrochim. Acta A61 (2005), 2526.
23. V. Krishnakumar and R.J. Xavier, Indian J. Pure Appl. Phys. 41 (2003), 597.
24. V. Krishnakumar, R.J. Xavier and T. Chithambarathanu, Spectrochim. Acta A62 (2005), 931.
25. M. Kurt and S. Yurdakul, J. Mol. Struct. 654 (2003), 1.
26. C. Lee, W. Yang and R.G. Parr, Phys. Rev. B 37 (1988), 785.
27. K.Y. Lee, S.H. Sung, S.H. Kim, Y.P. Jang, T.H. Oh and Y.C. Kim, Arch. Pharm. Res. 32(5) (2009), 677.
28. S.Y. Lee, Bull. Korean Chem. Soc. 19(1) (1998), 93.
29. S.Y. Lee and B.H. Boo, J. Phys. Chem. 100 (1996), 15073.
30. J.P. Merrick, D. Moran and L. Radom, J. Phys. Chem. A 111 (2007), 11683.
31. V.N. Pandey and G.N. Chatruvedi, Indian J. Med. Res. 57 (1969), 503.
32. R.G. Parr and W. Yang, Density Functional Theory of Atoms and Molecules, Oxford Univ. Press, New York, 1989.
33. M. Pfeier, F. Bauer, T. Stey, D. Leusser, D. Stalke, B. Engels, D. Moigno and W. Kiefer, J. Mol. Model. 6 (2000), 299.
34. J.A. Pople, H.B. Schlegel, R. Krishnan, D.J. Defrees, J.S. Binkley, M.J. Frisch, R.A. Whiteside, R.F. Hout and W.J. Hehre, Int. J. Quant. Chem. S15 (1981), 269.
35. J.A. Pople, A.P. Scott, M.W. Wong and L. Radom, Isr. J. Chem. 33 (1993), 345.
36. A. Prabha Misra, CDRI Lucknow, PhD thesis, 1999, p. 52, submitted to Banaras Hindu University, India.
37. K. Raj, T.M. Vijayavitthal and A.P. Bhaduri, Modified iridoid glycosides (part III). Synthesis of hepatoprotective evaluation of modified iridoid glycosides, Ind. J. Chem. B 35 (1996), 1056–1061.
38. G. Rauhut and P. Pulay, J. Phys. Chem. 99 (1995), 3093.
39. H.B. Schlegel, J. Comput. Chem. 3 (1982), 214.
40. P. Sinha, S.E. Boesch, C. Gu, R.A. Wheeler and A.K. Wilson, J. Phys. Chem. A 108 (2004), 9213.
41. N. Sundaraganesan, H. Saleem, S. Mohan, M. Ramalingam and V. Sethuraman, Spectrochim. Acta A62 (2005), 740.
42. N. Sunderganesan, H. Saleem, S. Mohan, M. Ramalingham and V. Sethuram, Acta Cryst. A62 (2005), 740.
43. T.M. Vijayavitthal, K. Raj and A.P. Bhaduri, Studies on the profile of immunostimulant activities of modified iridoid glycosides, Bioorgan. Med. Chem. 6 (1998), 605–611.
44. P.K. Visen, B. Saraswat, K. Raj, A.P. Bhaduri and M.P. Dubey, Prevention gelactosamine-induced Hepatic damage by natural product loganin from the plant Srychnos nux-vomica: Studies on isolated Hepatocytes and bile flow in rat, Phytother. Res. 12 (1998), 405–408.
45. T. Ziegler, Chem. Rev. 91 (1991), 651.

CHAPTER 8

THEORETICAL STUDY OF COPPER COMPLEXES: MOLECULAR STRUCTURE, PROPERTIES, AND ITS APPLICATION TO SOLAR CELLS

Jesus Baldenebro-Lopez[1, 2], Norma Flores-Holguin[1], Jose Castorena-Gonzalez[2], Jorge Almaral-Sanchez[2] and Daniel Glossman-Mitnik[1]

[1]NANOCOSMOS Virtual Lab, Centro de Investigación en Materiales Avanzados, S.C., Miguel de Cervantes 120, Complejo Industrial Chihuahua, 31190 Chihuahua, CHIH, Mexico

[2]Facultad de Ingeniería Mochis, Universidad Autónoma de Sinaloa, Prol. Ángel Flores y Fuente de Poseidón, S.N., 81223 Los Mochis, SIN, Mexico

ABSTRACT

We present a theoretical investigation of copper complexes with potential applications as sensitizers for solar cells. The density functional theory (DFT) and time-dependent DFT were utilized, using the M06 hybrid meta-GGA functional with the LANL2DZ (D95V on first row) and DZVP basis sets. This level of calculation was used to find the optimized molecular structure, the absorption spectra, the molecular orbitals energies, and the chemical reactivity parameters that arise from conceptual DFT. Solvent effects have been taken into account by an implicit approach, namely, the polarizable continuum model (PCM), using the nonequilibrium version of the IEF-PCM model.

INTRODUCTION

A survey of our energy supply options reveals that the sun is the only clean energy source in sufficiently abundant supply to meet global demand, yet less than 0.1% of our energy needs are met through the direct conversion of sunlight [1]. For this reason, considerable attention and extensive research efforts have been devoted over the last years to dye-sensitized solar cells (DSSCs) as valuable and low-cost alternatives to silicon and other inorganic semiconductor-based photovoltaic devices [2–4]. The DSSC devices are composed of a nanocrystalline TiO_2 film adsorbed by organic or inorganic dyes as the sensitizer. When the dye molecule is excited by the absorption of sunlight, an electron is readily injected into the conduction band of TiO_2. The electron is then transported to the external circuit, and the dye molecule is regenerated by redox reaction coupled to ion pair in electrolyte [5].

One of the key components of a DSSC is the sensitizer [6], which is responsible for light absorption and the generation of electric charges [7], thus exerting a significant influence on the power conversion efficiency (η) [8]. The most common sensitizers are Ru(II) complexes based on 2,2′-bipyridine (bpy)

Citation: Jesus Baldenebro-Lopez, Norma Flores-Holguin, Jose Castorena-Gonzalez, Jorge Almaral-Sanchez, and Daniel Glossman-Mitnik, "Theoretical Study of Copper Complexes: Molecular Structure, Properties, and Its Application to Solar Cells," International Journal of Photoenergy, vol. 2013, Article ID 613064, 7 pages, 2013. doi:10.1155/2013/613064

ligands, such as cis-di(thiocyanato) bis(bpy-4,4′-dicarboxylate) ruthenium(II) (N3 [9] and N719 [10]). These complexes have reached over 11% power conversion efficiency [11]; but the rarity and high cost of the Ru may limit their practical usage. Alonso-Vante and coworkers [12] discovered that copper(I) complexes have similar photophysical properties with Ru complexes, indicating that the iterative chemical optimization of common metal complexes sensitizers can be comparable to that of Ru complexes [13]. Copper(I) complexes display a wide variety of excited states and especially photophysical and photochemical processes; particularly, copper(I)-polypyridine complexes exhibit low-lying MLCT transitions that can participate, among others, in electron transfer processes [14]. Copper-based dye sensitizers might be good substitutes to meet the requirements for next-generation DSSC devices. However, only a few studies have reported work on DSSCs coupled with copper-based dye sensitizers [15–18]. In this research, we propose the study of five molecular systems of the type $[CuL2]^+$ with bipyridine ligands; in order to calculate from the theoretical point of view their geometric and spectroscopic properties, as well as chemical reactivity.

MOLECULAR MODELING

Molecular calculations were carried out with density functional theory (DFT) [19–21] and time-dependent (TD) DFT [22–24] in the Gaussian09 program [25]. The hybrid-meta-GGA functional considered in this study is the M06 [26] due to the ability to study properties of organometallic compounds; our group has done a preliminary study [27]. The LANL2DZ (D95V on first row) for C, H, O, and N atoms [28] and DZVP for Cu atom [29] basis sets were used for the geometrical optimizations, as well as electrical and optical properties of all of the copper-based dye sensitizers considered, both in the gas phase and in solution. Ultraviolet-visible spectra (UV-Vis) have been simulated using the TD-DFT approach and analyzed through the SWizard program [30, 31], through pseudo-Voigt band shape with relative weight of Gaussian functions; solvent effects have been taken into account by an implicit approach, namely, the polarizable continuum model (PCM) [32] using the nonequilibrium version of the IEF-PCM model [33]; the equations were solved for 20 excited states.

Using the DFT framework makes it possible to find the chemical reactivity descriptors values, such as electron affinity (A), ionization potential (I), electronegativity (χ), hardness (η), and electrophilic index (ω); all these values were obtained using system energy calculations. The electronegativity, hardness, and electrophilic index are defined as

$$\chi = -\frac{(I+A)}{2},$$

$$\eta = \frac{(I-A)}{2},$$

$$\omega = \frac{\mu^2}{2\eta}, \qquad (1)$$

where μ denotes the chemical potential and is calculated as $-\chi$.

RESULTS AND DISCUSSION

Optimized molecular structures of copper complexes in the presence of solvent are shown in Figure 1. A selection of geometric parameters of these five molecules is listed in Table 1 (for more details, see supplementary material available online at http://dx.doi.org/10.1155/2013/613064). The molecular structures in gas phase and in solution are qualitatively similar; however, the greatest change is observed in bond angles and dihedral angles. The 5, 5′ position of the carboxylic acid on the bipyridine ligands for CuL4 copper complex favors the increase of distance between the copper atom and the nitrogen atoms with respect to 4, 4′ positions of the other complexes, because of the position of the acceptor. Substitution of the carboxilic acid with methyl ester groups do not change radically the distance of the metal center to the ligand.

Table 1: Selected bond lengths (angstroms), bond angles, and dihedral angles (degrees) for copper complexes

Parameter	Gas phase					In presence of solvent				
	CuL4	CuL6	CuL7	CuL8	CuL9	CuL4[a]	CuL6[a]	CuL7[b]	CuL8[b]	CuL9[b]
Cu-N1	2.0459	2.0310	2.0318	2.0321	2.0320	2.0445	2.0250	2.0289	2.0299	2.0300
Cu-N2	2.0493	2.0282	2.0318	2.0321	2.0320	2.0364	2.0274	2.0290	2.0300	2.0301
Cu-N3	2.0475	2.0332	2.0318	2.0321	2.0320	2.0427	2.0243	2.0289	2.0300	2.0301
Cu-N4	2.0481	2.0284	2.0318	2.0321	2.0320	2.0386	2.0261	2.0289	2.0299	2.0300
C1-O1	1.2342	1.2314	1.2355	1.2383	1.2359	1.2468	1.2444	1.2408	1.2446	1.2453
C1-O2	1.3691	1.3675	1.3586	1.3647	1.3706	1.3509	1.3502	1.3555	1.3649	1.3636
N1-Cu-N2	81.67	81.87	81.92	81.56	81.58	81.87	81.85	81.97	81.72	81.77
N1-Cu-N3	129.67	124.35	124.78	124.99	124.98	127.57	125.02	125.37	125.56	125.53
N3-Cu-N4	81.47	81.80	81.91	81.56	81.58	81.87	81.93	81.97	81.72	81.77
N2-Cu-N4	129.16	124.99	124.78	124.99	124.98	132.13	125.26	125.37	125.56	125.53
N1-N2-N3-N4	−73.96	−81.11	−80.87	−80.97	−80.97	−73.13	−81.47	−79.85	−79.85	−79.84

[a]MeOH solution, [b]Chloroform solution.

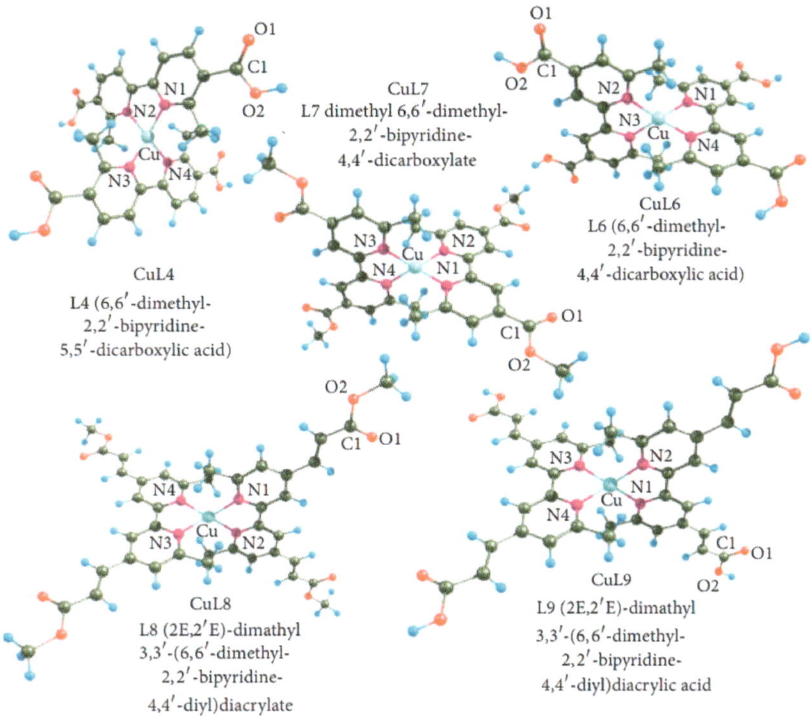

Figure 1: Optimized molecular structures of copper complexes in solution at the M06/ level of theory.

Absorption spectra for the proposed copper complexes are shown in Figure 2. These spectra were performed in the presence of solvent, under the same optimization conditions. The calculated value of the absorption wavelength (λ) is an important parameter, which indicates that these molecular systems should be considered for use as a functional material (as dye in this case) in a DSSC, the value of this parameter meets the requirement of the absorption property whose wavelength of maximum absorption (λ_{max}) should not be greater than 920 nm as established in the literature [34]. All studied sensitizers reach their λ_{max} above 500 nm, the largest value corresponds to the CuL4 sensitizer with 562 nm.

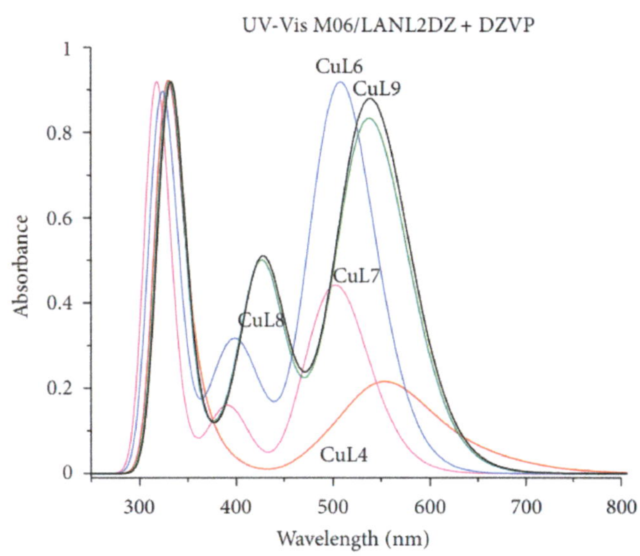

Complexes	λ Abs (nm)
CuL4	562
	495
	352
CuL6	506
	413
	381
	319
CuL7	500
	391
	315

Complexes	λ Abs (nm)
CuL8	535
	428
	358
	328
CuL9	536
	430
	329

Figure 2: Ultraviolet-visible (UV-Vis) spectra of copper complexes analyzed in this study.

Table 2 shows the results of TD-DFT calculations performed using the M06 functional and the LANL2DZ+DZVP basis sets; including the corresponding wavelengths (in nm), the energies (in eV), the oscillator strength (f), and the orbitals involved in the transitions. This information shows that the peak of the wavelength of maximum absorption is due to charge transfer between the metal atom and the ligand molecule, which may be explained more clearly in Figure 3. The excitation contains a mixture of HOMO→LUMO, HOMO − 1→LUMO, and HOMO − 1→LUMO + 1, this leads to consider them as degenerate orbitals.

Table 2: Copper complexes electronic excited states, showing wavelengths (nm), energies (eV), oscillator strength (), and the orbitals involved in the transitions. Only excited states with f > 0.02 are shown

Copper complex	λ (nm)	E (eV)	f	Assignment; H = HOMO, L = LUMO, and L + 1 = LUMO + 1
CuL4	562	2.21	0.048	H−1 → L (61%) H → L (26%) H−1 → L + 1 (10%)
	495	2.51	0.026	H−1 → L + 1 (75%) H → L + 1 (17%)
	352	3.52	0.057	H−1 → L + 3 (51%) H → L + 3 (29%) H−1 → L + 4 (11%) H → L + 4(7%)
CuL6	505	2.45	0.243	H → L (66%) H−1 → L + 1 (29%)
	413	3.00	0.030	H → L + 2 (54%) H−2 → L (21%) H−1 → L + 2 (16%) H → L + 3 (6%)
	381	3.25	0.028	H−1 → L + 3 (78%) H−2 → L + 1 (7%) H−3 → L (7%)
	319	3.89	0.092	H−4 → L (52%) H−3 → L (21%) H−6 → L (18%)
CuL7	500	2.48	0.267	H−1 → L + 1 (+52%) H → L (+47%)
	391	3.17	0.045	H → L + 3 (65%) H−1 → L + 2 (32%)
	315	3.94	0.285	H−4 → L + 1 (46%) H−6 → L (43%)
CuL8	535	2.32	0.391	H−1 → L + 1 (52%) H → L (45%)
	428	2.90	0.112	H → L + 3 (60%) H−1 → L + 2 (37%)
	358	3.46	0.025	H−1 → L + 4 (56%) H → L + 5 (40%)
	328	3.78	0.303	H−5 → L (47%) H−6 → L + 1 (40%) H−7 → L (5%)
CuL9	536	2.31	0.376	H−1 → L + 1 (52%) H → L (45%)
	430	2.88	0.102	H → L + 3 (61%) H−1 → L + 2 (37%)
	329	3.77	0.280	H−5 → L (48%) H−6 → L + 1 (39%)

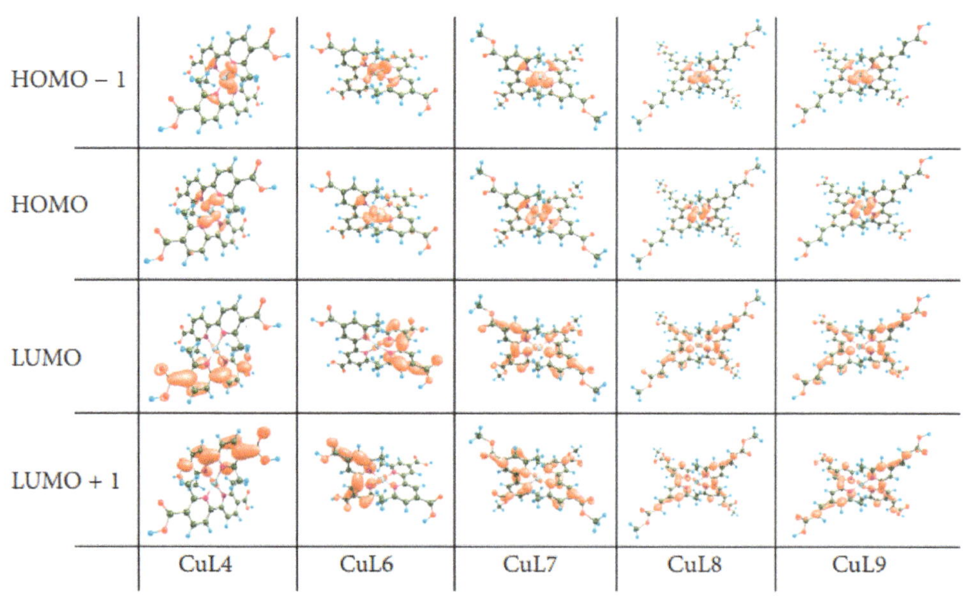

Figure 3: Molecular orbitals for Cu complexes calculated at the M06/level of theory.

The charge transfer efficiency from the sensitizer to the nanocrystalline oxide is extremely important in the solar cell design. Since the crucial electronic excitations occur from the highest occupied molecular orbitals (HOMO) to the lowest unoccupied molecular orbitals (LUMO), it is, therefore, important to form efficient charge-separated states with the HOMO localized on the donor subunit and the LUMO on the acceptor subunit. Figure 3 shows as the HOMO orbital density is located over around the copper atom in complexes; meanwhile, the density of the LUMO orbital is in the ligand, tending to move toward the carboxyl acid and methyl ester groups. This is beneficial in accordance with the above mentioned.

Figure 4 shows the HOMO-LUMO molecular orbitals' energetic position. These results are of great importance, since they can be used during synthesis to determine the solubility and chemical reactivity of the molecule, as reported in different studies [35, 36]. Additional factors to consider is the sensitizer LUMO level that is greater than the conduction band level of nanocrystalline oxides that are commonly used in such devices, also the HOMO level can be observed below the redox potential of the electrolyte [37]; both considerations are important to the design of the sensitizer. The CuL4 complex has the smaller energy gap (3.093 eV); however, the values for the other complexes studied are close to this. Molecular orbitals energy levels were calculated in the presence of solvent.

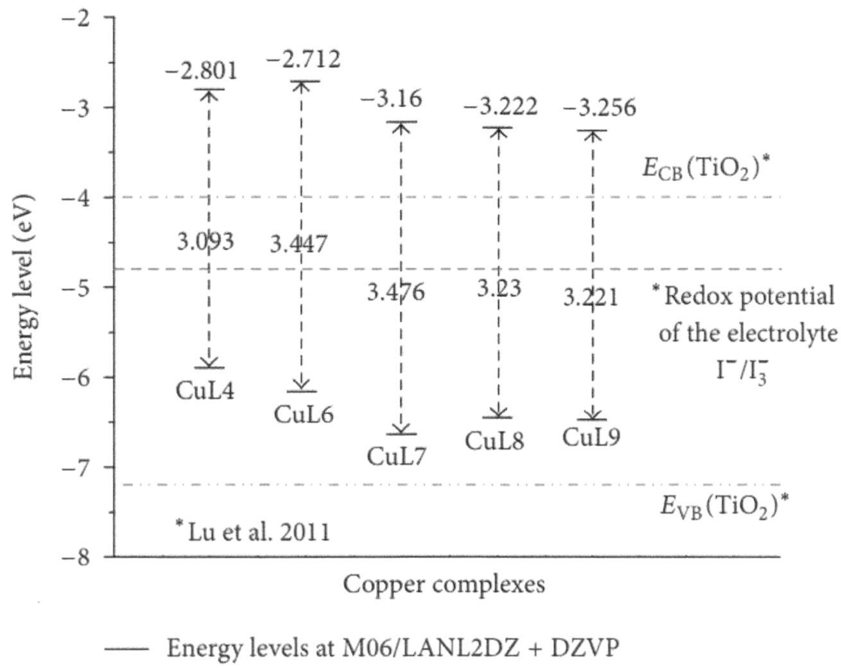

Figure 4: Molecular orbitals energy levels diagram.

Chemical reactivity parameters such as ionization potential (I), electron affinity (A), electronegativity (χ), chemical hardness (η), and electrophilicity index (ω) for the studied molecular systems (Table 3) were obtained by energy calculations (neutral and ionic state), taking into account the ground state geometry optimization.

Table 3: Chemical reactivity parameters using DFT descriptors

Copper complex	Chemical reactivity descriptors (eV)				
	I	A	χ	η	ω
CuL4	9.79	4.47	7.13	2.66	9.55
CuL6	10.15	4.53	7.34	2.81	9.58
CuL7	9.74	4.13	6.94	2.80	8.58
CuL8	9.35	4.26	6.80	2.54	9.10
CuL9	9.57	4.44	7.01	2.56	9.58

According to Table 3, complexes with carboxylic acid as anchor groups exhibit greater electron affinity, electronegativity, and electrophilicity index than those containing methyl ester groups. The trend in ionization potential for the copper complexes is CuL8 < CuL9 < CuL7 < CuL4 < CuL6. The results indicate that CuL6 and CuL7 copper complexes are harder than the rest of the systems studied; this can be interpreted as systems with greater resistance to change with the number of electrons.

CONCLUSIONS

In this work, the molecular structure and properties of five molecular systems proposed as sensitizers in dye-sensitized solar cells have been calculated. The applied methodology for this study is based on the density functional theory, using the M06 hybrid meta-GGA functional and the LANL2DZ and DZVP basis sets.

The molecular systems characterization includes the calculation of geometric parameters, ultraviolet-visible spectra, electronic excited states, molecular orbitals, and chemical reactivity parameters. The longer wavelength of maximum absorption and the smallest energy gap is for the CuL4 complex. For all cases, the HOMO orbital density is located over around the copper atom; meanwhile, the density of the LUMO orbital is in the ligand. Moreover, the molecular orbitals levels are appropriate considering the position of the conduction band of nanocrystalline oxides and the redox potential of the electrolyte. Analyzing these data makes it possible to find potential applications for these copper complexes in photovoltaic devices.

ACKNOWLEDGMENTS

This work has been supported by Consejo Nacional de Ciencia y Tecnología (CONACYT), Centro de Investigación en Materiales Avanzados, S.C. (CIMAV), Universidad Autónoma de Sinaloa, and Dirección de Investigación y Posgrado by PROFAPI 2012/040. J. Baldenebro-Lopez gratefully acknowledges a fellowship from CONACYT. D. Glossman-Mitnik and N. Flores-Holguin are researchers of CIMAV and CONACYT.

REFERENCES

1. P. G. Bomben, K. C. D. Robson, B. D. Koivisto, and C. P. Berlinguette, "Cyclometalated ruthenium chromophores for the dye-sensitized solar cell," Coordination Chemistry Reviews, vol. 256, pp. 1438–1450, 2012.
2. F. De Angelis, S. Fantacci, and R. Gebauer, "Simulating dye-sensitized TiO_2 heterointerfaces in explicit solvent: absorption spectra, energy levels, and dye desorption," Journal of Physical Chemistry Letters, vol. 2, no. 7, pp. 813–817, 2011.
3. Y. Shibano, T. Umeyama, Y. Matano, and H. Imahori, "Electron-donating perylene tetracarboxylic acids for dye-sensitized solar cells," Organic Letters, vol. 9, no. 10, pp. 1971–1974, 2007.
4. Y. Wang, M. Wu, X. Lin, A. Hagfeldt, and T. Ma, "Optimization of the performance of dye-sensitized solar cells based on Pt-like TiC counter electrodes," The European Journal of Inorganic Chemistry, vol. 2012, no. 22, pp. 3557–3561, 2012.
5. S.-H. Liu, H. Fu, Y.-M. Cheng et al., "Theoretical study of N749 dyes anchoring on the $(TiO_2)_{28}$ surface in DSSCs and their electronic absorption properties," The Journal of Physical Chemistry C, vol. 116, no. 31, pp. 16338–16345, 2012.
6. C. Dragonetti, A. Valore, A. Colombo et al., "A new thiocyanate-free cyclometallated ruthenium complex for dye-sensitized solar cells: beneficial effects of substitution on the cyclometallated ligand,"Journal of Organometallic Chemistry, vol. 714, pp. 88–93, 2012.
7. M. Katono, T. Bessho, M. Wielopolski et al., "Influence of the anchoring modes on the electronic and photovoltaic properties of D–π-A dyes," The Journal of Physical Chemistry C, vol. 116, no. 32, pp. 16876–16884, 2012.
8. X. Lu, Q. Feng, T. Lan, G. Zhou, and Z.-S. Wang, "Molecular engineering of quinoxaline-based organic sensitizers for highly efficient and stable dye-sensitized solar cells," Chemistry of Materials, vol. 24, no. 16, pp. 3179–3187, 2012.
9. M. K. Nazeeruddin, A. Kay, I. Rodicio et al., "Conversion of light to electricity by cis-X2bis(2,2′-bipyridyl-4,4′-dicarboxylate)ruthenium(II) charge-transfer sensitizers (X = Cl-, Br-, I-, CN-, and SCN-) on nanocrystalline TiO_2 electrodes," Journal of the American Chemical Society, vol. 115, no. 14, pp. 6382–6390, 1993.
10. M. K. Nazeeruddin, F. de Angelis, S. Fantacci et al., "Combined experimental and DFT-TDDFT computational study of photoelectrochemical cell ruthenium sensitizers," Journal of the American Chemical Society, vol. 127, no. 48, pp. 16835–16847, 2005.
11. G. C. Vougioukalakis, A. I. Philippopoulos, T. Stergiopoulos, and P. Falaras, "Contributions to the development of ruthenium-based sensitizers for dye-sensitized solar cells," Coordination Chemistry Reviews, vol. 255, no. 21-22, pp. 2602–2621, 2011.
12. N. Alonso-Vante, J.-F. Nierengarten, and J.-P. Sauvage, "Spectral sensitization of large-band-gap semiconductors (thin films and ceramics) by a carboxylated bis(1,10-phenanthroline)copper(I) complex," Journal of the Chemical Society, Dalton Transactions, no. 11, pp. 1649–1654, 1994.
13. J. Deng, Q. Xiu, L. Guo, L. Zhang, G. Wen, and C. Zhong, "Branched chain polymeric metal complexes containing Co(II) or Ni(II) complexes with a donor-π-acceptor architecture: synthesis, characterization, and photovoltaic applications," Journal of Materials Science, vol. 47, no. 7, pp. 3383–3390, 2012.
14. L. Barrientos, C. Araneda, B. Loeb, and I. G. Crivelli, "Synthesis, spectroscopic and electrochemical characterization of copper(I) complexes with functionalized pyrazino[2,3-f]-1,10-phenanthroline,"Polyhedron, vol. 27, no. 4, pp. 1287–1295, 2008.
15. S. Sakaki, T. Kuroki, and T. Hamada, "Synthesis of a new copper(I) complex,[Cu(tmdcbpy)2]+ (tmdcbpy = 4,4′,6,6′-tetramethyl-2,2′-bipyridine-5,5′- dicarboxylic acid), and its application to solar cells," Journal of the Chemical Society, Dalton Transactions, no. 6, pp. 840–842, 2002.
16. S. Hattori, Y. Wada, S. Yanagida, and S. Fukuzumi, "Blue copper model complexes with distorted tetragonal geometry acting as effective electron-transfer mediators in dye-sensitized solar cells," Journal of the American Chemical Society, vol. 127, no. 26, pp. 9648–9654, 2005.
17. T. Bessho, E. C. Constable, M. Graetzel et al., "An element of surprise-efficient copper-functionalized dye-sensitized solar cells," Chemical Communications, pp. 3717–3719, 2008.
18. Y.-Y. Lin, T.-H. Chu, S.-S. Li et al., "Interfacial nanostructuring on the performance of polymer/ TiO_2nanorod bulk heterojunction solar cells," Journal of the American Chemical Society, vol. 131, no. 10, pp. 3644–3649, 2009.
19. P. Hohenberg and W. Kohn, "Inhomogeneous electron gas," Physical Review, vol. 136, no. 3B, pp. B864–B871, 1964.
20. W. Kohn and L. J. Sham, "Self-consistent equations including exchange and correlation effects," Physical Review, vol. 140, no. 4A, pp. A1133–A1138, 1965.
21. R. G. Parr and W. Yang, Density-Functional Theory of Atoms and Molecules, Oxford University Press, New York, NY, USA, 1989.
22. K. Burke, J. Werschnik, and E. K. U. Gross, "Time-dependent density functional theory: past, present, and future," Journal of Chemical Physics, vol. 123, no. 6, 9 pages, 2005.
23. R. E. Stratmann, G. E. Scuseria, and M. J. Frisch, "An efficient implementation of time-dependent density-functional theory for the calculation of excitation energies of large molecules," Journal of Chemical Physics, vol. 109, no. 19, pp. 8218–8224, 1998.
24. R. Bauernschmitt and R. Ahlrichs, "Treatment of electronic excitations within the adiabatic approximation of time dependent density functional theory," Chemical Physics Letters, vol. 256, no. 4-5, pp. 454–464, 1996.
25. F. Gaussian 09 Revision A. 02, M. J., G. W.Trucks, H.

B. Schlegel et al., Gaussian, Inc., Wallingford CT, 2009.

26. Y. Zhao and D. G. Truhlar, "The M06 suite of density functionals for main group thermochemistry, thermochemical kinetics, noncovalent interactions, excited states, and transition elements: two new functionals and systematic testing of four M06-class functionals and 12 other functionals," Theoretical Chemistry Accounts, vol. 120, no. 1–3, pp. 215–241, 2008.

27. J. Baldenebro-López, J. Castorena-González, N. Flores-Holguín, J. Almaral-Sánchez, and D. Glossman-Mitnik, "Computational molecular nanoscience study of the properties of copper complexes for dye-sensitized solar cells," International Journal of Molecular Sciences, vol. 13, no. 12, pp. 16005–16019, 2012.

28. T. H. Dunning Jr. and P. J. Hay, Methods of Electronic Structure Theory, Plenum, New York, NY, USA, 1976.

29. C. Sosa, J. Andzelm, B. C. Elkin, E. Wimmer, K. D. Dobbs, and D. A. Dixon, "A local density functional study of the structure and vibrational frequencies of molecular transition-metal compounds," The Journal of Physical Chemistry, vol. 96, no. 16, pp. 6630–6636, 1992.

30. S. I. Gorelsky, "Swizard program," 2012, University of Ottawa, Ottawa, Canada, http://www.sg-chem.net/.

31. S. I. Gorelsky and A. B. P. Lever, "Electronic structure and spectra of ruthenium diimine complexes by density functional theory and INDO/S. Comparison of the two methods," Journal of Organometallic Chemistry, vol. 635, no. 1-2, pp. 187–196, 2001.

32. J. Tomasi, B. Mennucci, and R. Cammi, "Quantum mechanical continuum solvation models," Chemical Reviews, vol. 105, no. 8, pp. 2999–3094, 2005.

33. M. Cossi and V. Barone, "Time-dependent density functional theory for molecules in liquid solutions,"The Journal of Chemical Physics, vol. 115, no. 10, pp. 4708–4717, 2001.

34. M. A. Green, Solar Cells: Operating Principles, Technology and Systems Applications, Prentice-Hall, Englewood Cliffs, New Jersey, USA, 1982.

35. F. de Angelis, S. Fantacci, and A. Sgamellotti, "An integrated computational tool for the study of the optical properties of nanoscale devices: application to solar cells and molecular wires," Theoretical Chemistry Accounts, vol. 117, no. 5-6, pp. 1093–1104, 2007.

36. Y.-X. Weng, Y.-Q. Wang, J. B. Asbury, H. N. Ghosh, and T. Lian, "Back electron transfer from TiO_2nanoparticles to FeIII(CN)6 3−: origin of non-single-exponential and particle size independent dynamics," Journal of Physical Chemistry B, vol. 104, no. 1, pp. 93–104, 2000.

37. X. Lu, S. Wei, C.-M. L. Wu, S. Li, and W. Guo, "Can polypyridyl Cu(I)-based complexes provide promising sensitizers for dye-sensitized solar cells? A theoretical insight into Cu(I) versus Ru(II) sensitizers," Journal of Physical Chemistry C, vol. 115, no. 9, pp. 3753–3761, 2011.

CHAPTER 9

THE INFLUENCE OF MOLECULAR STRUCTURE MODIFICATIONS ON VIBRATIONAL PROPERTIES OF SOME BETA BLOCKERS: A COMBINED RAMAN AND DFT STUDY

A. Farcaş[1] C. Iacoviţă[2] E. Vinţeler[3] V. Chiş[3] R. Ştiufiuc[2] and C. M. Lucaciu[2a]

[1]Department of Mathematics and Computer Science, "Iuliu Hatieganu" University of Medicine and Pharmacy, L. Pasteur 6, 400349 Cluj-Napoca, Romania

[2]Department of Pharmaceutical Physics-Biophysics, "Iuliu Hatieganu" University of Medicine and Pharmacy, L. Pasteur 6, 400349 Cluj-Napoca, Romania

[3]Faculty of Physics, "Babes-Bolyai" University, Kogalniceanu 1, 400084 Cluj-Napoca, Romaniaa

ABSTRACT

We report results of a systematic Raman, SERS, and DFT study on four beta blocking molecules: Atenolol, Metoprolol, Propranolol, and, for the first time reported in the literature, Bisoprolol. The choice of these molecules was motivated by the structural similarities between Atenolol, Bisoprolol, and Metoprolol on one hand and by their differences relative to Propranolol. The density functional theory (DFT) approach, using the B3LYP method at the 6-311+G(d,p) level of theory, has been employed for geometry optimization and vibration bands assignments. The obtained results highlight the major role played by the central aromatic ring whose vibrations dominate the Raman spectra in all compounds. While the phenyl group vibrations dominate the Raman spectrum in the case of Atenolol, Bisoprolol, and Metoprolol, the spectrum of Propranolol presents high intensity vibrations of the naphthyl group. SERS performed on gold and silver colloids, at various pH conditions, revealed a higher sensitivity for Propranolol detection. The pH dependence of the spectrum indicates that the studied beta blockers attach themselves to the metal nanoparticles in a protonated form. The molecular adsorption geometry on metal nanoparticles surface has been evaluated by using the experimental SER spectra and the quantum chemical calculations.

INTRODUCTION

Raman spectroscopy and its more recent development Surface Enhanced Raman Spectroscopy (SERS) represent very powerful vibrational spectroscopy techniques allowing highly sensitive detection of low concentration analytes, in different media [1]. As a consequence of the fact that a single Raman scan can provide molecular specific spectral data in the 40–4000 cm^{-1} range, Raman spectroscopy gained a lot of popularity in different areas of pharmaceutical industry. Its ability to provide a high degree of specificity in analysis and its capacity to attain a high degree of lateral and depth resolution represent other powerful arguments for its future use in pharmaceutical industry. In the case of SERS, high amplifications are observed for the Raman modes which are perpendicular to the

Citation: A. Farcaş, C. Iacoviţă, E. Vinţeler, V. Chiş, R. Ştiufiuc, and C. M. Lucaciu, "The Influence of Molecular Structure Modifications on Vibrational Properties of Some Beta Blockers: A Combined Raman and DFT Study," Journal of Spectroscopy, vol. 2016, Article ID 3137140, 9 pages, 2016. doi:10.1155/2016/3137140

metallic surface, thus offering the possibility to investigate molecular orientation at the metallic surface and eventually the functional groups involved in the molecular attachment to the metallic surface. However, the complete interpretation and understanding of experimental Raman/SER spectra of molecules of pharmaceutical interest under different experimental conditions are not straightforward, the subject being under constant debate in the scientific literature.

The beta blockers are primarily used in the therapy of cardiovascular diseases [2]. Propranolol (PRNL) (1-isopropylamino-3-(1-naphthyloxy)-2-Propranolol) was the first beta blocking molecule used in cardiovascular therapy. It is a nonselective beta blocking agent, having a membrane stabilizing activity owing to the fact that it is the most hydrophobic compound among all beta blockers. Bisoprolol (BIS) (1-{4-{[2-(1-Methylethoxy) ethoxy] methyl}phenoxy}-3-[(1-methylethyl)amino]-2-propanol), Atenolol (ATE) ((±)-4-[2-Hydroxy -3-[(1-methylethyl) amino]propoxy]benzeneacetamide), and Metoprolol (MET) ((±)1-(Isopropylamino)-3 -[p-(β-methoxyethyl)phenoxy]-2-propanol (+)) represent other three drugs belonging to the class of beta1-adrenergic receptor blockers, able to specifically block adrenergic receptors from heart muscle and kidneys, with little activity against beta2-adrenergic receptors of the lungs and vascular smooth muscle, having a very similar chemical structure. As a common structural feature, all currently available beta blockers possess an amino-alkanol side chain and an aromatic group: phenyl in the case of ATE, BIS, and MET and naphthyl in the case of PRNL. The amino-alkanol side chain is characterized by the presence of an asymmetric carbon atom, resulting in the existence of a d- and an l-enantiomer (in the case of propanolamine type compounds, like the ones studied in this paper, the d-enantiomers show the (R)-configuration and the l-enantiomers show the (S)-configuration), the latter being more potent in blocking the adrenergic receptors. These drugs are used in clinical practice as racemates, although in vitro studies have demonstrated in the early 60s that l-enantiomers are orders of magnitude more potent than their d-enantiomer counterparts [3]. Over a long period of time it was considered that the d-enantiomers have no side effects, but during the years, more and more studies reported highly specific side effects of the d-enantiomers of beta blockers. Therefore, a "chiral switch," that is, replacing the currently clinically used racemic mixtures with optically pure l-enantiomers, was proposed [4]. On the other hand, beta blockers are prohibited in some sports according to the List of Prohibited Methods and Substances, released every year by the World Anti-Doping Agency [5]. As a consequence there is a need for a sensitive and reliable technique able to detect and measure the beta blocker concentration in body fluids.

The first use of Raman and SERS for beta blockers detection was reported by Rupérez and Laserna in 1996 [6]. More recently Levene et al. [7] were able to determine the best experimental parameters leading to the detection of PRNL down to a concentration of 2.36 ng/mL (7.97 nM), well within the physiological concentrations. In previous papers published in our group [8] the SERS, Raman, and DFT were synergistically employed for elucidating the adsorption geometry of PRNL molecules on metallic nanoparticles surface. The nanoparticles mediated enantioselective recognition of PRNL enantiomers through native cyclodextrin complexation by using SERS was also reported [9, 10]. Cozar et al. [11] reported vibrational studies together with DFT simulations for ATE and MET and proposed possible adsorption geometries for these molecules on metallic surface.

In this paper we present a systematic Raman, SERS, and DFT investigation of four widely used beta blockers: ATE, BIS, MET, and PRNL. By extending the object of this study to four beta blocking molecules having very similar structures it was possible to better understand the relationship between the chemical structures and the vibrational properties of the molecules in a holistic manner. The conventional Raman (solid phase) and SERS spectra of BIS are also reported for the first time. We investigated the structural and vibrational similarities between ATE, MET, and BIS and compared them with PRNL. The choice of the four molecules is motivated by the fact that the main structural difference between ATE, MET, and BIS on one hand and PRNL on the other hand is represented by the central aromatic group which is a phenyl in the case of the first three and a naphthyl in the case of PRNL. Thus, it was possible to detect the main changes in the Raman spectra induced by this replacement. For a better understanding of the vibrational spectra and for a correct assignment of the vibrational frequencies the density functional theory (DFT) approach using the B3LYP method at the 6-311+G(d,p) level of theory has been employed. The calculations have been performed for the ground state geometry of the most stable species. The SERS experiments have been performed on both silver and gold colloids in various pH conditions for assessing the availability of the SERS method for detecting the beta blockers and to compare the SERS sensitivity among the studied molecules. The molecular adsorption geometry on noble metal nanoparticles surface has been evaluated by using the experimental SER spectra and the quantum chemical calculations.

MATERIALS AND METHODS

All reagents used were of analytical grade. (R)-(+)-Propranolol hydrochloride, (S)-(−)-Propranolol hydrochloride, (R)-(+)-Atenolol, (S)-(−)-Atenolol, racemic Metoprolol, and Bisoprolol (analytical standard) were purchased from Sigma-Aldrich. Hydroxylammonium chloride and sodium hydroxide were purchased from Merck and VWR, respectively, whereas trinatriumcitrate-dihydrate and tetrachloroauric(III) acid trihydrate were from Roth. All reagents were used without further purification. Ultrapure water (18.2 MO, Barnstead EASYPure ROdi) was used for the preparation of aqueous solutions.

Preparation of Colloids

Silver colloids were prepared according to the Leopold and Lendl method [12]. An amount of 0.017 g of silver nitrate was dissolved in 90 mL of double distilled water. In a separate recipient, 0.017 g of hydroxyl-ammonium chloride was dissolved in 10 mL of water, followed by the addition of 250 mL of aqueous sodium hydroxide solution (0.1 M). This solution was then added rapidly to the silver nitrate solution under vigorous stirring. After few seconds, a grey-brown colloidal solution was obtained which was further stirred for 10 minutes. The pH value of the silver colloid, measured immediately after preparation, was 6.9. The UV-Vis spectrum of the silver colloidal solution exhibits a single narrow absorption peak at 418 nm, indicating the formation of spherical silver nanoparticles.

Gold colloid has been prepared by dissolving 0.0485 g of tetrachloroauric(III) acid trihydrate in 100 mL of water and boiled under vigorous stirring. A volume of 3 mL of 1% sodium citrate aqueous solution was then added dropwise while boiling and stirring. The solution was allowed to boil for further 20 minutes with continuous stirring. A significant color shift of the aqueous solution to a dark red color was observed [13]. By the end of the synthesis the total volume of the colloidal solution was adjusted to 100 mL by water addition. The pH value of the gold colloid, measured after cooling to room temperature, was 6. The UV-Vis spectrum of the gold colloidal solution presents a single narrow absorption peak at 522 nm, indicating the formation of spherical gold nanoparticles.

Methods and Instrumentation

Raman spectra (solid phase) and some SER spectra were recorded on a Renishaw micro-Raman spectrometer, using 3 wavelengths laser lines (532 nm, 633 nm, and 785 nm with powers up to 200 mW, 17 mW, and 300 mW, resp.). The system is integrated with a Leica optical microscope with 50x objective lens. The powder samples or a droplet of colloidal solution were placed on an aluminum plate in front of the lens. Each spectrum represents the average of minimum 5 recordings, the acquisition time for each spectrum being usually 5 s. A part of the SER spectra has been acquired in back-scattering geometry in the 200–2000 cm^{-1} range using a DeltaNu Advantage spectrometer (DeltaNu, Laramie, WY) equipped with a laser diode emitting at 785 nm. The laser power was 100 mW and the spectral resolution was 5 cm^{-1}. The spectra were recorded for 1 mL glass vials filled with 540 μL of colloid and 60 μL of the analyte. In some experiments, as indicated in the figure legends, 100 μL of 1 M NaCl was added in order to produce the aggregation of the nanoparticles and the creation of "hot spots." The pH of the solutions was adjusted with solutions of HCl and NaOH before the addition of the aggregating agent. Each SER spectrum is the average of five recordings taken with an acquisition time of 30 seconds. The UV-Vis absorption spectra of colloidal solutions were recorded using a T92+ UV-VIS Spectrophotometer from PG INSTRUMENTS, using standard 1 cm quartz cells at room temperature, over a spectral range between 190 nm and 900 nm with a spectral resolution of 2 nm.

DFT Computations

All the calculations related to molecular geometry optimization and generation of theoretical vibrational spectra were performed using Gaussian 09 software package [14] based on density functional theory (DFT) methods at the B3LYP/6-311+G(d,p) level of theory [15, 16]. No symmetry restrictions were applied during the optimization process. The calculated Raman activities (S_i) were converted into relative Raman intensities (I_i) using the relation given as follows [17, 18]:

$$I_i = \frac{\left[f\left(\upsilon_0 - \upsilon_i\right)^4 S_i\right]}{\left\{\upsilon_i\left[1 - \exp\left(-hc\upsilon_i/k_\beta T\right)\right]\right\}}. \quad (1)$$

Here f is a normalization factor for all peak intensities, υ_0 represents the exciting laser wavenumber, υ_i is the wavenumber of the ith vibrational mode, c stands for the speed of light, h and k_β are Planck's and Boltzmann's constants, and T represents the temperature.

Since the vibrational wavenumbers obtained by quantum chemical calculation are typically larger than their experimental counterparts, the computed wavenumbers for the frequencies larger than 1000 cm^{-1} have been scaled by 0.967 as proposed by Scott and Radom [19].

RESULTS AND DISCUSSIONS

The structures of the most stable conformers of the 4 beta blockers studied in this paper are presented in Figure1 together with the atoms numbering. The chemical structure and the optimized geometry of the molecules were calculated at B3LYP/6-311+G(d,p) level of theory. The vibrational frequencies were computed for the optimized geometries in order to avoid the occurrences of imaginary frequencies. Consequently, the optimized geometry corresponds to a local minimum on the potential-energy surface.

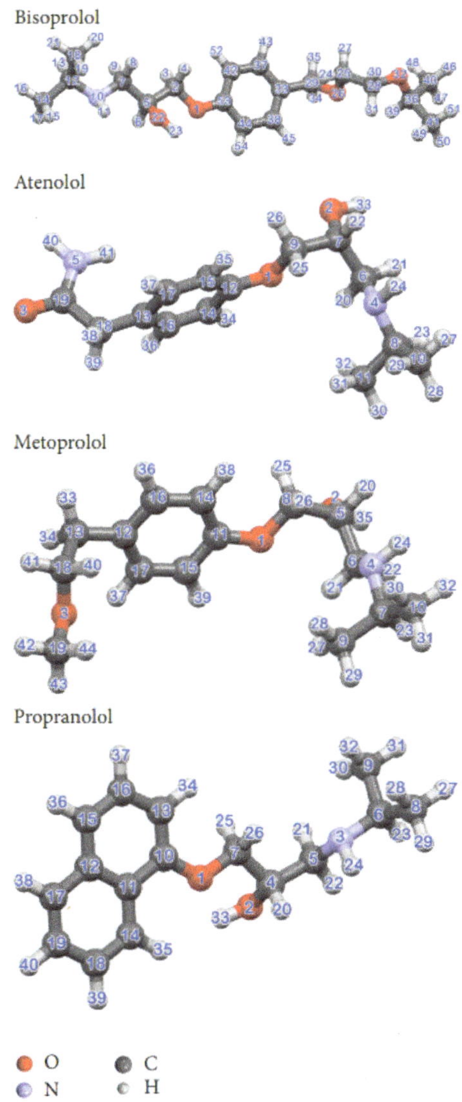

Figure 1: Optimized geometries of Atenolol, Bisoprolol, Metoprolol, and Propranolol from DFT calculations at B3LYP/6-311+G(d,p) level of theory.

At a closer look of their structure one can observe that the common structural features of these molecules are the amine alkanol side chain containing the asymmetric carbon atom, the amino groups, and the OH groups. The central part of the molecules consists of an aromatic group which is a phenyl one in the case of ATE, BIS, and MET and a naphthyl one in the case of PRNL. No other side chain exists in PRNL. ATE possesses an acetamide group bound to the central phenyl group, while BIS and MET possess propoxyethoxy and methoxyethyl groups, respectively.

The experimental Raman spectra of the 4 compounds are presented in Figure 2. One can easily see the likeness of the Raman vibration patterns in the case of ATE, BIS, and MET on one side and the complete different pattern recorded for PRNL. In the case of PRNL the Raman spectrum is dominated by two strong vibrations at 1385 cm^{-1} and 737 cm^{-1} while in the case of the other three beta blockers there are several high intensity common bands at ~1612 cm^{-1}, ~850 cm^{-1}, and ~638 cm^{-1}. These differences can be attributed to the substitution of the phenyl group of ATE, MET, and BIS with the naphthyl group in the case of PRNL, as we will demonstrate bellow.

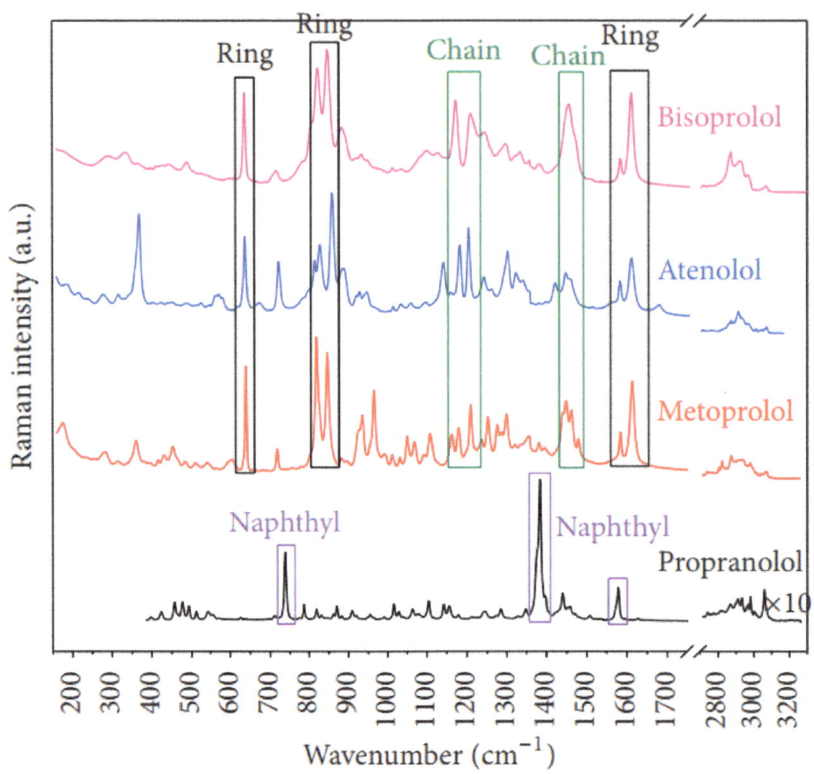

Figure 2: Raman spectra of Bisoprolol, Atenolol, Metoprolol, and Propranolol using a 785 nm excitation wavelength. The rectangles indicate the bands assigned to similar vibrations in Atenolol, Bisoprolol, and Metoprolol and the naphthyl group vibrations in Propranolol. The spectra were normalized for comparison.

For getting more insights of the relation between the molecular chemical structure and the Raman spectra, the simulated and experimental Raman spectra of the studied beta blockers are shown in Figure 3. The vibrational bands assignments obtained from DFT for all beta blocking molecules are indicated in Tables 1 and 2. For BIS (Figure 3(a)) we obtained a good correlation between the experimental and the simulated spectra. The most intense experimental Raman bands are at 638 cm^{-1} (652 cm^{-1} in DFT), assigned to the δ(CCC ring), 823 $cm^{-1}$and 849 cm^{-1} (835 cm^{-1} and 859 cm^{-1} in DFT), both assigned to ring breathing, 1612 cm^{-1} (1597 cm^{-1} in DFT), assigned to ν(CC ring), and 1173 cm^{-1} (1182 cm^{-1} in DFT), assigned to the δ(CC, ring). It is worthy to mention that all these vibrations involve the central phenyl ring of the molecule.

Table 1: Experimental (Exp) and simulated (Calc) vibrational band wavenumbers and Raman Intensities (RI) spectra of atenolol, bisoprolol and metoprolol

Atenolol				Metoprolol				Bisoprolol			
Calc/cm^{-1}	RI (%)	Exp/cm^{-1}	Assignments	calc/cm^{-1}	RI (%)	Exp/cm^{-1}	Assignments	calc/cm^{-1}	RI (%)	Exp/cm^{-1}	Assignments
282	20	279	$\omega(N_5H_2)$	263	13	281	$\delta(O_3C_{18}C_{19}) + \rho(C_{19}H_3)$				
								348	9	336	$\delta(O_{26}\text{-}CH_2)$
353	23	368	$\rho(N_5H_2) + \rho(C_{18}H_2)$	359	33	362	$\delta(C_{11}O_1) + \rho(C_{13}H_2)$				
476	14	479	$\rho(C_9H_2)$								
576	5	569	τ(CCC ring) + $\delta(O_3C_{19}\text{-}N_5H_2)$								
648	19	637	δ(CCC ring) + $\tau(N_5H_2)$	654	20	640	δ(CCC ring)	652	18	638	δ(CCC ring) + $\rho(C_{29}H_2)$
732	10	722	ω(CCC ring) + $\rho(N_5H_2)$	754	16	719	$\delta(N_4H_{24})$ + δ(CC ring)	725	7	716	δ(CCC ring) + $\rho(C_{29}H_2)$
837	51	828	$\rho(N_5H_2)$ + δ(CCC ring)	835	13	820	$\rho(C_5H_2) + \nu_s(O_3C_{18}C_{19})$	835	31	823	$\omega(C_{44}H_{54})$ + ring breathing
873	34	859	$\nu(C_7C_9)$ + ring breathing	855	87	848	ring breathing + $\delta(N_4H_{24})$	859	36	849	$\nu(C_{53}O_1)$ + ring breathing
								891	20	887	$\nu(C_5O_{22})$ + ν(NH)
934	8	928	$\rho(C_{18}H_{38}) + \omega(C_{19}O_3N_5)$	952	8	935	$\rho(C_{11}H_2) + \delta(C_{10}H_3)$	935	10	936	$\nu(C_{53}O_1) + \rho(C_2H_2)$
1055	12	1058	$\rho(N_5H_2)$	1041	12	1051	$\nu(O_1C_8)$	1003	16	1014	$\nu(C_{29}O_{26}) + \rho(C_{29}H_2)$
1146	8	1143	$\delta(H_{23}C_8\text{-}N_4H_{24}) + \rho(C_{10}H_2)$	1105	17	1109	$\nu_{as}(O_3C_{18}C_{19})$	1106	7	1098	$\delta(C_{36}C_{41}H_{50}) + \nu(C_{36}C_{40})$
1159	16		δ(CH ring) + $\tau(C_{18}H_2)$	1159	17	1164	δ(CH ring)	1155	13	1173	δ(CH ring)
1185	51	1183	$\nu(C_{13}C_{18}) + \omega(C_{18}H_2)$	1183	55	1180	$\delta(C_{13}H_{34}) + \delta(O_2H_{35})$	1182	57	1173	$\nu(C_{29}C_{33})$ + δ(CH ring)
				1203	8	1210	$\delta(O_2H_{35}) + \delta(C_{13}H_{34})$				
								1219	12	1210	$\nu_{as}(C_2O_1) + \tau(C_2H_2)$
1235	15	1205	$\nu(O_1C_{12})$ + δ(CCC ring)					1242	17	1245	$\tau(C_{29}H_2) + \rho(C_{25}H_2)$
				1256	20	1253	$\nu_{as}(O_1C_{11}C_8) + \delta(C_{17}C_{37})$				
1271	100	1243	$\omega(C_{18}H_2) + \delta(C_{19}N_5H_{40})$					1279	11	1298	$\tau(C_{28}H_2) + \rho(C_{25}H_2)$
1303	7	1301	$\delta(C_{19}N_5H_{40}) + \omega(C_{18}H_2)$	1325	64	1300	$\delta(C_5H_{20}) + \omega(C_6H_2)$	1304	4		$\delta(N_{10}H_{11}) + \delta(C_{12}H_{13})$
1329	5	1323	$\delta(C_8H_{23}) + \omega(C_{11}H_3)$	1364	5	1358	$\omega(C_{13}H_2) + \tau(C_{18}H_2)$	1346	29	1334	$\omega(C_{29}H_2) + \delta(C_{36}H)$
1414	6	1423	$\delta(C_{18}H_2) + \delta(N_5H_2)$								
1440	19	1448	$\delta(C_{11}H_3) + \delta(C_9H_2)$	1438	11	1450	$\delta_{as}(C_9H_3) + \delta_{as}(C_{10}H_3)$	1440	12	1455	$\delta(C_{29}H_2)$
				1469	12	1461	$\delta(C_{18}H_2) + \delta(C_{19}H_2)$				
1558	17	1584	ν(CC ring) + $\delta(N_5H_2)$	1560	14	1585	ν(CC ring)	1564	10	1585	ν(CC ring)
1597	95	1612	ν(CC ring)	1598	96	1614	ν(CC ring)	1597	75	1612	ν(CC ring) + δ(CH ring)
1694	6	1681	$\nu(O_3C_{19}) + \delta(N_5H_{40})$								

Table 2: Experimental and simulated vibrational bands in the Raman spectrum of propranolol

Calc/cm^{-1}	RI (%)	Exp/cm^{-1}	Assignments
430	5	422	$\delta(N_3H_{24}) + \rho(C_5H_2)$
453	10	456	symetric stretching naphtalene
470	8	477	$\delta(N_3C_5C_6) + \rho(C_7H_2)$
493	18	493	$\rho(N_3C_5C_6)$
536	11	512	δ(CCC) (naphtalene)
758	25	737	naphtalene ring breathing + $\delta(N_3H_{24})$
783	11	785	$\delta(N_3H_{24}) + \nu(N_3C_6C_5)$
828	6	818	$\nu_s(C_6C_9C_8) + \delta(N_3H_{24})$
884	6	908	δ(CC naphtalene) + $\delta(N_3H_{24})$
952	6	953	$\nu(O_1C_7) + \rho(C_7H_2)$
1006	12	1015	$\nu(C_{18}C_{19}) + \delta(C_{17}H_{38})$
1021	6	1027	$\delta(O_2H_{33}) + \rho(C_5H_2)$
1130	5	1155	δ(CH) (naphtalene) + $\nu(C_6N_3)$
1255	10	1245	$\tau(C_7H_2) + \delta(C_4H_{20})$
1352	100	1383	ν(CC naphtalene)
1423	26	1439	δ(CH) (naphtalene) + ν(CC naphtalene)
1560	38	1578	ν(CC naphtalene) + δ(CH naphtalene)

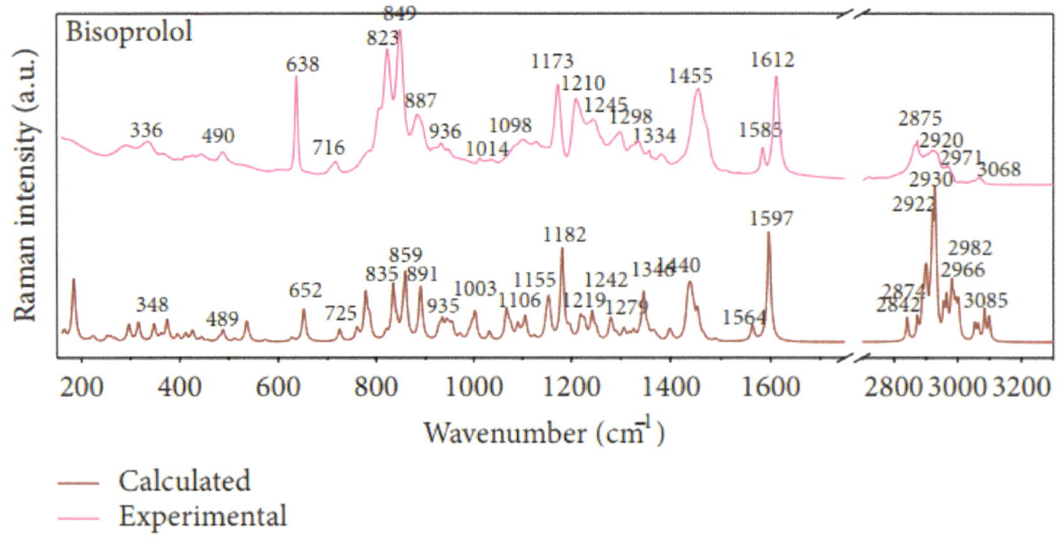
Bisoprolol
849
823
638
1173
1210
1612
1455
887
1245
1298
336
490
936
1098
716
1014
1334
1585
2875
2920
2971
3068
2930
2922
1597
1182
2982
859
2966
835
891
1242
1346
1440
652
1003
1155
348
725
935
1106
1219
1279
489
1564
2874
2842
3085
Raman intensity (a.u.)
200 400 600 800 1000 1200 1400 1600 2800 3000 3200
Wavenumber (cm^{-1})
Calculated
Experimental

(a)

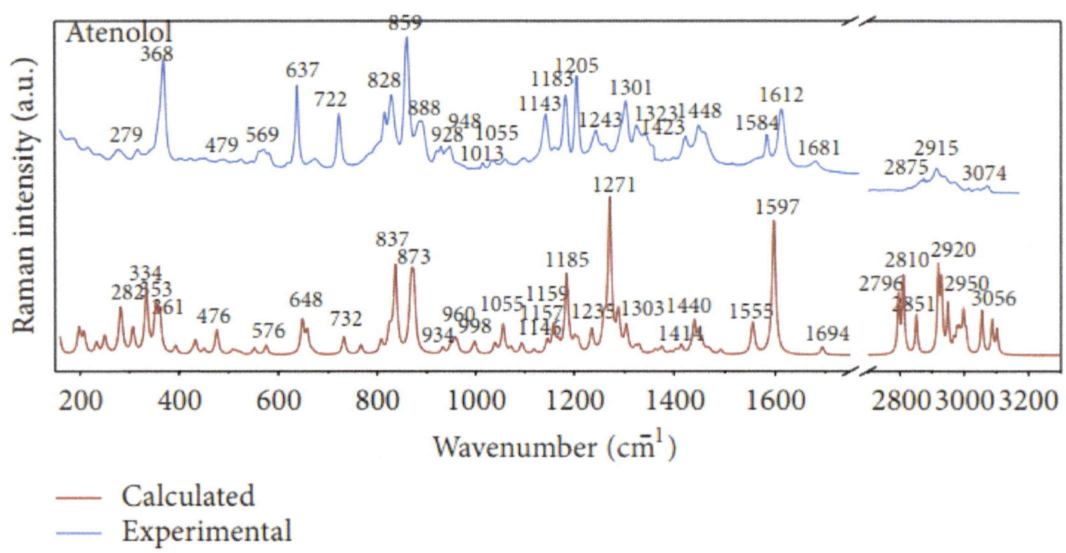
Atenolol
859
368
637
828
1205
1183
722
1301
888
1143
1612
1323
1448
948
1243
1584
928
1055
1423
279
479
569
1013
1681
2915
2875
3074
1271
1597
837
873
1185
2810
2920
334
282
353
648
1159
2796
2950
361
476
732
960
1055
1157
1235
1303
1440
1555
2851
3056
576
934
998
1146
1414
1694
Raman intensity (a.u.)
200 400 600 800 1000 1200 1400 1600 2800 3000 3200
Wavenumber (cm^{-1})
Calculated
Experimental

(b)

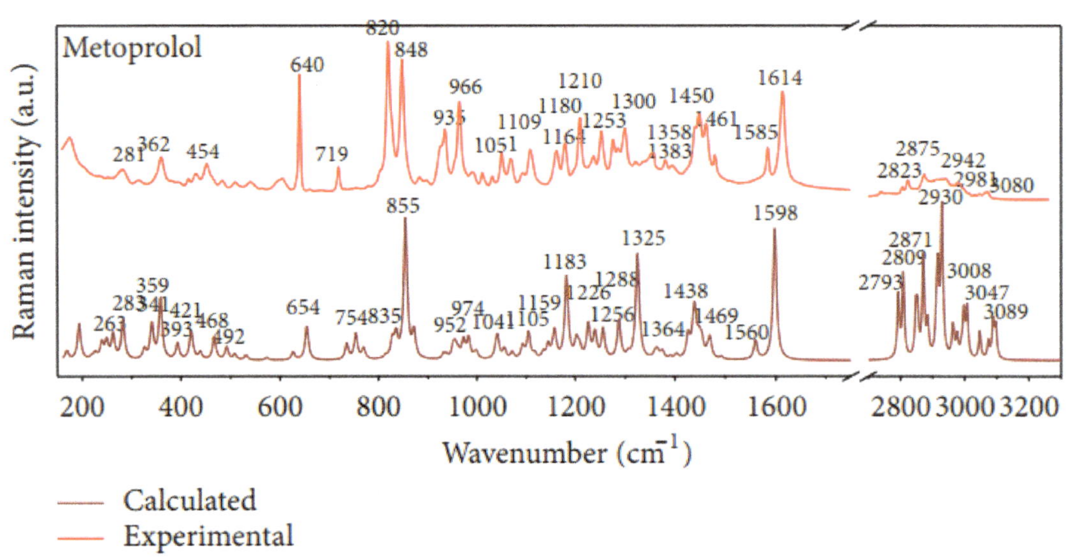
Metoprolol
820
848
640
966
1210
1614
1450
1180
1300
935
1109
1253
1461
1358
1051
1164
1585
281
362
454
719
1383
2875
2942
2823
2981
3080
2930
855
1598
1325
2871
2809
1183
3008
1288
1226
2793
1438
3047
359
283
341
421
468
654
754
835
974
1159
1256
3089
263
393
492
952
1041
1105
1364
1469
1560
Raman intensity (a.u.)
200 400 600 800 1000 1200 1400 1600 2800 3000 3200
Wavenumber (cm^{-1})
Calculated
Experimental

(c)

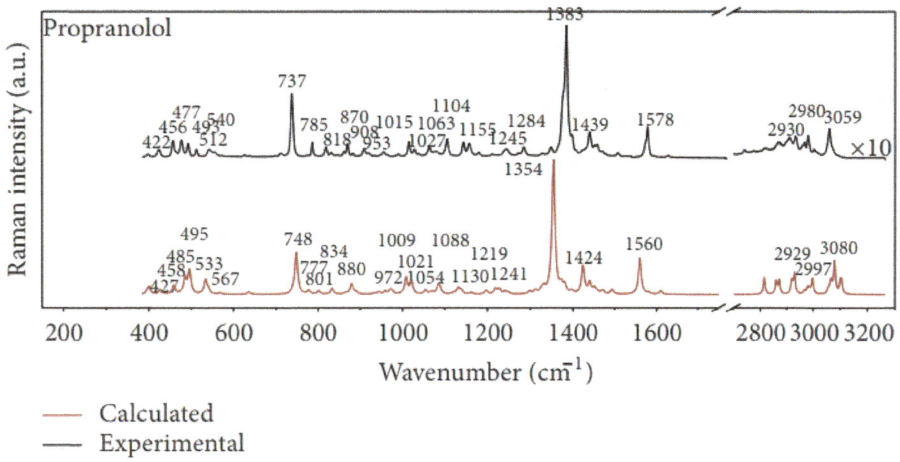

(d)

Figure 3: Experimental and simulated Raman spectra of the studied beta blocking molecules. In the case of Propranolol the experimental intensities at high wavenumbers were multiplied 10 times.

Other intense bands were observed at 1455 cm^{-1} (1440 cm^{-1} in DFT) assigned to $\delta(CH_2) + \delta$(CH ring), 1210 cm^{-1} (1219 cm^{-1} in DFT) assigned to $\nu_{as}(CO)$, 1337 cm^{-1} and 1298 cm^{-1} assigned to different CH_2 vibration modes, 1585 cm^{-1} assigned to the ν(CC ring) vibration, and 887 cm^{-1} assigned to ν(CO).

For ATE (Figure 3(b)) we experimentally measured the same intense bands as in the case of BIS, attributable to ring vibration modes at only slightly different wavenumbers (637 cm^{-1}, 828 cm^{-1}, 859 cm^{-1}, 1183 cm^{-1}, and 1611 cm^{-1}) (Table 1), and vibrations occurring at 1457 cm^{-1}, 1205 cm^{-1}, and 1301 cm^{-1} with the same assignments as in the case of BIS: $\delta(CH_2)$ + δ(CH ring), $\nu_{as}(CO)$, and different CH_2 vibration modes, respectively. The major differences in the Raman spectra of the two beta blockers were assigned to the acetamide group present in the ATE molecule. At 368 cm^{-1} an intense vibration band was assigned to the rocking mode of N_5H_2 group, specific for this molecule, an increase in the intensity of the vibrational band from 722 cm^{-1} (assigned in the case of BIS only to a CCC ring wagging mode) due to a supplementary contribution for the rocking of $N5H_2$ group, and a weaker but clearly noticeable vibration mode at 1680 cm^{-1}assigned to a $\nu(C_{19}O_3) + \delta(N_5H_{40})$. The peak from 1420 cm^{-1} can be assigned to a combined $\delta(C_{18}H_2)$ and$\delta(N_5H_2)$ vibration mode.

In the case of MET (Figure 3(c)) the major vibrations obtained in the Raman spectrum are again, like in the case of ATE and BIS, attributed to vibration modes of the ring atoms, that is, 848 cm^{-1} ring breathing, 640 cm^{-1} δ(CCC ring), and 1614 cm^{-1} and 1585 cm^{-1} δ(CC ring). Other intense bands were recorded at 1461 cm^{-1}assigned to $\delta(C_{18}H_2)$ and $\delta(C_{19}H_2)$ and similar to BIS and ATE the bands at 1210 cm^{-1}, 1180 cm^{-1}, and 1164 cm^{-1} assigned to $\delta(O_2H_{35})$, $\delta(C_{13}H_{34})$, and δ(CH ring), respectively. As it was the case for ATE and BIS, in the 900 cm^{-1}–1000 cm^{-1} region two supplementary medium intense bands are observed in the Raman spectrum of MET at 935 cm^{-1} and 966 cm^{-1}, which are assigned to $\rho(C_{11}H_2) + \delta(C_{10}H_3)$ and δ(CH ring) + $\delta(C_{18}H_2)$, respectively.

The experimental and simulated spectra for PRNL are very different as compared to ATE, BIS, and MET. The Raman spectrum (Figure 3(d)) is dominated by the 1383 cm^{-1} and 737 cm^{-1} bands assigned to the naphthyl ring vibration modes. Other medium intensity vibration modes were recorded at 1578 cm^{-1} (1560 cm^{-1} in the simulated spectrum) and at 1439 cm^{-1} (1423 cm^{-1} in the simulated spectrum) which are also associated with different modes of naphthyl group vibration. As a general observation we notice a very good correlation of the simulated spectrum with the experimental one both from the point of view of the wavenumbers and relative peak intensities, as can be seen in Figure 3(d) and in Table 2.

Finally, the mean of the absolute wavenumber deviations between experiment and theory for the most important bands are ±11 cm^{-1} for ATE, ±9 cm^{-1} for BIS, ±12 cm^{-1} for MET, and ±16 cm^{-1} for PRNL. Moreover, the root mean square standard deviation (RMSD) of residuals is 8 cm^{-1} in the case of ATE, 6 cm^{-1}for BIS, 9 cm^{-1} for MET, and 11 cm^{-1} in the case of PRNL. These discrepancies can be observed in Table 1where one can also find the assignments of the normal modes. The assignments of experimental vibrational bands were made by visual correlation with calculated vibrational bands using Gauss View 5.0, taking into account the wavenumber values and relative intensities.

The SER spectra of PRNL on both silver and gold colloids are represented in Figure 4 and they are fully consistent with data previously reported by Stiufiuc et al. [8]. We notice that the addition of PRNL to the colloidal solutions changes the color of both gold and silver colloids from red to blue and light grey to dark grey, respectively. This phenomenon can be attributed to the aggregation of the colloid due the presence of Cl^-anions since PRNL was used as a hydrochloride salt. This aggregation leads to the formation of "hot spots" having increased Raman enhancements factors and higher sensitivity. The general pattern of the SER spectra of PRNL is the same for both silver and gold colloids. However, small differences can be observed. The spectra are dominated by the naphthyl ring vibration at 1383 cm^{-1} and by the ring breathing at 737 cm^{-1} with the wavenumbers being very close to the pure Raman vibrations. This small difference in the wavenumbers (~2 cm^{-1}) indicates a weak interaction of the molecule with the metallic surface. Other medium intensity bands were recorded at 1440 cm^{-1} and 1577 cm^{-1}, assigned to the ring CC stretching and ring CH bending vibrations, respectively. However there is a difference in the relative intensity of these two bands for the two colloids; the 1577 cm^{-1} band is more intense for Au colloid, whereas the 1440 cm^{-1} peak is more intense for Ag colloid. These differences can be explained by small differences in the orientation of the molecule at the metallic surface and by different interaction strengths with the metals, thus emphasizing the role of the metallic surface in SERS measurements.

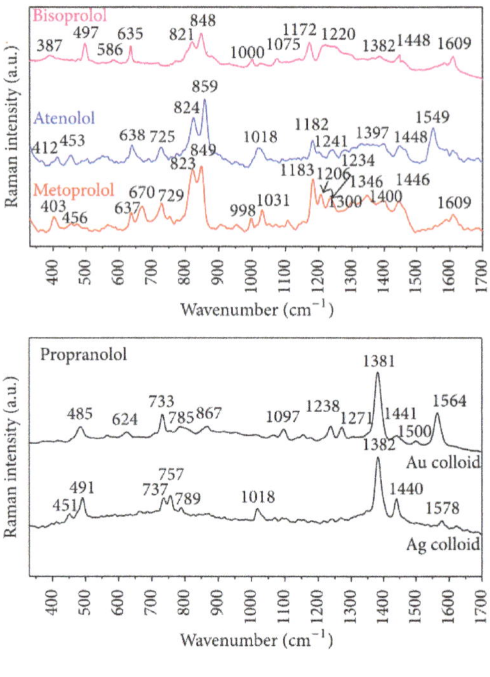

Figure 4: SER spectra for 10 mM Atenolol, Bisoprolol, and Metoprolol obtained on silver nanoparticles with 0.1 M NaCl used as aggregating agent (upper frame). SER spectra for 0.1 mM Propranolol on silver and gold nanoparticles (lower frame).

Stiufiuc et al. have demonstrated that the adsorption of PRNL on the metallic surface occurs through the oxygen atoms with the naphthalene ring lying perpendicular to it [8]. Their explanation was based on experimentally acquired SER spectra, Raman assignments, Molecular Electrostatic Potential (MEP) contour map, and SERS selection rules stating that the normal modes having a change in the polarizability component normal to the surface are enhanced [20, 21]. Our results, based on the used DFT method are in good agreement with these findings. The SER spectra of ATE, MET, and BIS, obtained on silver colloids, are presented in Figure 4 (upper frame). It is worth mentioning that for analytes concentrations of 10^{-3}–10^{-4} M no specific SER spectra could be recorded without the addition of an aggregating agent. ATE, BIS, and MET used in this study were free bases, whereas PRNL was a hydrochloride salt. The different behaviour of PRNL on one side and ATE, BIS, and MET, on the other side, can be attributed to the presence/absence in the colloidal solution of Cl^- anions.

In order to increase the Raman signal for MET, BIS, and ATE, NaCl was added up to a concentration of 10^{-4} M to the colloid-analyte mixture. Previously, several other salts (NaF, KBr) were tested, in order to assess the eventual fake vibration peaks belonging to the capping agents, used in the synthesis process, or to the impurities in the colloidal suspension strongly enhanced by colloid aggregation. In agreement with previous reports [22, 23], where either NaCl or LiCl were added to the colloid suspension, any additional vibrational bands were not detected in the SER spectra, apart from the well-known metal-Cl^- band, situated in the low wavenumber region (227 cm^{-1} for Ag-Cl and 250 cm^{-1} for Au-Cl). Therefore, the Cl^- anions are the most suitable to increase the Raman signal. The SER spectra of all three betal blockers are dominated by the breathing vibrations of the phenyl ring at 848 cm^{-1} for BIS, 859 cm^{-1} for ATE, and 849 cm^{-1} for MET. The small difference between the Raman vibrational bands and the SERS ones indicate a small interaction between the analyte and the metallic surface. Other vibrations belonging to the ring are also enhanced: 635 cm^{-1}, 821 cm^{-1}, and 1609 cm^{-1} for BIS, 638 cm^{-1} and 824 cm^{-1} for ATE, and 637 cm^{-1}, 823 cm^{-1}, and 1609 cm^{-1} for MET. It is interesting to note that the 1612 cm^{-1} CC ring stretching vibration is not enhanced in the SER spectrum of ATE but the 1549 cm^{-1} band, assigned to a combined CC ring stretching and the bending of N_5H_2has a medium intensity, while in the case of MET and BIS the vibrational bands at 1612 cm^{-1} are enhanced. This different behaviour of ATE with respect to the other two similar beta blockers (MET and BIS) can be explained by a different orientation of the molecule to the surface, the enhancement of the 1549 cm^{-1} band being due to the N_5H_2 bending (a group specific only for ATE). Also, as it can be observed from the SER spectrum of ATE, the amide I vibration band at 1681 cm^{-1} is lacking. This is an indication that the carbonyl group from acetamide lies in a configuration parallel to the metallic surface. ATE presents a clearly distinguishable vibration band at 725 cm^{-1}, corresponding to 722 cm^{-1} in the Raman spectrum with a higher intensity as compared to the other two beta blockers, attributed to a supplementary contribution due to a rocking vibration of the same N_5H_2 group.

Other group of medium intensity vibration bands were recorded at 1172 cm^{-1} for BIS and 1183 cm^{-1} for both ATE and MET. These bands were assigned to different CC aliphatic chain vibration modes. The weak vibrations at 1448 cm^{-1} for BIS and ATE and 1446 cm^{-1} for MET were assigned to CH bending vibration modes. The mechanism of molecular attachment to the metallic nanoparticles surface was proposed to take place through the amino group of the amino alkanol side chain up to a pH value of ~9.5 which is in fact the pKa for this amino group [8, 11]. Bellow this pH value the molecule is mostly protonated and can be electrostatically attracted by the nanoparticles. The pH dependence of the SER spectra for ATE, BIS, and MET (data not shown) demonstrates that as the pH is increased from acidic values the intensity of the SERS peaks increases up to pH = 9. For pH values greater than 11 the SER spectra could not be detected as the molecules became deprotonated. The same behaviour was proved experimentally for Propranolol [8]. The main information concerning the molecules orientation at the surface can be obtained from the SER spectra. The possible groups involved are either the protonated amino groups, the lone electron pairs of N and O atoms, or the π electrons of the phenyl ring. A flat orientation of the ring to the surface would imply a dramatic decrease in the SERS intensity corresponding to the ring vibrations, which is not the case. The fact that the ring vibration dominates the SERS spectra for the studied beta blockers implies that the aromatic rings possess a tilted position with respect to the metallic surface. The relatively low enhancement of the vibration bands in SERS for BIS, ATE, and MET indicated that the adsorption onto the silver surface is done through a weak physisorption process. The MEP maps data (not shown) indicates that the most negative part of the molecules is around the oxygen atoms; therefore we cannot preclude their involvement in the attachment.

CONCLUSIONS

A systematic Raman, SERS, and DFT study was performed on four beta blocker molecules. The used DFT theoretical methods allowed us to obtain the optimized geometries, the assignment of vibrational bands, and the MEP maps. The Raman spectra revealed the major role played by the central aromatic group, the ring vibrations dominating the overall spectra in the 200–2000 cm^{-1} region. We also noticed a great degree of "vibrational" similarity for BIS, ATE, and MET. The replacement of the phenyl ring (existing in ATE, BIS, and MET) with a phenyl one (for PRNL) leads to a dramatic change in the Raman pattern. In this case an increased domination of the ring vibrations has been recorded on the Raman spectrum. The SERS results show a higher sensitivity for PRNL as compared to BIS, ATE, and MET. The SERS mechanism rely mainly on molecular physisorption to the metallic surface through weak electrostatic forces acting between the protonated form of the molecule and the metallic nanoparticles. The molecular orientation relative to the surface is most probably determined by the oxygen atoms with the central ring having a tilted orientation relative to the nanoparticles surface.

CONFLICT OF INTERESTS

The authors declare that there is no conflict of interests regarding the publication of this paper.

ACKNOWLEDGMENT

This research was supported by CNCSIS-UEFISCDU, Project no. PN-II-ID-PCE-2011-3-0954. Dr. Anca Farcas acknowledges financial support from the POSDRU Grant no. POSDRU/159/1.5/S/136893, entitled "Parteneriat Strategic Pentru Cresterea Calitatii Cercetarii Stiintifice din Universitatile Medicale Prin Acordarea de Burse Doctorale si Postdoctoral—DocMed. Net_2.0".

REFERENCES

1. E. Smith and G. Dent, Modern Raman Spectroscopy: A Practical Approach, John Wiley & Sons, Chichester, UK, 2005.
2. N. Freemantle, J. Cleland, P. Young, J. Mason, and J. Harrison, "β blockade after myocardial infarction: systematic review and meta regression analysis," The British Medical Journal, vol. 318, no. 7200, pp. 1730–1737, 1999.
3. A. G. Wilson, O. G. Brooke, H. J. Lloyd, and B. F. Robinson, "Mechanism of action of βadrenergic receptor blocking agents in angina pectoris: comparison of action of propranolol with dexpropranolol and practolol," British Medical Journal, vol. 4, no. 5680, pp. 399–401, 1969.
4. K. Stoschitzky, W. Lindner, and G. Zernig, "Racemic beta-blockers—fixed combinations of different drugs," Journal of Clinical and Basic Cardiology, vol. 1, no. 1, pp. 14–18, 1998.
5. Beta-Blockers, August 2015, http://list.wada-ama.org/list/p2-beta-blockers.
6. A. Rupérez and J. J. Laserna, "Surface-enhanced Raman spectrometry of chiral β-blocker drugs on colloidal silver," Analytica Chimica Acta, vol. 335, no. 1-2, pp. 87–94, 1996.
7. C. Levene, E. Correa, E. W. Blanch, and R. Goodacre, "Enhancing surface enhanced Raman scattering (SERS) detection of propranolol with multiobjective evolutionary optimization," Analytical Chemistry, vol. 84, no. 18, pp. 7899–7905, 2012.
8. R. Stiufiuc, C. Iacovita, C. M. Lucaciu et al., "Adsorption geometry of propranolol enantiomers on silver nanoparticles," Journal of Molecular Structure, vol. 1031, pp. 201–206, 2013.
9. E. Bodoki, M. Oltean, A. Bodoki, and R. Ştiufiuc, "Chiral recognition and quantification of propranolol enantiomers by surface enhanced Raman scattering through supramolecular interaction with β-cyclodextrin," Talanta, vol. 101, pp. 53–58, 2012.
10. R. Stiufiuc, C. Iacovita, G. Stiufiuc, E. Bodoki, V. Chis, and C. M. Lucaciu, "Surface mediated chiral interactions between cyclodextrins and propranolol enantiomers: a SERS and DFT study," Physical Chemistry Chemical Physics, vol. 17, no. 2, pp. 1281–1289, 2015.
11. O. Cozar, L. Szabó, I. B. Cozar et al., "Spectroscopic and DFT study of atenolol and metoprolol and their copper complexes," Journal of Molecular Structure, vol. 993, no. 1–3, pp. 357–366, 2011.
12. N. Leopold and B. Lendl, "A new method for fast preparation of highly surface-enhanced raman scattering (SERS) active silver colloids at room temperature by reduction of silver nitrate with hydroxylamine hydrochloride," The Journal of Physical Chemistry B, vol. 107, no. 24, pp. 5723–5727, 2003.
13. K. C. Grabar, R. Griffith Freeman, M. B. Hommer, and M. J. Natan, "Preparation and characterization of Au colloid monolayers," Analytical Chemistry, vol. 67, no. 4, pp. 735–743, 1995.
14. M. J. Frisch, G. W. Trucks, H. B. Schlegel, et al., Gaussian 09, Revision A.02, Gaussian, Wallingford, Conn, USA, 2009.
15. A. D. Becke, "Density-functional thermochemistry. III. The role of exact exchange," The Journal of Chemical Physics, vol. 98, no. 7, pp. 5648–5652, 1993.
16. C. Lee, W. Yang, and R. G. Parr, "Development of the Colle-Salvetti correlation-energy formula into a functional of the electron density," Physical Review B, vol. 37, no. 2, pp. 785–789, 1988.
17. P. L. Polavarapu, "Ab initio vibrational Raman and Raman optical activity spectra," Journal of Physical Chemistry, vol. 94, no. 21, pp. 8106–8112, 1990.
18. G. A. Guirgis, Y. E. Nashed, and J. R. Durig, "Raman and infrared spectra, conformational stability, ab initio calculations and vibrational assignments for 2-chloroethyl silane," Journal of Molecular Structure,

vol. 510, no. 1–3, pp. 13–34, 1999.

19. A. P. Scott and L. Radom, "Harmonic vibrational frequencies: an evaluation of Hartree-Fock, Møller-Plesset, quadratic configuration interaction, density functional theory, and semiempirical scale factors,"The Journal of Physical Chemistry, vol. 100, no. 41, pp. 16502–16513, 1996.

20. J. A. Creighton, "Surface raman electromagnetic enhancement factors for molecules at the surface of small isolated metal spheres: the determination of adsorbate orientation from sers relative intensities,"Surface Science, vol. 124, no. 1, pp. 209–219, 1983.

21. M. Moskovits and J. S. Suh, "Surface selection rules for surface-enhanced Raman spectroscopy: calculations and application to the surface-enhanced Raman spectrum of phthalazine on silver," The Journal of Physical Chemistry, vol. 88, no. 23, pp. 5526–5530, 1984.

22. X. Dong, H. Gu, and F. Liu, "Study of the surface-enhanced Raman spectroscopy of residual impurities in hydroxylamine-reduced silver colloid and the effects of anions on the colloid activity," Spectrochimica Acta Part A: Molecular and Biomolecular Spectroscopy, vol. 88, pp. 97–101, 2012.

23. N. R. Yaffe and E. W. Blanch, "Effects and anomalies that can occur in SERS spectra of biological molecules when using a wide range of aggregating agents for hydroxylamine-reduced and citrate-reduced silver colloids," Vibrational Spectroscopy, vol. 48, no. 2, pp. 196–201, 2008.

CHAPTER 10

SYNTHESIS, SPECTROSCOPIC, MOLECULAR STRUCTURE, AND ANTIBACTERIAL STUDIES OF DIBUTYLTIN(IV) SCHIFF BASE COMPLEXES DERIVED FROM PHENYLALANINE, ISOLEUCINE, AND GLYCINE

Har Lal Singh and Jangbhadur Singh

Department of Chemistry, Faculty of Engineering & Technology, Mody University of Science and Technology, Laxmangarh, Sikar, Rajasthan 332311, India

ABSTRACT

New series of organotin(IV) complexes and Schiff bases derived from amino acids have been designed and synthesized from condensation of 1H-indole-2,3-dione, 5-chloro-1H-indole-2,3-dione, and α-amino acids (phenylalanine, isoleucine, and glycine). All compounds are characterized by elemental analyses, molar conductance measurements, and molecular weight determinations. Bonding of these complexes is discussed in terms of their UV-visible, infrared, and nuclear magnetic resonance (^{1}H, ^{13}C, and ^{119}Sn NMR) spectral studies. The results suggest that Schiff bases behave as monobasic bidentate ligands and coordinate with dibutyltin(IV) in octahedral geometry according to the general formula $[Bu_2Sn(L)_2]$. Elemental analyses and NMR spectral data of the ligands with their dibutyltin(IV) complexes agree with their proposed distorted octahedral structures. Few representative compounds are tested for their in vitro antibacterial activity against Gram-positive (B. cereus, Staphylococcus spp.) and Gram-negative (E. coli, Klebsiella spp.) bacteria. The results show that the dibutyltin complexes are more reactive with respect to their corresponding Schiff base ligands.

INTRODUCTION

Amino acids and their compounds with different metal ions play an important role in pharmaceutical industries [1–5]. A number of studies have indicated that biologically active compounds become more bacteriostatic and carcinostatic upon chelation. Some drugs show increased activity when administered as metal chelates and inhibit the growth of tumor. Moreover, the development in the field of bioinorganic chemistry has increased the interest in Schiff base complexes, since it has been recognized that many of these may serve as models for biologically important species [6–9]. During the past few decades, research about new organotin compounds has increased dramatically, most likely due to their diverse biological applications [10–13]. A huge interest in metal complexes of Schiff bases derived from amino acids and salicylaldehyde has emerged due to their structural, magnetic, and electrochemical properties, as well as their potential use as models for a number of important biological systems [14–16]. Among their several biological functions, they show antimicrobial, antimalarial,

Citation: Har Lal Singh and Jangbhadur Singh, "Synthesis, Spectroscopic, Molecular Structure, and Antibacterial Studies of Dibutyltin(IV) Schiff Base Complexes Derived from Phenylalanine, Isoleucine, and Glycine," Bioinorganic Chemistry and Applications, vol. 2014, Article ID 716578, 12 pages, 2014. doi:10.1155/2014/716578

antiproliferative, chemotherapeutic, radiopharmaceutical, insulin-mimetic, and fungicidal properties [17–23]. One of the most important bioinorganic chemistry research areas in organotin compounds is the investigation of their cytotoxic and antitumor activities [24]. Moreover, tin(IV) complexes characterized by the presence of one or more carbon-tin bonds have proved to be cytotoxic against the breast adenocarcinoma tumor and the colon carcinoma [25]. In general, the toxicity of organotin compounds seems to increase with the chain length of the organic alkyl groups, which are often more active than aryl ones [26, 27].

Although the major uses of these derivatives are well known for their versatile and significant important biological and pharmaceutical activities and as wood preservatives and pesticides [28–34], the organotin(IV) compounds possess potential applications in the field of industrial and medicinal chemistry. Organotin(IV) complexes have been extensively studied due to their coordination geometries as well as structural diversity. In view of this, the synthesis of organotin(IV) complexes of Schiff bases derived from the condensation of 1H-indole-2,3-dione and 5-chloro-1H-indole-2,3-dione with different amino acids (phenylalanine, isoleucine, and glycine) derivatives is reported herein. The characterization of the complexes was realized by elemental analysis and spectroscopic (UV, IR, ^{1}H, ^{13}C, and ^{119}Sn NMR) studies. Their antibacterial activities were screened against various Gram-positive and Gram-negative bacteria.

EXPERIMENTAL

All reagents were commercially available (Aldrich or Merck) and used as supplied. Solvents were purified and dried according to standard procedures and moisture was excluded from the glass apparatus using $CaCl_2$ drying tubes. Melting points were determined in open glass capillaries and were uncorrected. The ligands were prepared by the condensation of isatin and 5-chloroisatin with amino acids (phenylalanine, isoleucine, and glycine) as described earlier [35].

Analytical Methods and Spectral Measurements

Tin was estimated gravimetrically as SnO_2. CHN analyses were carried out on a Perkin Elmer 2400 Elemental Analyser. Molecular weight determinations were carried out by the Rast camphor method. Molar conductance measurements were made in anhydrous dimethylformamide at 25 ± 5°C using a Systronics conductivity bridge model 305. The electronic spectra were recorded in DMSO on a Thermo UV1 spectrophotometer. IR spectra were recorded on a Perkin Elmer Spectrum SP-2 Fourier transform spectrophotometer using KBr pellets (4000–400 cm^{-1}). ^{1}H and ^{13}C NMR spectra were recorded on a Bruker Avance-II (400 MHz) FTNMR spectrometer using DMSO-d_6 as solvent at 400 MHz and 100 MHz, respectively. TMS was used as internal reference for ^{1}H NMR and ^{13}C NMR. The ^{119}Sn NMR spectra with proton noise decoupling were recorded on a Bruker Avance-II spectrometer using dry DMSO as the solvent at 149.21 MHz and tetramethyltin (TMT) as an external standard. X-ray powder patterns were obtained with a SIEMENS D-5000 X-ray diffractometer using Cu Kα radiation (l = 1.5405984 Å) which was operated at 30 kV and 15 mA.

3D Molecular Modeling Analysis

The molecular modeling of a representative compound was carried out on a CS Chem3D Ultra Molecular Modeling and using MM2 analysis program [36]. It is an interactive graphics program that allows rapid structure building, geometry optimization with minimum energy, and molecular display. Correct sequence of atoms was obtained to get reasonable low energy molecular models to determine their molecular representation in three dimensions. Complications of molecular transformations could be explored using output obtained. An attempt to gain a better insight into the molecular structure of compounds, geometric optimization, and conformational analysis was performed using MM2 force field [37].

Syntheses of Organotin(IV) Complexes

Dibutyltin(IV) oxide (0.280 g; 1.125 mmol) was added to the calculated amount of the ligands (0.491−0.829 g; 2.250 mmol) in a 1 : 2 molar ratio in dry benzene (75 mL), with methanol (25 mL) mixture as reaction medium. The contents were refluxed on a fractionating column for about 5–7 hours. The water liberated in the reaction was removed azeotropically with benzene. On completion of the reaction, the resulting products were rendered free from solvent and then washed repeatedly with dry cyclohexane. The crystalline solids were separated out and purified by recrystallization from the same solvent. The products so formed were finally dried in vacuum at 40 ±5 °C for 2-3 hours. The purity of the complexes was checked by TLC using silica gel type G as adsorbent. Their physical properties and analytical data were recorded in Table 1.

Table 1: Analytical and physical data of the organotin(IV) complexes

C. number	Reactants		Molar ratio	Products & colour	M.P. °C (d)	Yield (%)	Elemental analysis found (calcd.)				Mol. Wt. Found (Calcd.)
	Metal	Ligands					%Sn	%C	%H	%N	
$Bu_2Sn(L^1)_2$	Bu_2SnO	$C_{17}H_{14}N_2O_3$	1 : 2	$C_{42}H_{44}N_4O_6Sn$ Red	164	94	14.35(14.49)	61.30(61.55)	5.44(5.41)	6.72(6.84)	824.55(819.53)
$Bu_2Sn(L^2)_2$	Bu_2SnO	$C_{14}H_{16}N_2O_3$	1 : 2	$C_{36}H_{48}N_4O_6Sn$ Red	262	79.14	15.84(15.80)	57.50(57.54)	6.38(6.44)	7.56(7.46)	742.19(751.50)
$Bu_2Sn(L^3)_2$	Bu_2SnO	$C_{10}H_8N_2O_3$	1 : 2	$C_{28}H_{32}N_4O_6Sn$ Red	174	76.81	18.43(18.57)	52.45(52.61)	5.00(5.05)	8.65(8.76)	630.88(639.29)
$Bu_2Sn(L^4)_2$	Bu_2SnO	$C_{17}H_{13}ClN_2O_3$	1 : 2	$C_{42}H_{42}Cl_2N_4O_6Sn$ Dark brown	210	72.39	13.03(13.36)	56.84(56.78)	4.73(4.76)	6.39(6.31)	872.54(888.42)
$Bu_2Sn(L^5)_2$	Bu_2SnO	$C_{14}H_{15}ClN_2O_3$	1 : 2	$C_{36}H_{46}Cl_2N_4O_6Sn$ Light brown	270	68.29	14.34(14.47)	52.59(52.71)	5.60(5.65)	6.70(6.83)	831.57(820.38)
$Bu_2Sn(L^6)_2$	Bu_2SnO	$C_{10}H_7ClN_2O_3$	1 : 2	$C_{28}H_{30}Cl_2N_4O_6Sn$ Dark brown	180–184	71.72	16.84(16.76)	47.41(47.49)	4.25(4.27)	7.85(7.91)	716.67(708.18)

Antibacterial Activity

Synthesized compounds were screened for their antibacterial activity against Gram-positive (B. cereus, Staphylococcus spp.) and Gram-negative (E. coli, Klebsiella spp.) bacteria at two concentrations 500 ppm and 200 ppm. The compounds were dissolved in DMSO. The in vitro antibacterial activity of the free ligands and their tin complexes was tested using the agar well diffusion method [38]. Using a sterile cork borer (5 mm in diameter) wells were made in each agar plate, more than 0.1 mL of the tested compounds were poured into three wells, and the dishes were incubated at 28 ± 2 °C for 24 hours. The growth of the microorganisms was inhibited by diffusion of the test solutions and then the inhibition zone around the well was measured. The antibacterial activity of each compound was compared with standard antibiotics such as streptomycin. DMSO was used as a control under the same conditions to test organisms and no activity was found.

RESULTS AND DISCUSSION

New organotin(IV) complexes were synthesized by the reactions of dibutyltin(IV) oxide with Schiff bases, which have been carried out in 1 : 2 molar ratios using anhydrous benzene and absolute methanol in a 3 : 1 ratio as solvent. These reactions proceed with the liberation of water, which was azeotropically removed. The results of elemental analysis (C, H, and N) with molecular formula and the melting points are presented in Table 1. The results obtained are in good agreement with those calculated for the suggested formula. The scheme of the organotin(IV) complexes preparation is given below (Scheme 1).

$$Bu_2SnO + 2\overset{\frown}{N\ OH} \xrightarrow[\text{Reflux 5–7 h}]{\text{Methanol + benzene}} Bu_2Sn(\overset{\frown}{N\ O})_2 + H_2O$$

where $\overset{\frown}{N\ OH}$ represents the donor system of the Schif bases

Scheme 1: Representative equation illustrating the formation of dibutyltin(IV) complexes.

The above reactions were found to be quite facile and could be completed in 5–7 h of refluxing. All these complexes are intensively coloured and are solids. They are soluble in methanol, DMF, and DMSO. The compounds were dissolved in DMF and molar conductance 10^{-3}M of solution at 25°C was measured. The molar conductance values of the complexes fall in the range from 10.32 to 20.14 $\Omega^{-1}cm^2mol^{-1}$ indicating that these compounds are nonelectrolytes.

Electronic Spectra

The spectra of the ligands and their complexes were recorded in dry DMSO. The various bands observed were assigned to interligand and charge transfer of n–π^* transitions according to their energies and intensities. Two intense maxima are observed in the complexes at 210–232 and 375–382 nm ranges which may be assigned to n–π^* transition of the carboxylate [39] and of the C=N chromophore. The appreciable shifting observed in the n–π^* transition (~370 nm) is due to the polarization in the C=N bond caused by tin-ligand electron interaction. This clearly indicates the coordination of the azomethine nitrogen with the tin atom. A band in the region 355–338 nm in the spectra of the Schiff bases and complexes is likely to be the secondary band of the benzene ring coupled with the intramolecular charge transfer transition taking place within the ligand moiety. Furthermore, sharp bands were observed in the region 246–332 nm in the spectra of the complexes which could be assigned to the charge transfer transition from ligand to tin atom [40].

Infrared Spectra

The IR spectra of these compounds have been recorded in the range of 4000–400 cm^{-1}. Tentative assignments have been made on the basis of earlier publications and the important data are listed in Table 2. The absorptions of interest in the spectra of the complexes are ν(COO), ν(Sn–C), and ν(Sn–O). The position and the intensities of these peaks are expected to be changed on chelation. New peaks and quasi-peaks are also a guide to chelation. The IR spectra of all the ligands show the absence of bands at 3450 and 1750 cm^{-1} due to $\nu(NH_2)$ group of amino acids and ν(C=O) of isatin. Instead, a new prominent band at 1635±5 cm^{-1} due to azomethine ν(C=N) linkage appeared in all the ligands [12, 34], indicating that condensation between ketone moiety of isatin and that of amino group of amino acids has taken place resulting in the formation of the desired ligands (L^1H–L^6H). Moreover, comparison of the IR spectra of the ligands with their organotin(IV) complexes showed a major shift to lower wave numbers by 10–15 cm^{-1} in azomethine ν(C=N) at 1620 ± 5 cm^{-1} suggesting the involvement of the azomethine nitrogen with the organotin(IV) ion [9, 11, 32]. The appearance of a new band of medium intensity in the region ~552–540 cm^{-1} in all of the derivatives studied, which may be assigned to ν(SnN), further confirms the coordination of the amino nitrogen with the tin moiety.

The absence of the ν(COOH) in all the organotin(IV) compounds in the 2750–3100 cm^{-1} region and the presence of ν(Sn–O) in the 432–420 cm^{-1} range indicate deprotonation of the carboxylic acid group and consequent coordination of the carboxylate group with the tin metal as expected. IR stretching frequencies $\nu_{as}(OCO)$ and $\nu_s(OCO)$ for carboxylates have been utilized as a characteristic tool to confirm the mode of coordination through carboxylate oxygen and also to identify the nature (monodentate, bidentate, or bridging/chelating) of bonding of the carboxylate group. Unless the carboxylate group is branched at the α-carbon or the organic substituents at the tin atom are bulky, carboxylate groups in organotin(IV) carboxylates generally adopt a bridge structure in the solid state [41].

Table 2: Important IR spectral data (cm^{-1}) of Schiff bases and their corresponding organotin(IV) complexes

Compounds	ν(OH)ν	ν(C=N–)	ν(C=O)	ν_{as}(COO)	ν_s(COO)	Δν	ν(Sn←N)	ν(Sn–O)	ν(Sn–C)
L^1H	3090–2740 br	1620 s	1728 s	—	—	—	—	—	—
$Bu_2Sn(L^1)_2$	—	1610 s	1722 s	1590 vs	1325 s	265	540 w	430 m	622 m
L^2H	3100–2740 br	1625 s	1730 s	—	—	—	—	—	—
$Bu_2Sn(L^2)_2$	—	1612 s	1726 s	1605 m	1333 m	272	545 m	426 s	640 m
L^3H	3090–2750 br	1630 s	1720 s	—	—	—	—	—	—
$Bu_2Sn(L^3)_2$	—	1616 s	1722 s	1601 s	1332 m	268	552 m	425 m	630 m
L^4H	3110–2750 br	1622 s	1735 s	—	—	—	—	—	—
$Bu_2Sn(L^4)_2$	—	1605 s	1732 s	1608 s	1330 m	278	545 s	420 w	635 m
L^5H	3080–2730 br	1630 s	1735 s	—	—	—	—	—	—
$Bu_2Sn(L^5)_2$	—	1612 s	1734 s	1595 vs	1326 s	269	540 w	425 w	625 m
L^6H	3109–2730 br	1622 s	1725 s	—	—	—	—	—	—
$Bu_2Sn(L^6)_2$	—	1610 s	1728 s	1610 vs	1340 vs	270	544 w	432 m	632 m

br: broad, vs: very sharp, v: sharp, m: medium, and w: weak.

The IR stretching frequencies of the carboxylate groups are very important for determining their structures; namely, when there are interactions between the carbonyl oxygen atoms of the carboxylate groups and the tin atom, the asymmetric absorption vibration frequencies ν_{as}(OCO) of the carboxylate groups decrease and the symmetric absorption frequencies ν_s(OCO) increase. In the IR spectra of the title compounds, the carboxylate bands are observed in the characteristic regions for ν_{as}(OCO) between 1610 and 1590 cm^{-1} and for ν_s(OCO) between 1340 and 1325 cm^{-1}(Table 2). In the diorganotin(IV) complexes, the differences [νΔ(OCO)] between ν_{as}(OCO) and ν_s(OCO) are more than 200 cm^{-1} indicating the covalent nature of the metal-oxygen bond. Ionic bonding and also bridging or chelation, therefore, can be excluded, and unidentate coordination of the carboxylic groups bonding to the metal must therefore be assumed. Thus, for diorganotin compounds the νΔ(OCO) values were found to be more than 200 cm^{-1}, which indicates that the carboxylate groups are chelated and bonded to the metal in a unidentate manner. Moreover, for complexes Δv below 200 cm^{-1} would be expected for bridging or chelating carboxylate, but greater than 200 cm^{-1} for the monodentate bonding carboxylate anions. Further evidence for the coordination with tin via oxygen atom was revealed by the presence of the ν(Sn–O) [33] stretching bands in the spectra of complexes in the region of 432–420 cm^{-1}. Besides this, several new bands in the complexes observed at ~630 cm^{-1}, ~540 cm^{-1}, and ~430 cm^{-1} may be ν(Sn–C) [34], ν(Sn←N) [12], and ν(Sn–O) [38], respectively. The vacant 5d orbital of tin atoms tends toward high coordination with ligands containing lone pairs of electrons. The intensity observed for the symmetric Sn–CH_2 stretching vibration at ~1440 cm^{-1} indicates a bent C–Sn–C moiety for the dibutyltin complexes. These data are consistent with a distorted octahedral configuration for the tin atom.

^{1}H NMR Spectra

The chemical shifts (δ ppm) of various protons in the complexes in DMSO-d_6 solution are given in Table 3. The conclusions drawn from ^{1}H NMR spectral studies lend further support to the mode of bonding discussed above. The absence of a signal at δ 12.00–11.00 ppm due to the C(O)OH proton suggests the deprotonation of the carboxylic oxygen atom of the ligands on complexation. The ligands give a complex multiplet signal in the region δ 7.10–7.85 ppm for the aromatic protons and these remain almost at the same position in the spectra of the metal complexes. The appearance of signals due to NH protons at the same positions in the ligand and its complexes shows the noninvolvement of this group in coordination. The butyl protons attached to tin appear as multiplets due to –CH_2–CH_2–CH_2– protons in theδ 1.65–1.15 ppm region and as a clear triplet or broad signal due to terminal methyl protons in the δ 0.80–0.85 ppm region with $^3J_{HH}$ = 7.35 Hz. The >CH– protons in the complexes of the ligands give signals in the δ 4.38–3.99 ppm region. The most important piece of information, obtained from ^{1}H NMR values in these compounds, demonstrates that diorganotin(IV) complexes show the coordination number greater than four, probably six, in noncoordinating solvent. The number of protons of various groups calculated from the integration curves and the number of those calculated for the expected molecular formula agree with each other.

Table 3: ^{1}H NMR spectral data[a] of the ligands and their corresponding organotin(IV) complexes

Compounds	Chemical Shift (δ, ppm)
L^1H	11.50 (s, 1H, COOH), 4.25 (t, 1H, N-CH-CH_2-), 3.08 (d, 2H, -CH_2-Ph), 8.01 (s, 1H, NH), 7.20–7.80 (m, 9H, aromatic).
$Bu_2Sn(L^1)_2$	4.21 (s, 1H, N-CH-CH_2), 3.10 (d, 2H, -CH-CH_2-Ph), 8.03 (s, 1H, NH), 7.10–7.75 (m, 9H, aromatic); 1.60–1.15 (br m. 6H, -CH_2-CH_2-CH_2-); 0.82 (br t, 3H, CH_3 of butyl groups).
L^2H	11.39 (s, 1H, COOH), 3.96 (d, 1H, N-CH-CH-), 2.10–2.16 (m, 1H, CH_3-CH-CH_2-), 1.50–1.60 (m, 2H, -CH-CH_2-CH_3), 8.05 (s, 1H, NH), 0.85 (d, 3H, -CH-CH_3), 0.95 (t, 3H, -CH_2-CH_3), 7.25–7.82 (m, 4H, aromatic).
$Bu_2Sn(L^2)_2$	3.98 (d, 1H, N-CH-CH), 2.08–2.16 (m, 1H, CH_2-CH-CH_3), 0.80 (d, 3H, -CH-CH_3), 1.48–157 (m, 2H, CH-CH_2-CH_3), 0.99 (t, 3H, -CH_2-CH_3); 8.10 (s, 1H, NH), 7.22–7.80 (m, 4H, aromatic); 1.65–1.21 (br m. 6H, -CH_2-CH_2-CH_2-); 0.80 (br t, 3H, CH_3 of butyl groups).
L^3H	11.28 (s, 1H, COOH), 4.30 (s, 2H, N-CH_2-), 8.02 (s, 1H, NH), 7.28–7.80 (m, 4H, aromatic).
$Bu_2Sn(L^3)_2$	4.38 (s, 2H, N-CH_2-), 8.05 (s, 1H, NH), 7.25–7.78 (m, 4H, aromatic); 1.62–1.18 (br m. 6H, -CH_2-CH_2-CH_2-); 0.84 (br t, 3H, CH_3 of butyl groups).
L^4H	11.42 (s, 1H, COOH), 3.97 (d, 1H, N-CH-CH_2-), 3.15 (d, 2H, -CH_2-Ph), 8.08 (s, 1H, NH), 7.20–7.85 (m, 8H, aromatic);
$Bu_2Sn(L^4)_2$	4.02 (t, 1H, N-CH-CH_2), 3.08 (d, 2H, -CH_2-Ph), 8.10 (s, 1H, -NH-), 7.20–7.75 (m, 8H, aromatic); 1.60–1.15 (br m. 6H, -CH_2-CH_2-CH_2-); 0.80 (br t, 3H, CH_3 of butyl groups).
L^5H	11.42 (s, 1H, COOH), 3.99 (d, 1H, N-CH-CH), 2.08–2.14 (m, 1H, CH_3-CH-CH_2), 1.42–1.58 (m, 2H, CH-CH_2-CH_3), 8.10 (s, 1H, NH), 0.82 (d, 3H, -CH-CH_3), 0.98 (t, 3H, -CH_2-CH_3), 7.20–7.80 (m, 3H, aromatic).
$Bu_2Sn(L^5)_2$	4.01 (d, 1H, N-CH-CH), 2.10–2.20 (m, 1H, CH_3-CH-CH_2), 0.80 (d, 3H, -CH-CH_3), 1.45–1.58 (m, 2H, -CH-CH_2-CH_3), 1.02 (t, 3H, CH_2-CH_3), 8.05 (s, 1H, NH), 7.20–7.82 (m, 3H, aromatic); 1.64–1.18 (br m. 6H, -CH_2-CH_2-CH_2-); 0.82 (br t, 3H, CH_3 of butyl groups).
L^6H	11.35 (s, 1H, COOH), 4.32 (s, 2H, -CH_2-), 8.12 (s, 1H, NH), 7.20–7.76 (m, 3H, aromatic).
$Bu_2Sn(L^6)_2$	4.40 (s, 2H, N-CH_2-), 8.10 (s, 1, NH), 7.28–7.76 (m, 3H, aromatic); 1.65–1.21 (br m. 6H, -CH_2-CH_2-CH_2-); 0.85 (br t, 3H, CH_3 of butyl groups).

[a]Chemical shift (δ) in ppm: Multiplicity is given as: s = singlet, d = doublet, t = triplet, q = quartet, m = complex pattern, br = broad.

^{13}C NMR Spectra

The ^{13}C NMR spectral data along with assignment of characteristic peaks of ligands and its diorganotin(IV) complexes are presented in Table 4. The signals due to the carbon atoms attached to the carboxylate oxygen and the azomethine groups in the ligands appear at ~δ 176 ppm and ~δ 163 ppm, respectively. However, in the spectra of the corresponding tin complexes, these appear at ~δ 185 ppm (carboxylate group) and at ~δ 155 ppm (azomethine group), respectively. The considerable shifts in the resonances of the carbon atoms attached to oxygen and nitrogen indicate the involvement of oxygen and nitrogen atoms in coordination. The alkyl carbons of Schiff base and organotin(IV) complexes appear well within the expected ranges and their assignments are given in Table 4. The heteronuclear coupling constant $^1J(^{13}C–^{119}Sn)$ in the studied organotin(IV) complexes is determined, which provides vital information about the coordination environment and the geometry of the complexes. In all the complexes, the alkyl group are observed with coupling constant of the values $^1J(^{13}C–^{119}Sn)$ are in the range 690.56–759.86 Hz. These values lie in a typical range for six-coordinated organotin(IV) complexes [42].

Table 4: ^{13}C NMR spectral data of the ligands and their corresponding organotin(IV) complexes

Compounds	Chemical shift in (δ ppm) COOH	CH	C=N	CH_2/CH_3	Aromatic carbons	Sn-$^{\alpha}CH_2$-$^{\beta}CH_2$-$^{\gamma}CH_2$-$^{\delta}CH_3$
L^1H	176.8	67.8	163.4	38.2	149.2, 135.8, 133.0, 128.6, 127.5, 126.3, 124.6, 122.9, 120.1	—
$Bu_2Sn(L^1)_2$	185.4	65.7	155.2	39.4	150.0, 136.1, 132.6, 128.2, 127.1, 126.5, 124.4, 123.3, 120.2	C-α, 22.2; C-β, 25.6; C-γ, 22.4; C-δ, 13.4
L^2H	178.5	65.7	162.4	16.3, 21.3	145.8, 133.2, 130.5, 126.2, 124.0, 119.3	—
$Bu_2Sn(L^2)_2$	184.9	66.2	154.6	17.4, 22.1	146.2, 132.9, 130.8, 126.54, 124.3, 120.1	C-α, 21.4; C-β, 25.3; C-γ, 21.6; C-δ, 13.5
L^3H	176.1	—	163.6	52.4	148.5, 131.0, 129.8, 125.3, 122.9, 119.1	—
$Bu_2Sn(L^3)_2$	183.7	—	154.5	58.3	149.0, 132.5, 129.6, 125.7, 123.4, 119.6	C-α, 22.1; C-β, 25.2; C-γ, 22.3; C-δ, 14.1
L^4H	178.5	65.7	162.4	39.1	150.2, 136.1, 132.8, 129.5, 127.3, 126.4, 123.2, 122.4, 119.8	—
$Bu_2Sn(L^4)_2$	184.5	66.2	154.5	38.7	149.5, 135.7, 132.1, 130.4, 126.9, 126.2, 123.5, 122.1, 120.3	C-α, 22.4; C-β, 26.7; C-γ, 22.8; C-δ, 14.6
L^5H	177.5	66.2	163.1	17.3, 20.5	147.7, 132.8, 131.3, 126.6, 123.8, 120.3	—
$Bu_2Sn(L^5)_2$	184.1	67.4	153.8	18.0, 20.2	148.1, 132.4, 129.7, 126.5, 124.0, 120.2	C-α, 22.0; C-β, 26.1; C-γ, 22.5; C-δ, 13.4
L^6H	176.9	—	162.9	55.4	149.5, 132.8, 128.9, 125.7, 123.0, 120.3	—
$Bu_2Sn(L^6)_2$	182.8	—	155.4	56.1	148.2, 133.1, 128.5, 125.8, 122.7, 120.5	C-α, 21.2; C-β, 25.3; C-γ, 22.24; C-δ, 13.7

^{119}Sn NMR Spectra

The ^{119}Sn NMR chemical shifts are very sensitive to coordination number of tin and are generally shifted upfield on bonding to Lewis base. It has been reported [43, 44] that the ^{119}Sn chemical shifts in the ranges from 200 to −60, from −90 to −190, and from −210 to −400 ppm are associated with four-, five-, and six-coordinated alkyltin(IV) compounds, and these ^{119}Sn shifts are higher with aryltin(IV) compounds.

In the ^{119}Sn NMR spectra of $Bu_2Sn(L^1)_2$, $Bu_2Sn(L^2)_2$, and $Bu_2Sn(L^3)_2$recorded in DMSO-d_6, the ^{119}Sn chemical shifts are observed at δ −265.4, −270.8, and −266.1 ppm, respectively, lying in the range of six-coordinated tin. The $^1J(^{119}Sn–^{13}C)$ values allow the calculation of the bond angle for the C–Sn–C fragment, which is in the order of ~117°, suggesting that the tin atom has a slightly distorted octahedral geometry (Figure 1); this behavior is in agreement with the values reported for hexacoordinated tin compounds [45].

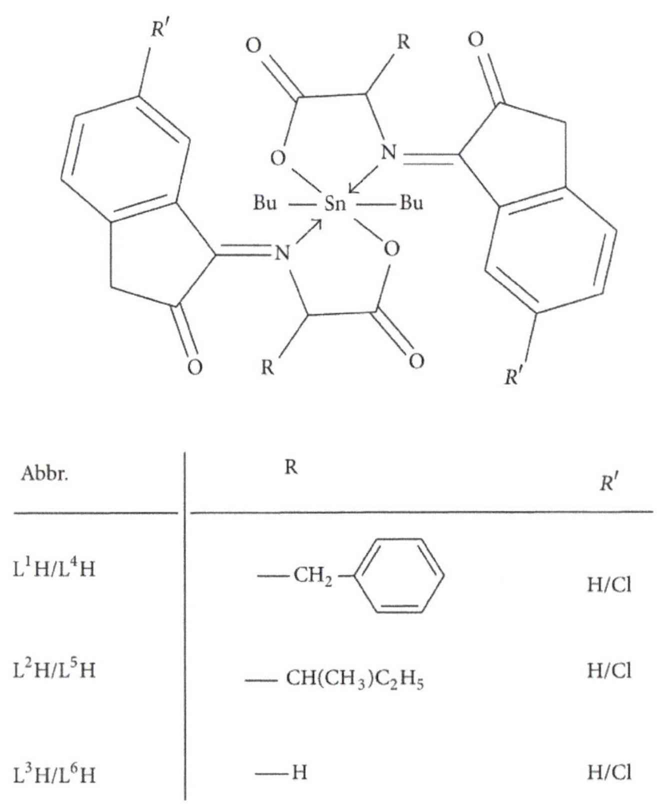

Figure 1: Structure of diorganotin(IV) complexes.

X-Ray Powder Diffraction Studies

The X-ray diffraction studies have been performed on Philips PW 1710 automated X-ray powdered diffractometer. The experimental conditions employed in reading the pattern were as follows: the operating target voltage was 30 kV; the tube current was 15 mA. The X-ray from copper target was filtered with nickel, and monochromatic Kα line of wavelength 1.540598 Å was obtained. Filtration reduces noise due to white radiation and increases resolution also.

The X-ray diffraction of L^1H, L^2H, L^3H, and their metal complexes indicates the crystalline nature of the L^1H, L^2H, and L^3H; $Bu_2Sn(L^3)_2$ is shown to have amorphous nature. All the reflection has been indexed for D (particle size) using Scherrer's equation. Average crystallite size values of 22.484 nm and 32.439 nm were shown for L^1H and L^2H, respectively. These values of particle size and 2θ for each peak have been calculated with the help of the cell parameters and corresponding FWHM values. The lattice spacing, FWHM, and particle size for L^1H and L^2H have been found out and are given in Table 5. The diffractogram of compounds has been shown in Figures 2 and 3.

Table 5: X-ray powder diffraction data for compounds (L^1H) and (L^2H)

S. number	2θ	"d" (Å)	Intensity (count)	FWHM	D^* (nm)
			L^1H		
1	13.517	6.54547	385.6	0.5425	25.381
2	14.936	5.92655	482.3	0.4308	31.912
3	16.873	5.25031	216.5	0.7307	18.770
4	21.492	4.13128	323.9	0.7307	18.643
5	22.046	4.02874	453.9	0.6139	22.169
6	22.521	3.94474	289.4	0.6139	22.151
7	24.025	3.70113	301.8	0.6139	22.091
8	24.805	3.58649	162.8	0.6139	22.059
9	25.067	3.54964	205.2	0.6139	22.047
10	26.359	3.37851	150.7	0.6139	21.991
11	27.324	3.26131	228.2	0.486	27.722
12	28.440	3.13581	217.5	0.7166	18.756
13	32.024	2.79258	112.5	0.7166	18.598
			L^2H		
1	11.495	7.69156	128	0.3461	39.860
2	13.137	6.73368	236.3	0.3806	36.191
3	14.870	5.95279	807.9	0.2844	48.343
4	16.683	5.30969	228.6	0.3597	38.139
5	21.263	4.17518	222.4	0.3597	37.885
6	22.035	4.03068	286.1	0.5986	22.736
7	23.706	3.75021	263.9	0.5544	24.476
8	24.902	3.57273	521.9	0.3499	38.695
9	26.232	3.39454	156.1	0.3499	38.593
10	27.114	3.28608	717.3	0.2874	46.899
11	31.637	2.82587	144.2	0.4665	28.597
12	37.570	2.3921	84.3	0.6166	21.289

$^*D$: crystallite size (in Å) and d: lattice spacing.

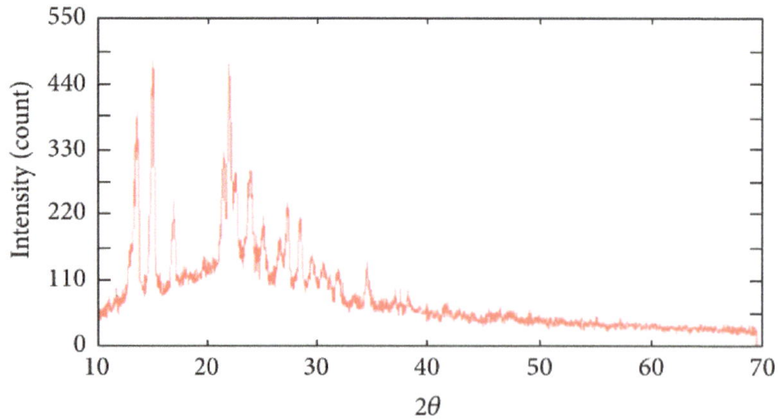

Figure 2: X-ray powder diffraction patterns of the compound [L^1H].

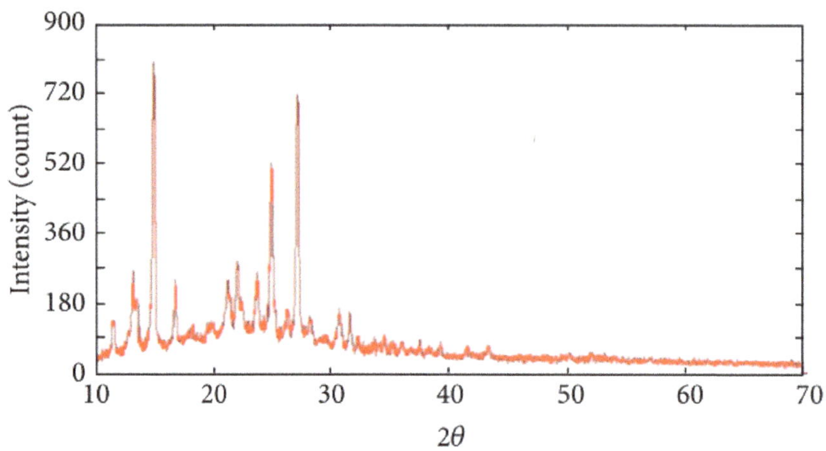

Figure 3: X-ray powder diffraction patterns of the compound [L^2H].

Molecular Structure

Molecular modeling helped to demonstrate the significant features, that is, molecular geometries, bond energy, and torsion angles of organometallic frameworks theoretically. Bond lengths, bond angles, and atomic coordinates depend on the hybridization of an atom and mode of bonding. Thus, molecular modeling was the blue print of three dimensional arrangements of atoms. Beside this, if deviations in bond distance, bond angles, or torsion angles are evidenced, specific electronic interactions can be detected and confirmed to the earlier spectral lines of evidence [46]. The bonding capabilities of atoms have impact on bond lengths and bond angles of concerned functional groups. Therefore, molecular models of complexes were demonstrated. Physical dimension of the molecules helped out to demonstrate the changes occurring during their topological assemblies. Energy minimization is used to find the structure with lowest energy using molecular mechanics analysis program [47, 48]; data analysis of bond lengths and bond angles is presented in Tables 6 and 7.

Table 6: Important bond lengths of compound [$Bu_2Sn(L^2)_2$]

S. number	Atoms	Actual (Å)	Optimal (Å)
1	Sn(1)–C(44)	2.2339	2.1620
2	Sn(1)–C(47)	2.2064	2.1620
3	N(9)–Sn(1)	2.1050	—
4	N(5)–Sn(1)	2.1010	—
5	Sn(1)–C(44)	2.2339	2.1620
6	Sn(1)–O(6)	2.0644	—
7	O(6)–C(7)	1.3666	1.3380
8	Sn(1)–O(2)	2.0581	—
9	O(2)–C(3)	1.3639	1.3380
10	Sn(1)–C(47)	2.2064	2.1620
11	N(5)–C(14)	2.1680	1.2600
12	C(4)–N(5)	1.4923	1.4700

Table 7: Important bond angles of compound [$Bu_2Sn(L^2)_2$]

S. number	Atoms	Actual (°)
1	C(47)–Sn(1)–C(44)	117.1540
2	C(47)–Sn(1)–N(5)	85.7227
3	C(47)–Sn(1)–N(9)	124.3910
4	C(47)–Sn(1)–O(6)	68.5077
5	C(47)–Sn(1)–O(2)	141.0197
6	C(44)–Sn(1)–N(5)	80.4837
7	C(44)–Sn(1)–N(9)	78.7094
8	C(44)–Sn(1)–O(6)	138.6752
9	C(44)–Sn(1)–O(2)	88.6745
10	N(5)–Sn(1)–N(9)	148.9524
11	N(5)–Sn(1)–O(6)	139.5356
12	N(5)–Sn(1)–O(2)	69.4566
13	N(9)–Sn(1)–O(6)	67.5865
14	N(9)–Sn(1)–O(2)	87.2268
15	O(6)–Sn(1)–O(2)	112.0005

The molecular structure is shown in Figure 4. The deprotonated ligand is coordinated as bidentate ligand via the carboxylate oxygen and azomethine nitrogen atoms. The six coordination number is completed by two carbon atoms of butyl groups. The organic molecule acts as an anionic bidentate with the ON donors placed in the same side. Since the synthesized compounds are related and differ only in substituted R group, one compound $Bu_2Sn(L^2)_2$ was theoretically studied. Based upon spectroscopic data, Sn(IV) compounds with O_2N_2 ligands generally adopt distorted octahedral geometry.

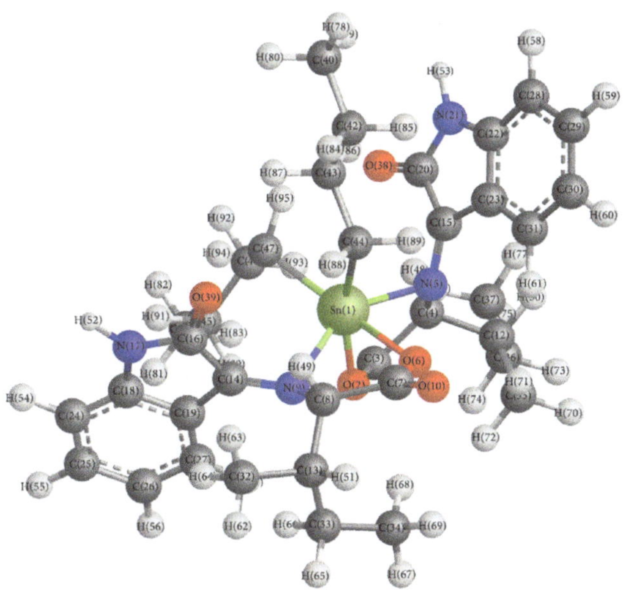

Figure 4: 3D molecular structure of $Bu_2Sn(L^2)_2$.

The optimized structure for the compound $Bu_2Sn(L^2)_2$ is shown in Figure 4 and pertinent bond parameters are given in Tables 6and 7. The bond angles around tin atom, for example, C(47)–Sn(1)–C(44) angle of 117.15°, N(5)–Sn(1)–N(9) angle of 148.95°, and O(6)–Sn(1)–O(2) angle of 112.00°, in $Bu_2Sn(L^2)_2$ are the representative of the distorted octahedral structure. The two Sn–O bond distances are close to being identical values. The calculated Sn–O bond distances of 2.0581/2.0644 Å in $Bu_2Sn(L^2)_2$ are also close to the already reported Sn–O distances in $\{CH_2N(Me)CH(Me)CH(Ph)O\}_2Sn$ (2.048/2.078) [49]. The two different Sn–N distances of Sn1–N5 and Sn1–N9 in compounds $Bu_2Sn(L^2)_2$ are 2.1010/2.1050 Å, which are similar to the already reported structures, $SalenH_2Sn$ 2.0535(9) and 2.0369(8) Å) [50].

In Vitro Antibacterial Activity

Antibacterial screening of the synthesized compounds was carried out using four bacterial strains of Gram-positive (B. cereus, Staphylococcus spp.) and Gram-negative (E. coli, Klebsiella spp.) bacteria. The zones of inhibition values in Table 8 represent the mean value of the three readings with standard deviation. Streptomycin was used as a reference compound for antibacterial activities. These bacterial strains are used because they are known as common pathogens of human beings. The antimicrobial studies suggested that the Schiff bases are biologically active and their metal complexes showed significantly enhanced antibacterial activity against microbial strains in comparison to the free ligands, and also crystalline materials are more active than the amorphous materials. Tested compounds showed zone of inhibition ranging from 14.2 mm to 32.1 mm against the Gram-positive bacteria and between 10.1 mm and 25.7 mm against Gram-negative bacteria. The ligands (HL) show zone of inhibition ranging from 14.2 mm to 21.5 mm against Gram-positive bacteria and from 10.1 mm to 22.6 mm against Gram-negative bacteria. It has been observed that the metal complexes showed increased zone of inhibition against the bacterial strains compared to ligands. On the basis of zone of inhibition produced against the test bacterium, compound $Bu_2Sn(L^3)_2$ was found to be the most effective against B. cereus, Staphylococcus spp., E. coli, and Klebsiella spp.with zone of inhibition of 32.1 mm, 31.2 mm, 16.4 mm, and 18.5 mm, respectively (Table 8). This also showed that the antibacterial activity of ligands is greatly enhanced when it is coordinated with metal ions. Although it is difficult to make out an exact structure-activity relationship between the antimicrobial activity and the structure of these complexes, it can possibly be concluded that the chelation as well as the addition of a substrate enhances the activity of the complexes. The increase in the activity of the complexes compared to that of the ligands can be explained on the basis of Overton's concept [51] and Tweedy's chelation theory [52]. Chelation considerably reduces the polarity of the metal ion because of partial sharing of its positive charge with the donor group and possible π-electron delocalization over the whole chelate ring. Such chelation could enhance the lipophilic character of the metal atom, which subsequently favours its permeation through the lipid layers of the cell membrane. In general, metal complexes are more active than ligands and they may serve as principal cytotoxic species. Thus, exhibiting their broad spectrum nature can be further used in pharmaceutical industry for mankind as an antimicrobial agent after testing its toxicity to human beings.

Table 8: Antibacterial activity of ligands and their organotin(IV) complexes

Schiff bases/complexes	Zone of inhibition (mm)							
	Gram-negative				Gram-positive			
	E. coli		*Klebsiella* spp.		*B. cereus*		*Staphylococcus* spp.	
	500 ppm	200 ppm	500 ppm	200 ppm	500 ppm	200 ppm	500 ppm	200 ppm
L^2H	18 ± 0.6	15 ± 0.1	22 ± 0.6	17 ± 0.1	14 ± 0.2	12 ± 0.9	23 ± 0.2	20 ± 0.7
$Bu_2Sn(L^2)_2$	25 ± 0.7	22 ± 0.4	30 ± 0.3	26 ± 0.8	20 ± 0.7	18 ± 0.4	30 ± 0.1	25 ± 0.5
L^3H	12 ± 0.1	10 ± 0.3	15 ± 0.1	13 ± 0.6	20 ± 0.1	18 ± 0.6	21 ± 0.5	18 ± 0.6
$Bu_2Sn(L^3)_2$	16 ± 0.4	15 ± 0.6	18 ± 0.5	16 ± 0.4	32 ± 0.1	28 ± 0.3	30 ± 0.2	27 ± 0.2
L^5H	14 ± 0.3	12 ± 0.9	15 ± 0.2	12 ± 0.1	24 ± 0.6	20 ± 0.1	19 ± 0.1	17 ± 0.1
$Bu_2Sn(L^5)_2$	20 ± 0.5	18 ± 0.1	20 ± 0.5	18 ± 0.5	28 ± 0.3	23 ± 0.4	29 ± 0.4	25 ± 0.6
L^6H	11 ± 0.3	13 ± 0.2	10 ± 0.1	8 ± 0.5	20 ± 0.5	17 ± 0.8	18 ± 0.6	15 ± 0.2
$Bu_2Sn(L^6)_2$	16 ± 0.2	15 ± 0.5	11 ± 0.8	10 ± 0.9	24 ± 0.9	19 ± 0.5	20 ± 0.1	18 ± 0.4
Streptomycin	28 ± 0.1	23 ± 0.3	26 ± 0.3	20 ± 0.5	26 ± 0.8	22 ± 0.1	32 ± 0.1	26 ± 0.3
DMSO	0	0	0	0	0	0	0	0

CONCLUSION

A new series of organotin complexes were prepared in good yields. Based on various physiochemical and structural investigations, it was concluded that the ligands act as bidentate (no donor) forming distorted octahedral complexes with $Bu_2Sn(IV)$ ion. Furthermore, the current study strongly demonstrates that these complexes are more effective antibacterial agents than the parent ligands. Selectivity is observed in the activities of some compounds over particular microorganisms, which is very important for the future pharmaceutical applications in order to avoid the side effects, so we can conclude that the compounds synthesized and tested look very promising.

CONFLICT OF INTERESTS

The authors declare that there is no conflict of interests regarding the publication of this paper.

ACKNOWLEDGMENTS

The authors are thankful to Professor J. V. Desai, Dean of the Faculty of Engineering & Technology, Mody University of Science and Technology, Laxmangarh, Sikar, for providing necessary facilities to carry out this research work. They are also thankful to Dr. Tejpal Dewa, Department of Microbiology, University of Delhi, for providing antimicrobial screening facilities.

REFERENCES

1. Z. Moradi-Shoeili, Z. Amini, D. M. Boghaei, and B. Notash, "Synthesis, X-ray structure and ascorbic oxidation properties of ternary α-amino acid Schiff base-bipy Cu(II) complexes as functional models for ascorbic oxidase,"Polyhedron, vol. 53, pp. 76–82, 2013.
2. A. R. Patil, K. J. Donde, S. S. Raut, V. R. Patil, and R. S. Lokhande, "Synthesis, characterization and biological activity of mixed ligand Co(II) complexes of schiff base 2-amino-4-nitrophenol-n-salicylidene with some amino acids,"Journal of Chemical and Pharmaceutical Research, vol. 4, no. 2, pp. 1413–1425, 2012.
3. Z. S. M. Al-Garawi, I. H. R. Tomi, and A. H. R. Al-Daraji, "Synthesis and characterization of new amino acid-schiff bases and studies their effects on the activity of ACP, PAP and NPA enzymes (In Vitro)," E-Journal of Chemistry, vol. 9, no. 2, pp. 962–969, 2012.
4. F. Ciolan, L. Patron, G. Marinescu, N. Cioatera, and M. Mureseanu, "Synthesis and characterization of Zn(II) binuclear complexes with three schiff bases derived from some amino acids and 2,2′-(propane-l,3- diyldioxy) dibenzaldehyde," Revista de Chimie, vol. 62, no. 12, pp. 1145–1149, 2011.
5. P. R. Reddy, A. Shilpa, N. Raju, and P. Raghavaiah, "Synthesis, structure, DNA binding and cleavage properties of ternary amino acid Schiff base-phen/bipy Cu(II) complexes," Journal of Inorganic Biochemistry, vol. 105, no. 12, pp. 1603–1612, 2011.
6. M. Nath, S. Goyal, S. Goyal, G. Eng, and N. Ogwuru, "Synthesis, spectral and thermal studies of organotin(IV) complexes of N-acetylamino acids," Synthesis and Reactivity in Inorganic and Metal-Organic Chemistry, vol. 28, no. 10, pp. 1619–1641, 1998.
7. A. S. P. Azzouz and R. T. Ali, "Synthesis of Schiff bases derived from benzaldehyde and salicylaldehyde with some amino acids by a new develop method," National Journal of Chemistry, vol. 37, pp. 158–168, 2010.
8. M. Shakir, N. Shahid, N. Sami, M. Azam, and A. U. Khan, "Synthesis, spectroscopic characterization and comparative DNA binding studies of Schiff base complexes derived from l-leucine and glyoxal," Spectrochimica Acta A: Molecular and Biomolecular Spectroscopy, vol. 82, no. 1, pp. 31–36, 2011.
9. H. L. Singh, J. B. Singh, and H. Sachedva, "Synthesis, spectroscopic and antimicrobial studies of lead (II) complexes of schiff bases derived from amino acids and isatins," Spectroscopy Letters, vol. 46, no. 4, pp. 286–296, 2013.
10. R. M. Wang, J. J. Mao, J. F. Song, C. X. Huo, and Y. F. He, "Antioxidant activity of bovine serum albumin binding amino acid Schiff-bases metal complexes,"Chinese Chemical Letters, vol. 18, no. 11, pp. 1416–1418, 2007.

11. A. G. Davies, M. Gielen, K. H. Pannell, and E. R. T. Tiekink, Tin Chemistry: Fundamentals, Frontiers and Applications, John Wiley & Sons, New York, NY, USA, 2008.

12. H. L. Singh and J. Singh, "Synthesis, characterization and biological activity of dibutyltin(IV) Schiff base complexes derived from substituted isatin and amino acids," Natural Science, vol. 4, no. 3, pp. 170–178, 2012.

13. X. Shang, X. Meng, E. C. B. A. Alegria et al., "Syntheses, molecular structures, electrochemical behavior, theoretical study, and antitumor activities of organotin(IV) complexes containing 1-(4-chlorophenyl)-1-cyclopentanecarboxylato ligands," Inorganic Chemistry, vol. 50, no. 17, pp. 8158–8167, 2011.

14. D. Point, W. C. Davis, S. J. Christopher et al., "Development and application of an ultratrace method for speciation of organotin compounds in cryogenically archived and homogenized biological materials," Analytical and Bioanalytical Chemistry, vol. 387, no. 7, pp. 2343–2355, 2007.

15. R. Singh, S. Joshi, and A. K. Agrawal, "Synthesis and structural studies of complexes of N-salicylidine amino acids (glycine, L-leucine with salicyldehyde) derived Schiff base with some bivalent transition metal ions,"International Journal of ChemTech Research, vol. 2, no. 1, pp. 728–732, 2010.

16. M. Tariq, N. Muhammad, M. Sirajuddin et al., "Synthesis, spectroscopic characterization, X-ray structures, biological screenings, DNA interaction study and catalytic activity of organotin(IV) 3-(4-flourophenyl)-2-methylacrylic acid derivatives," Journal of Organometallic Chemistry, vol. 723, pp. 79–89, 2013.

17. J. Zuo, C. Bi, Y. Fan et al., "Cellular and computational studies of proteasome inhibition and apoptosis induction in human cancer cells by amino acid Schiff base-copper complexes," Journal of Inorganic Biochemistry, vol. 118, pp. 83–93, 2013.

18. A.-J. Guo, X.-S. Xu, Y.-H. Hu, M.-Z. Wang, and X. Tan, "Effects of ternary complexes of copper with salicylaldehyde-amino acid Schiff base coordination compounds on the proliferation of BGC823 cells," Chinese Journal of Cancer, vol. 29, no. 3, pp. 277–282, 2010.

19. N. Korkmaz, A. G. Gökçe, S. T. Astley, M. Aygün, D. Astley, and O. Büyükgüngör, "A highly novel dinickel complex: a hydrogen-bonded anti-skew carboxylate bridge and a 2D supramolecular structure," Inorganic Chemistry Communications, vol. 12, no. 12, pp. 1204–1208, 2009.

20. P. Conti, M. de Amici, S. J. di Ventimiglia et al., "Synthesis and anticonvulsant activity of novel bicyclic acidic amino acids," Journal of Medicinal Chemistry, vol. 46, no. 14, pp. 3102–3108, 2003.

21. M.-Z. Wang, Z.-X. Meng, B.-L. Liu, G.-L. Cai, C.-L. Zhang, and X.-Y. Wang, "Novel tumor chemotherapeutic agents and tumor radio-imaging agents: potential tumor pharmaceuticals of ternary copper(II) complexes," Inorganic Chemistry Communications, vol. 8, no. 4, pp. 368–371, 2005.

22. Y. Zhang, X. Wang, and L. Ding, "Interaction between tryptophan-vanillin Schiff base and herring sperm DNA," Journal of the Serbian Chemical Society, vol. 75, no. 9, pp. 1191–1201, 2010.

23. M. Wang, Z. Meng, and J. Fu, "Synthesis and biodistribution of six novel^{99m}Tc complexes of 2-hydroxybenzaldehyde-amino acid Schiff bases," Applied Radiation and Isotopes, vol. 64, no. 2, pp. 235–240, 2006.

24. M. S. Ahmad, M. Hussain, M. Hanif, S. Ali, and B. Mirza, "Synthesis, chemical characterization and biological screening for cytotoxicity and antitumor activity of organotin (IV) derivatives of 3,4-methylenedioxy 6-nitrophenylpropenoic acid," Molecules, vol. 12, no. 10, pp. 2348–2363, 2007.

25. M. Gielen, "Tin-based antitumour drugs," Coordination Chemistry Reviews, vol. 151, pp. 41–51, 1996.

26. D. de Vos, R. Willem, M. Gielen, K. E. van Wingerden, and K. Nooter, "The development of novel organotin anti-tumor drugs: Structure and activity,"Metal-Based Drugs, vol. 5, no. 4, pp. 179–188, 1998.

27. D. Schonfelder and U. Thust, "Chemical structure-biological activity relationships: quantitative approaches," in Proceedings of the 3rd Congress of the Hungarian Pharmacological Society, F. Darvas, Ed., Pergamon Press, 1980.

28. W. Rehman, M. K. Baloch, and A. Badshah, "Synthesis, spectral characterization and bio-analysis of some organotin(IV) complexes,"European Journal of Medicinal Chemistry, vol. 43, no. 11, pp. 2380–2385, 2008.

29. Q. Li, M. F. C. Guedes da Silva, and A. J. L. Pombeiro, "Diorganotin(IV) derivatives of substituted benzohydroxamic acids with high antitumor activity," Chemistry, vol. 10, no. 6, pp. 1456–1462, 2004.

30. L. Pellerito and L. Nagy, "Organotin(IV)$^{n+}$ complexes formed with biologically active ligands: equilibrium and structural studies, and some biological aspects," Coordination Chemistry Reviews, vol. 224, no. 1-2, pp. 111–150, 2002.

31. E. M. Hodnett and W. J. Dunn III, "Structure-antitumor activity correlation of some schiff bases," Journal of Medicinal Chemistry, vol. 13, no. 4, pp. 768–770, 1970.

32. M. Nath, R. Jairath, G. Eng, X. Song, and A. Kumar, "Synthesis, spectral characterization and biological studies of some organotin(IV) complexes of l-proline, trans-hydroxy-l-proline and l-glutamine," Spectrochimica Acta A: Molecular and Biomolecular Spectroscopy, vol. 62, no. 4-5, pp. 1179–1187, 2005.

33. H. L. Singh and J. B. Singh, "Synthesis and characterization of new lead(II) and organotin(IV) complexes of Schiff bases derived from histidine and methionine," International Journal of Inorganic Chemistry, vol. 2012, Article ID 568797, 12 pages, 2012.

34. H. L. Singh, M. Sharma, and A. K. Varshney, "Studies of coordination compounds of organotin(IV) with Schiff bases of amino acids," Synthesis and Reactivity in Inorganic and Metal-Organic Chemistry, vol. 30, no. 3, pp. 445–456, 2000.

35. H. L. Singh and J. B. Singh, "Synthesis and characterization of new lead(II) complexes of Schiff bases derived from amino acids," Research on Chemical Intermediates, vol. 39, no. 5, pp. 1997–2009, 2013.

36. Chem, "3D Ultra molecular modeling and analysis," Cambridge, UK, http://www.cambridgesoft.com/.

37. N. L. Allinger, Molecular Structure: Understanding Steric and Electronic Effects from Molecular Mechanics, John Wiley & Sons, 2010.

38. H. L. Singh, J. B. Singh, and K. P. Sharma, "Synthetic, structural, and antimicrobial studies of organotin(IV) complexes of semicarbazone, thiosemicarbazone derived from 4-hydroxy-3- methoxybenzaldehyde,"Research on Chemical Intermediates, vol. 38, no. 1, pp. 53–65, 2012.

39. F. Ahmad, M. Parvez, S. Ali, M. Mazhar, and A. Munir, "Synthesis and spectral studies of tri- and diorganotin(IV) complexes with 5-benzoyl-α-methyl-2-thiopheneacetic acid: crystal structure of $[(CH_3)_3Sn(C_{14}H_{11}O_3S)]$,"Syn thesis and Reactivity in Inorganic and Metal-Organic Chemistry, vol. 32, no. 4, pp. 665–687, 2002.

40. M. Nath, C. L. Sharma, and N. Sharma, "Dibutyltin(IV) complexes of schiff bases derived from aminoacids," Synthesis and Reactivity in Inorganic and Metal-Organic Chemistry, vol. 21, pp. 807–824, 1991.

41. G. Roge, F. Huber, H. Preut, A. Silvestri, and R. Barbieri, "Triorganotin and triorganolead derivatives of N-acetylamino-acids," Journal of the Chemical Society, Dalton Transactions, no. 4, pp. 595–600, 1983.

42. M. S. Singh, M. D. Raju, K. Tawade, and A. K. Singh, "Synthesis and spectroscopic study of some organotin(IV) derivatives of benzilmonothiosemicarbazone," Main Group Metal Chemistry, vol. 21, no. 9, pp. 489–493, 1998.

43. S. G. Teoh, S. H. Ang, E. S. Looi, C. A. Keok, S. B. Teo, and J. P. Declercq, "Synthesis and crystal structure of di-n-butylbis(2-amino-5-chlorobenzoato) tin(IV)," Journal of Organometallic Chemistry, vol. 523, no. 1, pp. 75–78, 1996.

44. B. Sen Saraswat and J. Mason, "Modes of coordination of 1,2-aminothiols in organotin(IV) complexes, as demonstrated by ^{119}Sn, ^{15}N and ^{13}C NMR spectroscopy," Polyhedron, vol. 5, no. 9, pp. 1449–1458, 1986.

45. J. Holeček, M. Nádvorník, K. Handlíř, and A. Lyčka, "^{13}C and ^{119}Sn NMR spectra of Di-n-butyltin(IV) compounds," Journal of Organometallic Chemistry, vol. 315, no. 3, pp. 299–308, 1986.

46. P. Comba and T. W. Hambley, Molecular Modeling of Inorganic Compounds, Wiley-VCH, Weinheim, Germany, 1995.

47. A. P. Murthy, "Electrochemical reductive cleavage of carbon-halogen bonds-application of Marcus theory in the analysis of nonlinear activation—driving force relationships," International Archives of Science and Technology, vol. 6, pp. 10–15, 2006.

48. P. Kumar, P. Mishra, R. P. Singh, and R. P. Singh, "Synthesis, spectroscopic, XRPD and computational study of transition metal-organic frameworks (TMOFs) derived from (6R)-6-(α-phenyl-D-glycylamino) penicillanic acid,"Russian Journal of Inorganic Chemistry, vol. 54, pp. 1301–1309, 2009.

49. T. Zöller, L. Iovkova-Berends, T. Berends, C. Dietz, G. Bradtmöller, and K. Jurkschat, "Intramolecular N→Sn coordination in tin(II) and tin(IV) compounds based on enantiopure ephedrine derivatives," Inorganic Chemistry, vol. 50, no. 17, pp. 8645–8653, 2011.

50. D. A. Atwood, J. A. Jegier, K. J. Martin, and D. Rutherford, "Tetradentate $-N_2O_2$ ligand complexes of Tin(II). X-ray crystal structure of [N,N′-(1,2-ethylene) bis (salicylaldamine)]tin(II), $(SaleanH_2Sn)$," Journal of Organometallic Chemistry, vol. 503, no. 1, pp. C4–C7, 1995.

51. E. Abele, "Activation of silicon bonds by fluoride ion in the organic synthesis in the new millennium: a review," Main Group Metal Chemistry, vol. 28, no. 2, pp. 45–69, 2005.

52. B. G. Tweedy, "Possible mechanism for reduction of elemental sulfur by monilinia fructicola," Phytopathology, vol. 55, pp. 910–914, 1964

CHAPTER 11 APPLICATION OF SPECTROSCOPIC METHODS FOR STRUCTURAL ANALYSIS OF CHITIN AND CHITOSAN

Jolanta Kumirska[1], Małgorzata Czerwicka[1], Zbigniew Kaczyński[1], Anna Bychowska[1], Krzysztof Brzozowski[1], Jorg Thöming[2] and Piotr Stepnowski[1]

[1]Faculty of Chemistry, University of Gdansk, Sobieskiego 18/19, PL-80-952 Gdansk, Poland

[2]UFT-Centre for Environmental Research and Sustainable Technology, University of Bremen, Leobener Straße UFT, D-28359 Bremen, Germany

ABSTRACT

Chitin, the second most important natural polymer in the world, and its N-deacetylated derivative chitosan, have been identified as versatile biopolymers for a broad range of applications in medicine, agriculture and the food industry. Two of the main reasons for this are firstly the unique chemical, physicochemical and biological properties of chitin and chitosan, and secondly the unlimited supply of raw materials for their production. These polymers exhibit widely differing physicochemical properties depending on the chitin source and the conditions of chitosan production. The presence of reactive functional groups as well as the polysaccharide nature of these biopolymers enables them to undergo diverse chemical modifications. A complete chemical and physicochemical characterization of chitin, chitosan and their derivatives is not possible without using spectroscopic techniques. This review focuses on the application of spectroscopic methods for the structural analysis of these compounds.

INTRODUCTION

Chitin, the second most abundant natural polymer in the world, functions as a natural structural polysaccharide [1]. A major component of the carapaces, crusts and shells of crustaceans such as shrimps, crabs and lobsters, it is also an ingredient of cell walls in fungi and yeast [2]. Its estimated production is 10^{10}–10^{12} tonnes per year [3]. Chitin is a linear polymer consisting mainly of β-(1→4)-linked 2-acetamido-2-deoxy-β-d-glucopyranose units and partially of β-(1→4)-linked 2-amino-2-deoxy-β-d-glucopyranose. In this form, chitin is insoluble in water and common organic solvents, dissolving only in specific solvents such as *N,N*-dimethylacetamide (DMAc)-LiCl [4], hexafluoroacetone or hexafluoro-2-propanol [5]. When the degree of *N*-acetylation (defined as the average number of *N*-acetyl-d-glucosamine units per 100 monomers expressed as a percentage) is less than 50%, chitin becomes soluble in aqueous acidic solutions (pH < 6.0) and is then called chitosan [6,7]. Thus, chitosan is a collective name for a group of fully and partially deacetylated chitins, but a rigid

Citation: Jolanta Kumirska, Małgorzata Czerwicka, Zbigniew Kaczyński, Anna Bychowska, Krzysztof Brzozowski, Jorg Thöming and Piotr Stepnowski, Application of Spectroscopic Methods for Structural Analysis of Chitin and Chitosan, doi: 10.3390/md8051567

nomenclature with respect to the degree of *N*-deacetylation between chitin and chitosan has not been defined [8]. The structures of "ideal" chitin and "ideal" chitosan, and the "real" structures of these compounds are shown in Figure 1.

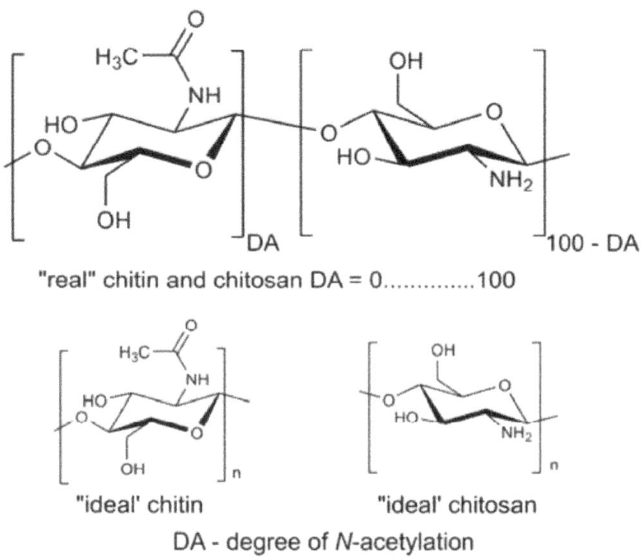

Figure 1 The chemical structures of chitin and chitosan.

According to the nomenclature proposed by the European Chitin Society (EUCHIS) [9], chitin and chitosan should be classified on the basis of their solubility and insolubility in 0.1 M acetic acid; the soluble material is named chitosan, whereas chitin is insoluble. Chitin and chitosan have a molecular mass of up to several million g/mol. Commercially available chitosan has an average molecular weight ranging between 3,800 and 500,000 g/mol and its degree of *N*-acetylation is 2% to 40% [8,10].

Chitin and chitosan are of commercial interest because of their high nitrogen content (6.89%) and their excellent properties such as biocompatibility, biodegradability, non-toxicity and adsorptive abilities [2,8,11]. These compounds are of very low toxicity: LD_{50} of chitosan in laboratory mice is 16 g/kg body weight, which is similar to that of sugar and salt. In rats, chitosan is safe up to 19% in the diet [12]. As a material that is highly insoluble and has a low chemical reactivity, chitin has limited applicability. Recently, however, chitosan has come back into the spotlight because of its numerous and extensive applications–in biomedicine, waste water treatment, food, cosmetics and the fiber industry [7,8,11,13–17]. Most of the characteristic properties of chitosan are due to the high content of primary amino groups with a pKa of 6.3. At low pH, the positive charge on the-NH_3^+ groups converts chitosan to a water-soluble cationic polyelectrolyte; when pH increases to above 6.0 the positive charge on the amino groups is lost and chitosan becomes insoluble [7]. The soluble-insoluble transition of chitosan occurs around pH 6.0–6.5 at the pKa of its primary amino groups. The pKa depends closely on the degree of *N*-acetylation (DA), hence the solubility of chitosan is dependent on DA and the method of *N*-deacetylation [18]. The amino groups are also responsible for several straightforward chemical modifications of chitosan, which predisposes its ongoing development for many applications [19–22]. The improved solubility in water and organic solvents of chemically modified chitin and chitosan has been reported by many scientists [23–25].

Chitin is isolated from the exoskeletons of crustaceans, molluscs, insects and certain fungi, but the main commercial sources of chitin are crab and shrimp shells [2,26]. Depending on the source, chitin occurs in two allomorphs, the α and β-forms, and additionally as γ-chitin, which appears to be a combination of the α and β structures rather than a different allomorph [3]. α-Chitin is by far the most abundant and is usually isolated from the exoskeleton of crustaceans, particularly from shrimps and crabs. β-Chitin can be extracted from squid pens, and γ-chitin from fungi and yeast [27]. β-Chitin is easily converted to α-chitin by alkaline treatment followed by flushing in water [28].

Several techniques to extract chitin from different sources have been published [26,29–31]. Crustacean shells consist of proteins (30–40%), calcium carbonate (30–50%), chitin (20–30%) and pigments (astaxanthin, canthaxanthin, lutein or β-carotene). These proportions vary from species to species and from season to season [32]. The most common method for chemically isolating chitin from crustacean shells involves a number of major steps: the washing, grinding and sieving of raw shells, followed by their demineralization (elimination of calcium carbonate in dilute acidic acid) and deproteinization in aq NaOH or KOH. The use of enzymatic hydrolysis for deproteinization [29] and microorganisms for both demineralization and deproteinization has been also reported [29,33].

Industrially chitin is converted into the more readily applicable chitosan by partial or complete deacetylation of chitin in both the solid (heterogeneous process) and dissolved (homogeneous process) states under alkaline conditions or by enzymatic hydrolysis (using a chitin deacetylase). The source of natural chitin used to produce chitosan affects the production parameters and chitosan preparations. It has been shown that β-chitin is more reactive in *N*-deacetylation than α-chitin [34]. The difference in the semicrystalline morphology of chitin means, for example, that chitosans obtained in a solid-state reaction have a heterogeneous distribution of *N*-acetyl groups along the molecular chains [35]. Differences in the chitosan production process (e.g., temperature, alkali concentration, ratio of alkali solutions to the shells) also mean that chitosan preparations consist of a mixture of chitosans varying in molecular weight and degree of *N*-acetylation. Those produced from chitin may also contain impurities such as heavy metals, protein residues and acid/alkaline residues.

For most of the commercially available chitosans, parameters like weight-average molecular weight (M_W), polydispersity (M_W/M_N), degree of *N*-acetylation (DA), pattern of acetylation (PA) and impurity content (protein, heavy metal) are usually unknown [32,36]. Knowledge of the microstructure of chitosan samples is thus essential for an understanding of the structure–property–activity relationships in them, and special emphasis in this respect is placed on the chitosans used in biomedical applications [32,36,37]. The complete structural characterization of synthesized, biologically active and/or water-soluble derivatives of chitin and chitosan is also very important.

The present paper reviews the applications of spectroscopic methods for the structural analysis and physicochemical characterization of chitin, chitosan and their derivatives. The most important techniques are X-ray spectroscopy, infrared (IR) and UV-Vis-spectroscopy, mass spectrometry (MS), and nuclear magnetic resonance spectroscopy (NMR). The usefulness of these methods for determining and confirming molecular structures, for monitoring reactions and controlling the purity of these compounds is discussed.

APPLICATION OF SPECTROSCOPIC METHODS FOR ANALYZING THE STRUCTURE AND DETERMINING THE PHYSICOCHEMICAL PROPERTIES OF CHITIN, CHITOSAN AND THEIR DERIVATIVES

X-ray Spectroscopy

X-ray spectroscopy is unarguably the most versatile and widely used means of characterizing materials of all forms [38]. There are two general types of structural information that can be studied by X-ray spectroscopy: electronic structure (focused on valence and core electrons, which control the chemical and physical properties, among others) and geometric structure (which gives information about the locations of all or a set of atoms in a molecule at an atomic resolution). This method encompasses several spectroscopic techniques for determining the electronic and geometric structures of materials using X-ray excitation: X-ray absorption spectroscopy (XAS), X-ray emission spectroscopy (XES), X-ray photoelectron spectroscopy (XPS) and X-ray Auger spectroscopy. Which type of X-ray spectroscopy is employed depends on whether the target information is electronic, geometric or refers to oxidation states: for instance, XAS (first developed by de Broglie) to probe empty states and the shapes of molecules or local structures [39], and XPS (first developed by Siegbahn) to investigate occupied electronic states [40]. X-ray spectroscopy is thus a powerful and flexible tool and an excellent complement to many structural analysis techniques such as UV-Vis, IR, NMR or Raman.

Typical Conditions of X-ray Measurements

In an X-ray diffraction measurement, a crystal is mounted on a goniometer and gradually rotated while being bombarded with X-rays, producing a diffraction pattern of regularly spaced spots known as reflections [41]. The two-dimensional images taken at different rotations are converted into a three-dimensional model of the density of electrons within the crystal using the mathematical method of Fourier transforms, combined with the chemical data obtained for the sample. If single crystals of sufficient size cannot be obtained, various other X-ray methods, like fiber and powder diffraction, can be applied to record less detailed information.

Clark and Smith in 1937 were the first to make crystal studies of chitin and chitosan using X-ray diffraction (XRD) [42]. They carried out those investigations using a commercial copper-target diffraction tube operated at 30 kV and 25 mA as the X-ray sources, which generated principally Cu-K$_\alpha$ lines. The diffraction patterns were recorded on a flat film perpendicular to the beam with the sample 5.0 cm from the film. In later research, the conditions for X-ray measurements of chitin and chitosan were mostly modifications of the ones used by Clark and Smith [42]. For instance, X-ray diffraction measurements were done at 100% relative humidity in a helium atmosphere [43] to avoid the X-ray scattering that led to a "dirty" background on the X-ray film. Present-day X-ray analyses of chitin, chitosan and their derivatives are carried out on advanced X-ray diffractometers.

Apart from traditional X-ray diffraction (XRD), other X-ray techniques for determining chitin/chitosan and their derivatives have been applied: X-ray photoelectron spectroscopy (XPS), the second most popular X-ray spectroscopy technique–for determining the bonding energies of C, O and N atoms on the surface of chitosan and its metal chelate, and for other chitin and chitosan investigations [44–49]; X-ray emission spectroscopy (XES)–perfect for studying the chemical bonding in chitosan and cross-linked chitosan derivatives [50]; X-ray absorption spectroscopy (XAS)–for determining the coordination number of Fe atoms in chitosan-metal complexes [51].

X-ray Spectra of Chitin and Chitosan

Numerous X-ray spectroscopic studies of chitin and chitosan have yielded the diffractive patterns of these compounds [52–59]. However, different sources have characterised these patterns with differently indexed crystalline peaks: these can be labeled either using d, the centre-to-centre spacing of the crystallites, or Miller indices, e.g., (020). It is also very common to describe the diffraction pattern using the values of angles.

Typical spectra of chitin fiber and chitosan fiber are shown in Figure 2 [56]. The spectrum of the former exhibits broad peaks at d = 0.34, 0.45, 0.50 and 1.09 nm with a shoulder at 0.71 nm; in the latter the spectral peaks are at d = 0.45, 0.88 and 2.93 nm.

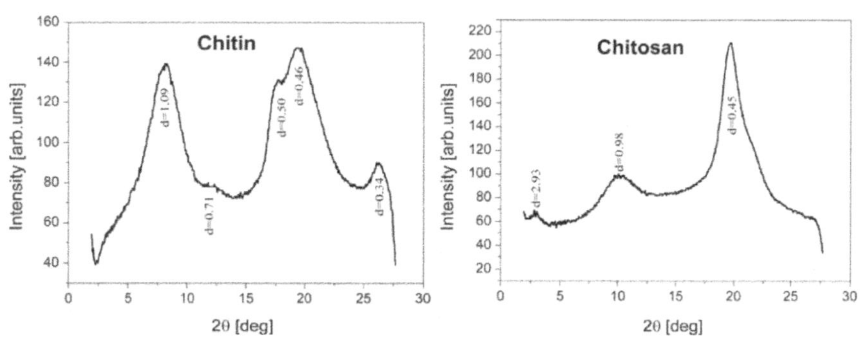

Figure 2 X-ray diffraction spectra of chitin and chitosan fibers. Reprinted from Carbohydrate Polymers 56, 2004, Muzzarelli, C., Francescangeli, O., Tosi, G., Muzzarelli, R.A.A., Susceptibility of dibutyryl chitin and regenerated chitin fibers to deacetylation and depolymerization by lipases, 137–146, Copyright (2010), with permission from Elsevier [56].

The Miller indices of the diffraction peaks characteristic of chitin were (020), (110), (120), (101) and (130) [57,58], although Muzzarelli *et al.* reported the lack of a strong (020) peak for chitin. [59]. In turn, the typical chitin diffraction pattern, given in angle form, showed strong reflections at 2 around 9–10° and 2 of 20–21° and minor reflections at higher 2 values, e.g., at 26.4° and higher [60].

X-ray Analysis of Chitin and Chitosan Polymorphs

As already mentioned, the crystal structures of chitosan have been examined since the work of Clark and Smith in 1937 [42]. A great number of diffraction experiments have been undertaken in an attempt to elucidate the molecular geometry of chitosan [42,61–71]. The first X-ray studies showed that the chitosan molecule can adopt at least two different conformations in crystals–a 2-fold [42,61–63] and an eight-fold right-handed helical structure [64,65]. Apart from these two helical conformations, other polymorphs of chitosan have been characterized [66–71]. In 1994 Yui *et al.* determined the detailed crystal structure of the anhydrous form of chitosan by combining X-ray diffraction analysis with a stereochemical model refinement [67]. Chitosan chains crystallized in an orthorhombic unit cell of the following dimensions: a = 0.828 nm, b= 0.862 nm and c (fiber axis) = 1.043 nm. The X-ray diffraction pattern was recorded on the imaging plate, and the intensity of each reflection was estimated by two-dimensional measurement and subsequent background removal. The results showed that the chain conformation of this anhydrous form of chitosan was an extended two-fold helix stabilized by intramolecular O3···O5 hydrogen bonds and that the adjacent parallel chains were connected by O6···N2 hydrogen bonds.

At the same time Mazeau *et al.* examined the conformation and packing of a chitosan polymorph crystallized at a high temperature on the basis of diffraction data [68]. In this kind of polymorph, the chitosan chains form orthorhombic crystals with a = 0.807 nm, b = 0.844 nm and c (chain axis) = 1.034 nm. This chitosan molecule adopted a two-fold helical conformation stabilized by two intramolecular hydrogen bonds, a strong one between O5′ and O3, and a weaker one between O5′ and O6. These two anhydrous forms of chitosan have relatively similar structures.

In 1997 Yui's group published another paper on the X-ray fiber diffraction method used to determine the crystal structure of a hydrated form of chitosan [69]. The results indicated that the hydrated form of chitosan molecules had a two-fold helical symmetry reinforced by a O3····O5 hydrogen bond with a repeating period of 10.34 Å; this is a structure typical of β (1→4) linked polysaccharides such as cellulose, mannan and chitin.

The following crystalline polymorphs of chitosan have so far been found using X-ray diffraction measurements: the most abundant "tendon-chitosan" [42,69], as well as the "annealed" [63], "1–2", "L-2" [66], "form I" and "form II" [61] and "8-fold right-handed" forms [64,65]. Apart from the last one, all the polymorphs of chitosan molecules have the extended 2-fold helix configuration; the 8-fold polymorph is unstable and is easily converted into the 2-fold helix [72].

Depending on the source, chitin can occur in the α-, β- and γ-forms. The differences among them depend on the arrangement of chains in the crystalline regions [52]. The most abundant and stable form is α-chitin, which Minke and Blackwell studied in detail using XRD in 1978 [45]. These authors determined the α-chitin structure on the basis of intensity data from deproteinized lobster tendon. They discovered that α-chitin chains form orthorhombic crystals with a = 0.474 nm, b = 1.886 nm and c (fiber axis) = 1.032 nm. Additionally, the chains form hydrogen-bonded sheets linked by C=O...H–N bonds approximately parallel to the a axis, and each chain is stabilized by an C(3′)O–H····OC(5) intramolecular hydrogen bond, as in cellulose. These data also indicated that a statistical mixture of CH_2OH orientations was present, equivalent to half an oxygen on each residue, each forming inter- and intramolecular hydrogen bonds. As a result, Minke and Blackwell found that the α-chitin structure contained two types of amide groups, differing in their hydrogen bonding. In addition, the inability of this polymorph to swell in the presence of water was explained by the extensive intermolecular hydrogen bonding. The modes of hydrogen bonding in α- and β-chitin are illustrated in Figures 3 and 4 [18].

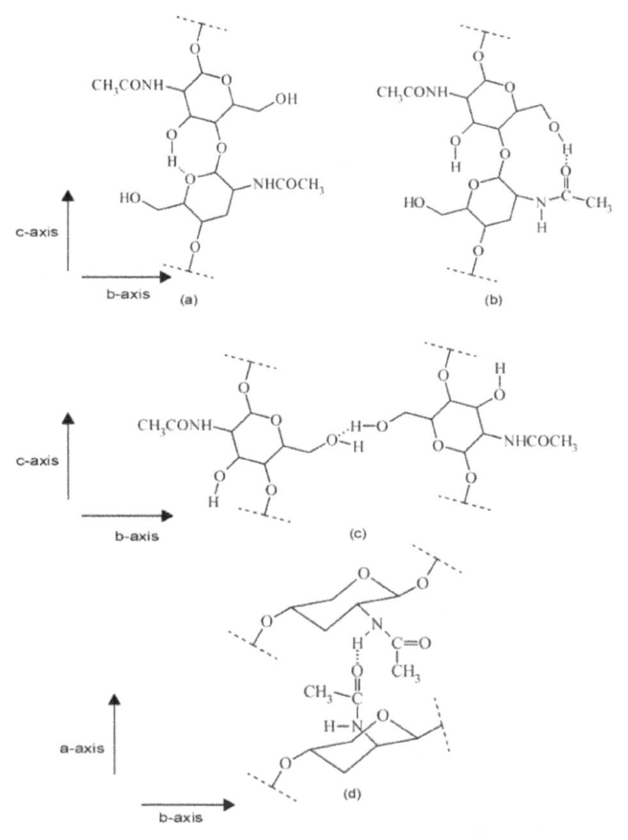

Figure 3 Modes of hydrogen bonding in α-chitin: (a) intrachain C(3′) OH···OC(5) bond; (b) intrachain C(6′$_1$)OH···O=C(7$_1$) bond; (c) interchain C(6′$_1$)O···HOC(6$_2$) bond; (d) interchain C(2$_1$)NH···O=C(7$_3$) bond. Adapted with permission from Biomacromolecules. 2000, 1, 609–614. Copyright 2010, American Chemical Society [18].

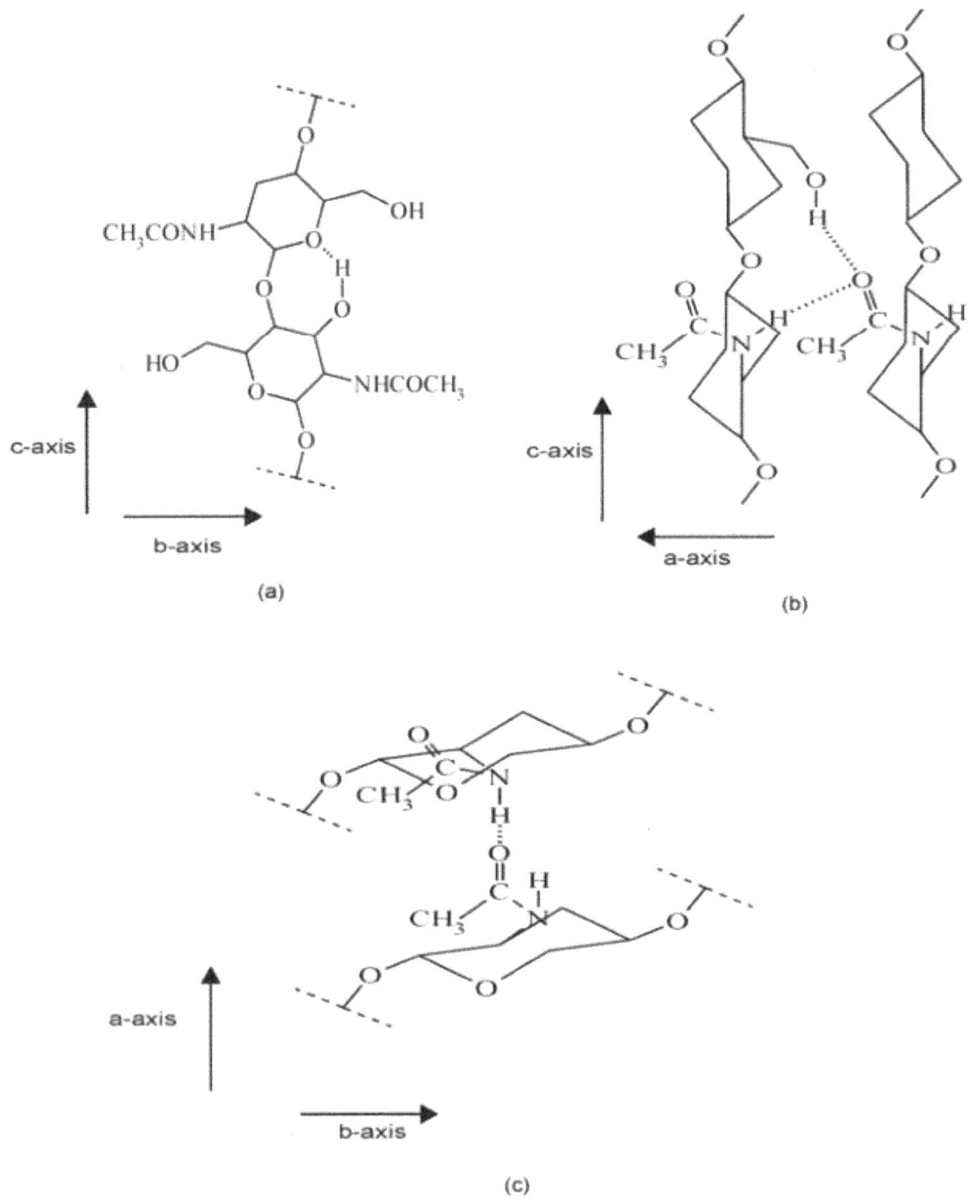

Figure 4 Modes of hydrogen bonding in β-chitin: (a) intrachain C(3′)OH···OC(5) bond; (b) interchain $C(2_1)NH\cdots O{=}C(7_3)$ bond and $C(6'_1)OH\cdots O{=}C(7_3)$ bond (*ac* plane projection). Adapted with permission from Biomacromolecules. 2000, 1, 609–614. Copyright 2010, American Chemical Society [18].

As mentioned above, the differences among chitin polymorphs are due to the arrangement of the chains in the crystalline regions: α-chitin has a structure of antiparallel chains [45], β-chitin has intrasheet hydrogen-bonding by parallel chains [52], and γ-chitin, being a combination of α- and β-chitin [52], has both parallel and antiparallel structures. Because of these differences each chitin polymorph has different properties specific to it. For example, β-chitin is more soluble in and more reactive towards solvents and has a greater affinity towards them; it is also more susceptible to swelling than α-chitin. β-Chitin is also more amenable to*N*-deacetylation than α-chitin. In this case, comparing the diffraction profiles of α- and β- chitin before and after *N*-deacetylation using X-ray spectroscopy seems to be a relatively good solution for distinguishing these forms of chitin. Abdou *et al.* used X-ray diffraction to study two chitin polymorphs, the α- and β-forms, obtained from different sources [60]. The chitin samples were converted into the more soluble chitosan by steeping them in solutions of NaOH of various concentrations and for extended periods of time. The X-ray diffraction patterns of the α-chitin samples and the corresponding hydrolyzed chitosans showed strong reflections at 2 around 9–10° and 2 of 20–21° and minor reflections at higher 2 values at e.g., 26.4° and higher. The chitin bands were sharper than the chitosan bands, even though there was only a slight decrease in the crystallinity percentage. In turn, the X-ray diffraction patterns of the β-chitin samples and their corresponding hydrolyzed chitosans showed that the band at 2 = 9.9° decreased significantly after deacetylation, and that this was followed by a dramatic decrease in the crystallinity percentage. It was therefore concluded that β-chitin is much more amenable to *N*-deacetylation than the α-form. The X-ray diffraction patterns of α- and β-chitin and the corresponding hydrolysed chitosans make them easily distinguishable. The crystallinity index (*CI*) can also be calculated on the basis of X-ray diffractograms. This takes different values for different forms of chitin; for example, Lima showed that *CI* for α-chitin is 28.3% but that for β-chitin it is 20.8% [73].

Many other studies of chitin polymorphs have revealed differences in crystallinity peaks between α-, β- and γ-chitins obtained from various sources [52–55]. For example, Jang *et al.* found crystalline peaks at 9.6, 19.6, 21.1 and 23.7° for α-chitin, at 9.1 and 20.3° for β-chitin, and at 9.6 and 19.8° for γ-chitin [52]. Similarly, Cárdenas *et al.* [53] reported that WAXD patterns of α-chitin (chitins from shrimp, lobster, prawn and king crab) and β-chitin (chitin from squid) exhibited their major characteristic peak at 19.2–19.3° and 18.8° respectively. Kim [54] found that β-chitin from squid pen exhibited crystalline peaks at 9.8° and 19.3°. Yen and Mau [55] found that fungal chitin (γ-chitin) showed two crystalline reflections at 5.4–5.6° and 19.3–19.6°. Irrespective of their origin, the three types of chitin consistently display a major peak at ~19° in their crystallinity structure.

Physicochemical Characterization of Chitin and Chitosan Using X-Ray Diffraction

In 1937, Clark and Smith, in their pioneer X-ray diffraction studies of chitin/chitosan and their derivatives [42], were the first to investigate the physicochemical properties of chitin. They reported the action of hydrochloric acid, lithium thiocyanate and nitric acid on chitin. Their data showed that even at room temperature the ether linkages of chitin were hydrolysed in hydrochloric acid; concurrently, but more slowly, the amide groups were also hydrolysed. In addition, those authors investigated the dispersion of chitin in lithium thiocyanate; at a temperature of 200 °C chitin formed a definite addition compound with lithium thiocyanate, but at lower temperatures only intramicellar swelling was observed. In turn, chitin nitrate was roughly as soluble in hydrochloric acid as the original chitin–no substantial hydrolysis of the acetyl groups had occurred.

Subsequent studies revealed that the properties of chitin and chitosan depended mostly on the degree of *N*-acetylation, molecular weight, polydispersity and crystallinity [31]. Commonly used to measure crystallinity, XRD is also applied to determine the degree of *N*-acetylation of chitin and chitosan [74].

In 1990, Focher *et al.* [75] used XRD to study chitin and postulated the following equation for determining the crystallinity index (*CI*):

$$CI\,(\%) = [(I_{110} - I_{am})/I_{110}] \times 100 \quad (1)$$

where I_{110} (arbitrary units) is the maximum intensity of the (110) peak at around 2 = 19°, and I_{am} (arbitrary units) is the amorphous diffraction at 2 = 12.6°. This expression had in fact been employed three years earlier by Struszczyk to determine the *CI* of chitosan [76]. Currently this equation is routinely applied during investigations of chitin, chitosan and their derivatives [73,77,78]. In most cases, *CI* provides information about the crystal state, but it is also very useful for distinguishing α-chitin from β-chitin [73].

On the basis of X-ray powder diffractograms of chitin and chitosan with different degrees of *N*-acetylation, Zhang noted two maximum peaks of the following intensities: one at the (020) reflection and the other at the (110) reflection [74]. He therefore postulated a crystallinity index (*CI*) expressed by two equations:

$$CI_{020} = [(I_{020} - I_{am})/I_{110}] \times 100 \quad (2)$$

$$CI_{110} = [(I_{110} - I_{am})/I_{110}] \times 100 \quad (3)$$

Further chitin and chitosan studies indicated that crystallinity could also be assigned from an X-ray diffractogram by dividing the area of the crystalline peaks by the total area under the curve (background area) [60,79,80]. In these calculations, the crystallinity percentage supplied information on relative crystallinity.

A lot of studies have been carried out in which X-ray measurements were applied to determine which parameters affect the crystalline structure of chitin and chitosan and how they do so. In 1991 Ogawa reported an increase in crystallinity with a decrease in the M_w of chitosan [71]. One year later, he also determined how chitosan polymorphism and its crystallinity in the membrane depended on the membrane preparation procedure and the molecular weight of the chitosan [43]. XRD measurements demonstrated that when an acetic acid solution of chitosan was dried in air and then soaked in an alkaline solution, both hydrated and anhydrous polymorphs of chitin were present in the resulting membrane. On the other hand, when a highly concentrated solution of chitosan in aqueous acetic acid was neutralized with an alkaline solution, no anhydrous polymorphs were detected in the membrane because drying was incomplete. In another paper from 1993 [81], Ogawa compared the crystallinities of partially *N*-deacetylated chitin (PDC) and partially *N*-acetylated chitosan (PAC) samples with a similar degree of *N*-acetylation and their behavior by heating them in water. He discovered that the *N*-acetylation of pure chitosan is far superior to the solid-state *N*-deacetylation of chitin for producing a less crystalline sample, and in particular, for obtaining a less anhydrous crystal.

The effect of DA on solubility in relation to the crystal structure of deacetylated chitin was also discussed by Cho *et al.* in 2000 [18]. Wide-angle X-ray diffractometry (WAXD) revealed that chitin with ca 72% DA retained the crystal structure of α-chitin with significantly reduced crystallinity and crystallite perfection. The water-soluble chitin with *circa* 51% DA had a new crystal structure resembling that of β-chitin rather than that of either α-chitin or chitosan, suggesting that homogeneous deacetylation converted the crystal structure of chitin from the α- to the β-form.

Using X-ray powder diffraction Zhang *et al.* [74] also tried to look for a relationship between the crystalline state and DA of chitin. Figure 5 presents the XRD patterns of chitin and chitosan with different degrees of *N*-acetylation [74]. These authors noted that the peak of maximum intensity at the (020) reflection diminished together with the decrease in DA and moved to a higher angle. The second intensive peak at the (110) reflection also diminished with the decrease in DA. Consequently, CI_{020} decreased linearly with the decrease in DA. This linear relationship between CI_{020} and the degree of *N*-acetylation suggested the possible use of XRD for determining DA of macromolecular chitin and chitosan.

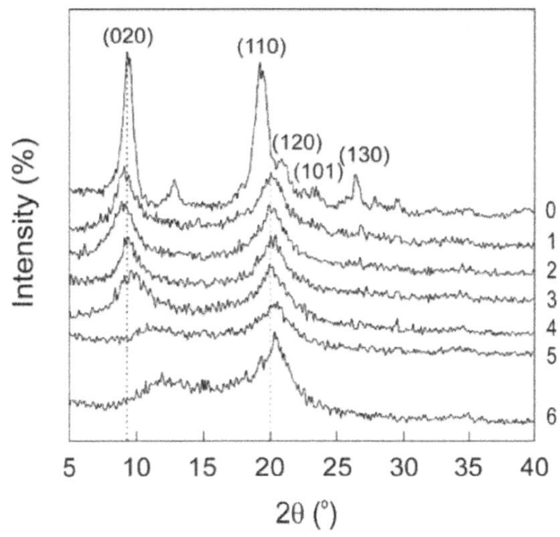

Figure 5 Comparison of X-ray powder diffractograms of chitin and chitosan with different degrees of *N*-acetylation. Figures 0–6 imply different DA (%): 0-83.1, 1-40.6, 2-36.5, 3-41.3, 4-28.6, 5-13.0, 6-7.2. Reprinted from Carbohydrate Research 340, 2005, Zhang, Y., Xue, C., Xue, Y., Gao, R., Zhang, X., Determination of the degree of deacetylation of chitin and chitosan by X-ray powder diffraction, 1914–1917, Copyright (2010), with permission from Elsevier [74].

As demonstrated, molecular weight and DA are the major parameters significantly influencing the crystal structure of chitin and chitosan, although it has also been reported that the crystallinity index depends on other factors. For example, Seoudi reported that *CI* decreased after chitin was treated with HCl and NaOH [82], and Wada and Saito [57] found that when the biopolymer was heated from room temperature to 250 °C the α-chitin remained structurally stable. Moreover, the influence of alkali-freezing treatment on the solid state structure of chitin was examined by Feng [58]. XRD revealed that during this treatment, the crystal space parameters of chitin changed, and the order of the hydrogen bonds in chitin was modified.

X-ray diffraction was used to measure the elastic moduli E_l of the crystalline regions of α-chitin and chitosan [83]. This parameter provided important information on the molecular conformation in the crystal lattice and the mechanism of deformation in the crystalline regions. The data enabled the elastic moduli E_l of the crystalline regions in the direction parallel to the chain axis at 20 °C to be assigned as 41 GPa for α-chitin and 65 GPa for chitosan. These E_l values, which are low compared to those for cellulose I (138 GPa), were due to the contracted skeletons of α-chitin and chitosan in the crystal lattice. In addition, the elastic moduli calculated from X-ray data showed that the molecular chain of α-chitin in the crystal lattice was mechanically stable from − 190 °C to 150 °C.

X-ray Analysis of Chitosan Salts

Chitosan has a regular distribution of aliphatic primary amino groups along its chain. These produce salts when the molecule reacts with inorganic or organic acids. Ogawa and Inukai used XRD to study several crystallized inorganic acid salts of chitosan [64]. On the basis of X-ray fiber diffraction, these authors suggested that chitosan acid salts took up two different conformations. One, which they called "type I salt", retained the extended two-fold helix of the unreacted chitosan molecule, although they were different crystals. The second one, the "type II salt", had an eight-fold helical conformation in the crystal. The salts forming with HNO_3, HBr and HI took up the former structure, and those with HF, HCl, and H_2SO_4 the latter one. Crystals of type I salts were anhydrous, whereas those of type II were hydrated. Despite the different anion sizes, all the type II salts gave fiber patterns that were very similar to each other. They crystallized in a monoclinic unit cell with a helical repetition of 4.073 nm. The chirality of the eight-fold helix was right-handed, since their fiber patterns were very similar to those obtained by Cairns *et al.* [65].

Apart from inorganic salts, chitosan can easily form organic salts, for example, when it reacts with L-ascorbic acid. In 1996 Ogawa carried out an X-ray study of the chain conformation of the ascorbic acid salt of chitosan [84]. He determined that both l- and d-ascorbate chitosan salts retained the extended 2-fold helical conformation of the unreacted chitosan chain and that both crystals were anhydrous. In view of these results, he classified them as type I salts of the chitosan acid salts [64], although during the preparation of the l- andd-ascorbates of chitosan, that author discovered that d-ascorbic acid displayed a higher reactivity towards chitosan than the l-isomer. Those differences in reactivity between l- and d-ascorbic acids could be connected with the optical resolution of ascorbic acid.

Chitosan salts were also examined in 1999 by Kawada *et al.* [72]. They studied the spontaneous water-removing action of acid by preparing chitosan salts of many different (monocarboxylic, inorganic and organic) acids, and examined their structures using X-ray diffraction. The results indicated that the temperature required for salt formation depended on the hydrophobicity of the acid; for instance, the chitosan formic acid salt could be prepared at room temperature, whereas the formation of the propionic acid salt had to be carried out at 4 °C. Moreover, type II salts of monocarboxylic acids, the hydrated crystals of chitosan, could be dehydrated even at room temperature without any loss of orientation or decomposition of the chitosan specimen.

X-ray Analysis of Chitosan Derivatives

Chitin and chitosan are a family of polymers with highly variable chemical and physical properties. These compounds and their derivatives have at least 200 potential and current applications [32] in the biomedical, food, biotechnological, agricultural and cosmetics industries.

Muzzarelli and co-workers found that chitosan exhibited good adsorption selectivity towards some transition and post-transition metal ions from aqueous solution [85]. A chitosan-metal complex dissociates easily when the pH is lowered; therefore, chitosan is very often used in the recovery of useful transition metals from waste. Ogawa and Oka reported on the behavior of chitosan-metal complex formations examined by X-ray diffraction [86]. The unit cells of all the salt complexes studied were orthorhombic, although the cupric salt complexes showed some unindexed reflections. Their lattice parameters and the number of water molecules in the cell depended on the counteranions of the metal salt and not on the metal ion. The ratio of glucosamine residues to metal salts was 2:1. On the basis of the fiber diffraction patterns of various chitosan-transition metal salt complexes, these authors postulated a coordination mode which they named the "pendant model". This model had already been put forward by Ogawa *et al.* back in 1984, but unequivocal experimental evidence to support it was lacking [70]. In this "pendant model" metal anions were coordinated to the amino groups of the chitosan chain like a pendant. Additionaly there was another contrasting model called "the bridge model", in which metal ions were coordinated with four nitrogen atoms of the intra- and inter-chitosan chains [87]. In spite of these differences, chitosan exhibits a high affinity for metal ions, a property that has been used to recover transition metals from waste water. At present, the adsorption properties of metal ions on chitin and chitosan derivatives are still routinely examined by X-ray diffraction [49,88–90].

The majority of current studies on chitin and chitosan are seeking to discover new derivatives with unusual properties and different potential applications [56,91–97]. X-ray measurements are still very often applied to characterize most of these new derivatives. For example, chitosan-based nanocomposite films, containing chitin nanocrystals as functional components, were successfully prepared and cross-linked using glutaraldehyde [97]. XRD showed that chitin nanocrystals retained their crystalline morphology in the nanocomposites before and after cross-linking, and that chitosan also retained its amorphous characteristics in the nanocomposites. Another example, novel chitosan/gelatin membranes were prepared using a suspension of chitosan hydrogel mixed with gelatin [95]. XRD studies showed that the chitosan and gelatin in these membranes are compatible and interact well with each other. In addition, the incorporation of gelatin reduced the crystallinity of chitosan. Finally, natural rubber/chitosan blends were studied by XRD analysis [91], the measurements indicating that vulcanization enhanced the crystallinity.

Other X-ray Techniques Used in Chitin and Chitosan Analysis

As mentioned at the beginning of this section, X-ray spectroscopic techniques other than the traditional X-ray diffraction measurements have been successfully used to analyze chitin, chitosan and their derivatives.

X-ray photoelectron spectroscopy (XPS) is usually used to determine the bonding energies of C, O and N atoms on the surface of chitosan and its metal chelates, although this is not its only use [44–49]. For example Matienzo and Winnacker [46] presented high-resolution C1s, N1s and O1s XPS spectra for a chitosan film coated on an Al-silicon surface. The films were treated in either an oxygen plasma environment or under UV/ozone irradiation. XPS data showed that hydroxyl and amine entities participated only minimally in the modification. In addition, deposition of chitosan films onto Al-coated silicon wafers produced films with a more ordered chitosan structure. Surface analysis of modified films by XPS also indicated that neither the hydroxyl groups nor the amine segments appeared to participate in surface degradation reactions by either UV/ozone or oxygen plasma during the exposure times chosen for those studies. XPS also provided information regarding the forms of species absorbed on the polymer [48]. For example, a study of the interactions of Cu^{2+}, Mo^{4+} and Cr^{3+} with chitosan beads, cross-linked chitosan beads and chitosan flakes revealed that the adsorption of Mo^{4+} and Cr^{3+} on chitosan flakes and beads was followed by reduction of the Mo^{4+} and Cr^{3+}. Another, example of the use of XPS was presented by Veleshko *et al.* [47]. They examined the complexation between the uranyl group U(VI) and chitosan by means of X-ray photoelectron spectra. The results showed that the interaction of chitosan with the uranyl group yielded complexes containing the nitrogen atom of the amino group and, very probably, oxygen atoms from the chitosan ring and free hydroxyl groups in the equatorial plane.

X-ray emission spectroscopy (XES), also known as X-ray fluorescence spectroscopy (XFS), is a very sensitive probe for examining the local electronic structure and chemical bonding of the emitting atoms. This X-ray technique was used by Kurmaev *et al.* in 2002 [98] to study the chemical bonding in chitosan and chitosan cross-linked with ethylene glycol diglycidyl ether (EGDE). These authors concluded that the changes in the width of resonantly excited O Kα XES were due to site-selective excitation of oxygen atoms belonging to different functional groups (OH and –O–). Comparison of the nitrogen Kα spectra of unmodified and cross-linked chitosan proved that the preferred structural model was the one according to which EGDE was linked via the hydroxyl group.

Infrared Spectroscopy

Infrared (IR) spectroscopy is one of the most important and widely used analytical techniques available to scientists working on chitin and chitosan. It is based on the vibrations of the atoms of a molecule. The infrared spectrum is commonly obtained by passing infrared electromagnetic radiation through a sample that possesses a permanent or induced dipole moment and determining what fraction of the incident radiation is absorbed at a particular energy [99]. The energy of each peak in an absorption spectrum corresponds to the frequency of the vibration of a molecule part, thus allowing qualitative identification of certain bond types in the sample. An IR spectrometer usually records the energy of the electromagnetic radiation that is transmitted through a sample as a function of the wavenumber or frequency. Nowadays, the total spectrum is analyzed by an interference process and converted into the frequency or wavenumber range by means of a mathematical process known as the Fourier transform. Fourier-transform infrared (FTIR) spectroscopy has dramatically improved the quality of infrared spectra and minimized the time required to obtain data [99,100]. Progress in modern infrared spectroscopy is reviewed in literature [101,102].

Typical Conditions for the FTIR Spectroscopic Analysis of Chitin, Chitosan and Their Derivatives

FTIR spectra are usually recorded in the middle infrared (4000 cm^{-1} to 400 cm^{-1}) with a resolution of 4 cm^{-1} in the absorbance mode for 8 to 128 scans at room temperature. The samples for FTIR analysis are prepared by grinding the dry blended powders with powdered KBr, often in the ratio of 1:5 (Sample: KBr) and then compressed to form discs. Spectra are sometimes measured using a deuterated triglycerinesulphate detector (DTGS) with a specific detectivity of 1×10^9 $cmHz^{1/2}$ w^{-1} [103] or on films using an attenuated total refraction (ATR) method in an IR spectrometer [104–106]. Diffuse Reflectance Infrared Fourier-Transform (DRIFT) spectroscopic analysis is also applied [107].

Physicochemical Characterization of Chitin and Chitosan Using Infrared Spectroscopy

As already mentioned, natural chitin occurs mainly as α- and β-chitin. The description and interpretation of the infrared spectra of these two forms of chitin have been published by many scientists [108–110]. By way of example, the spectra of α- and β-chitin, and the *Ianthella basta* scaffold after NaOH treatment and H_2O_2purification are shown in Figure 6 [110].

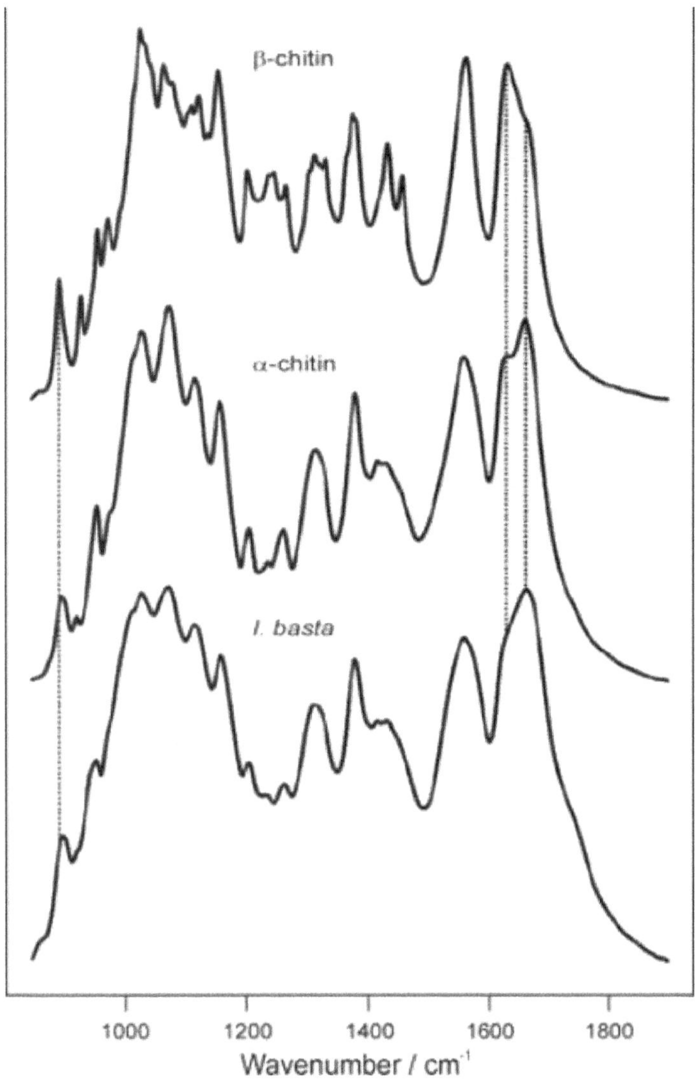

Figure 6 FTIR spectra of β-chitin from *T. rotula*, α-chitin from crabs and *I. basta*chitin after NaOH and H_2O_2 treatment. Dashed vertical lines are drawn to mark characteristic differences between α- and β-chitin. Reprinted from Journal of Structural Biology 168, 2009, Brunner, E., Ehrlich, H., Schupp, P., Hedrich, R., Hunoldt, S., Kammer, M., Machill, S., Paasch, S., Bazhenov, V.V., Kurek, D.V., Arnold, T., Brockmann, S., Ruhnov, M., Born, R., Chitin-based scaffolds are an integral part of the skeleton of the marine demosponge lanthella basta, 539–547, Copyright (2010), with permission from Elsevier [110].

The spectra of α- and β-chitin display a series of narrow absorption bands, typical of crystalline polysaccharide samples. The C=O stretching region of the amide moiety, between 1700 and 1500 cm^{-1}, yields different signatures for α- and β-chitin. For α-chitin, the amide I band is split into two components at 1660 and 1630 cm^{-1} (due to the influence of hydrogen bonding or the presence of an enol form of the amide moiety [109–111]), whereas for β-chitin it is at 1630 cm^{-1} (Figure 6). The amide

II band is observed in both chitin allomorphs: at 1558 cm^{-1} for α-chitin and 1562 cm^{-1} for β-chitin [110]. Another characteristic marker is the CH deformation of the β-glycosidic bond. This band shifts from 890 cm^{-1} in β-chitin to 895 cm^{-1} in α-chitin. Infrared spectra of β-chitin reveal two additional bands for CH_x deformations at about 1455 and 1374 cm^{-1} and a greater number of narrower bands in the C–O–C and C–O stretching vibration region (1200–950 cm^{-1}) not observed in α-chitin. As shown in Figure 6, the FTIR spectrum of the chitin isolated from *I. basta*confirmed the finding that this chitin resembles α-chitin more closely than β-chitin [110], demonstrating that FTIR can be used to determine chitin allomorphs.

FTIR spectroscopy has been used to characterize not only isolated chitin [110] but also the source of chitin, e.g., in two species of black coral, *Antipathes caribbeana* and *A. pennacea* [112]. Although FTIR absorption spectra of the natural samples (without deproteinization) showed similar distribution patterns for both species of coral, and confirmed the presence of chitin in both species, small differences were observed (e.g., the intensity of the IR absorption bands in *A. caribbeana* was stronger). The absence of a free hydroxyl in the hydroxymethyl groups CH_2OH in *A. caribbeana* (determined by FTIR analysis) indicated that the chitin chains were organized in sheets, where they were hydrogen-bonded to adjacent chains, a situation that favors a denser fiber packing of chitin. This means that the FTIR measurements permitted an explanation of why natural *A. caribbeana* coral was harder to pulverize and required a longer deproteinization time than *A. pennacea*. The presence of chitin in polyplacophoran sclerites was also confirmed by IR [113].

FTIR spectroscopy has also been used to compare the yield and purity of chitin isolated from pupae of the silkworm (*Bombyx mori*) using two methods of extraction: an open reactor and 1 h of an acidic reaction, and extraction in a closed reactor within 24 h of a basic reaction [114]. The efficiency of chitosan production by the *N*-deacetylation of chitin was also investigated by IR in this work. During the *N*-deacetylation of chitin, the band at 1655 cm^{-1} gradually decreased, while that at 1590 cm^{-1} increased, indicating the prevalence of NH_2 groups. The spectra of chitin and chitosan obtained from the *N*-deacetylation of chitin using a solution of NaOH (40 wt%) in the presence of $NaBH_4$ for 5 h are presented in Figure 7.

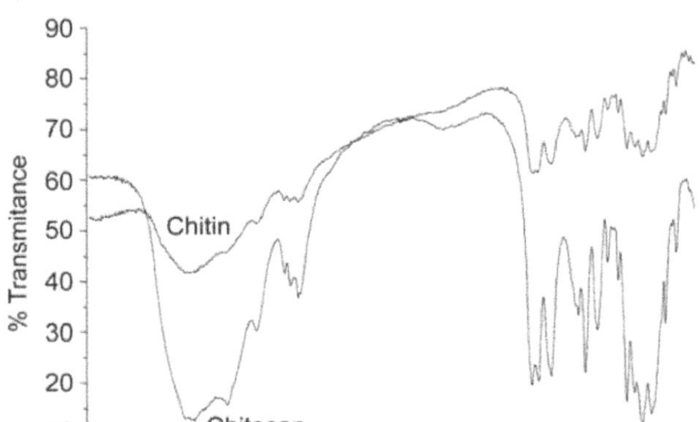

Figure 7 FTIR of chitin (A) and chitosan (B) produced from silkworm pupae; range: 1400–1700 cm^{-1}. Reprinted from Carbohydrate Polymers 64, 2006, Paulino, A.T., Simionato, J.I., Garcia, J.C., Nozaki J., Characterization of chitosan and chitin produced from silkworm crysalides, 98–103, Copyright (2010), with permission from Elsevier [114].

The band at 1590 cm^{-1} displayed a greater intensity than the one at 1655 cm^{-1} and demonstrated the effective deacetylation of chitin. FTIR analyses were also used to find the optimal conditions for the *N*-deacetylation of chitin whiskers (the alkali concentration and the treatment time) using a microwave technique [115].

Prashanhi *et al.* [116] applied IR spectroscopy to observe the changes occurring in the crystallinity and polymorphic nature of chitosan as a function of the *N*-deacetylation of chitin under different conditions: uncontrolled conditions (chitosan A), under an N_2 atmosphere (chitosan B), and with thiophenol (chitosan C). The FTIR spectra of chitosans A, B, and C were similar to each other, but there were subtle differences in the absorption intensities. Apart from the expected decrease in the band at 1665 cm^{-1} (amide I), the vibrational mode of amide II at 1550 cm^{-1} for chitin appeared at 1604 cm^{-1}, 1598 cm^{-1} and 1592 cm^{-1} for chitosan A, B and C respectively [116]. In none of these spectra were there any sharp absorptions at *circa*3500 cm^{-1}, which confirms that the hydroxyl groups in positions C2 and C6 of the chitosans are involved in intra- and intermolecular hydrogen bonds. The region above 3000 cm^{-1} was centred at 3395 cm^{-1} in chitosan A, at 3407 cm^{-1} in chitosan B and at 3419 cm^{-1} in chitosan C; the shift to the higher frequency demonstrated a higher-order structure for these three chitosans. The CH_2 stretching bands of chitosan B around 1425 cm^{-1} were more intense than those of chitosans A and C. Furthermore, the FTIR spectra exhibited a progressive weakening of the bands at 3265 cm^{-1} and 3100 cm^{-1} during *N*-deacetylation, and the A_{1382}/A_{2920} cm^{-1} ratios of 0.65, 0.56 and 0.46 indicated a higher order structure of the chitosans prepared with thiophenol than those prepared under an N_2 atmosphere. The ratio of the band intensities at 1379 and 2900 cm^{-1} was also used to estimate the crystallinity of chitin and chitosan by Focher *et al.* [75] and Wu *et al.* [117], whereas Prashanth and Tharanathan used the sharp absorption peak around 618 cm^{-1}[118].

The impurities in chitosan preparation were determined by FTIR analysis [119]. The FTIR spectra of control chitosan and low-molecular-weight water-soluble chitosan (LMWSC) prepared in this study were compared to establish the contaminants and prove the synthesis of LMWSC. In the LMWSC spectrum the carboxyl group absorption band derived from lactic acid and impurities formed during enzyme degradation disappeared or significantly decreased.

FTIR spectroscopy has been employed to measure the critical concentration of two chitooligosaccharides form a lyotropic liquid crystalline phase in formic acid (C_l) [120]. Strong interactions between sugar chains and solvent were revealed by the widening of bands attributed to the –OH, –NH, –NHCO– of the chitooligosaccharide, including the C=O of formic acid. FTIR measurements of the shift of seven bands–1. $-NH_2$, –OH (3390–3418 cm^{-1}), 2. C=O of formic acid (1716–1724 cm^{-1}), 3. amide I (1626–1633 cm^{-1}), 4. amide II (1520–1531 cm^{-1}), 5. 6. C_3–OH (double peaks, 1178–1189 cm^{-1} and 1148–1153 cm^{-1}), 7. C_6–OH (1073–1074 cm^{-1})–enabled the C_1 values of these chitooligosaccharides to be established.

Determination of the Degree of N-Acetylation of Chitin and Chitosan Using Infrared Spectroscopy

The degree of *N*-acetylation is one of the most important chemical parameters capable of influencing the performance of chitosan and chitin in many of their applications [2,7,121]. Of the various analytical techniques developed for DA determination [122], infrared spectroscopy is at the centre of attention. A convenient and relatively quick technique, it allows the DA values of chitin/chitosan to be determined on the basis of absorption ratios, also in the solid state [123]. Several procedures using different absorption ratios have already been proposed for determining DA for chitin and chitosan samples [124–130]. A review article summarizing the latest literature information on DA determination by IR spectroscopy for chitin and chitosan was published by Kasaai [131]. In that paper, various IR procedures were compared for their performances and limitations, advantages and disadvantages, and different factors affecting the experimental results were discussed, as were the validity data of DA measurements by FTIR spectroscopy. In view of this, the present review will discuss only general information on DA determination by IR.

DA can be determined by IR techniques in the following ways:

- Determination of the A_M/A_R ratio, where A_M is the intensity of the characteristic band of *N*-acetylation, which is a measure of the *N*-acetyl or amine content, and A_R is the intensity of a reference band that does not change with different DA values. The DA parameter of unknown samples can be established by comparing the determined A_M/A_R values with similar ratios of a few reference samples of known DA.
- Drawing a calibration curve by plotting the absorption ratio of chitin/chitosan samples of known DA *versus* their DA as established by IR or a reference method such as ^{1}H NMR spectroscopy. The DA values of unknown samples can then be estimated from the calibration curve.
- Statistical evaluation of several absorption band ratios [131,132].

IR techniques require choosing an appropriate band measure, an appropriate reference band, and drawing a good base line, necessary for measuring the intensity of absorption. The amide I bands at 1655 cm^{-1}(sometimes together with the amide I band at 1630 cm^{-1}) or the amide II band at 1560 cm^{-1} are used as the characteristic band(s) of *N*-acetylation. Among the postulated internal reference bands are the OH stretching band at 3450 cm^{-1} [125,133], the C-H stretching band at 2870–2880 cm^{-1} [134], the –CH_2 bending centred at 1420 cm^{-1} [129], the amide III band at 1315–1320 cm^{-1} [135], the anti-symmetric stretching of the C-O-C bridge at around 1160 cm^{-1} [136], the skeletal vibrations involving the C-O-C stretching bands at 1070 or 1030 cm^{-1} [127,128] and the band at 897 cm^{-1} (C-O-C bridge as well as glycosidic linkage) [137]. The different baselines suggested in the literature [129] are presented in Figure 8.

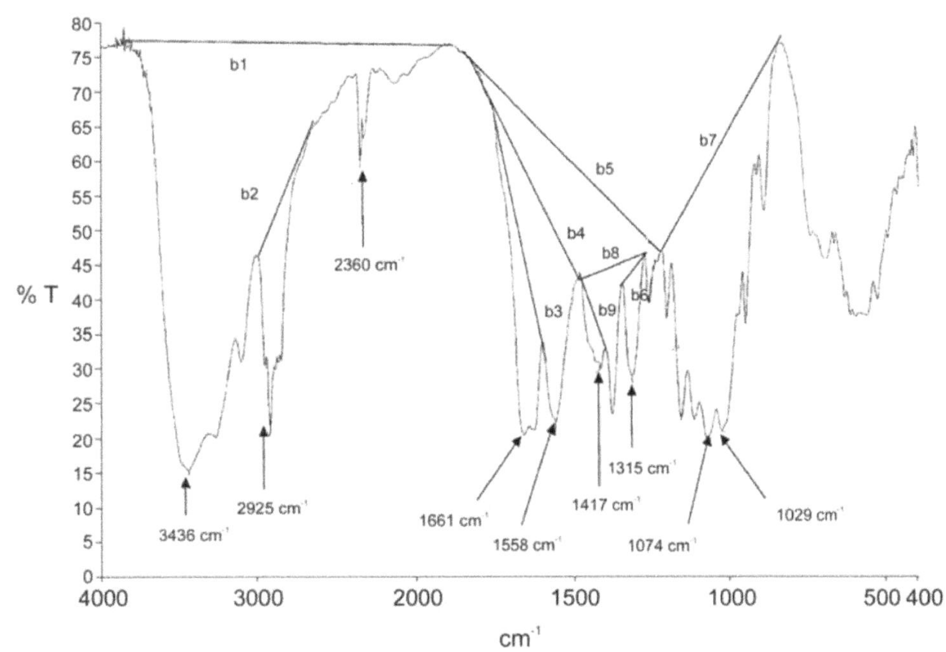

Figure 8 IR spectrum of chitin. Representation of the different baselines mentioned in the literature. Reprinted from Polymer 42, 2001, Brugnerotto, J., Lizardi, J., Goycoolea, F.M., Argüelles-Monal, W., Desbrières, J., Rinaudo, M., An infrared investigation in relation with chitin and chitosan characterization, 3569–3580, Copyright (2010), with permission from Elsevier [129].

Many different absorption band ratios, such as A_{1560}/A_{2875}, A_{1655}/A_{2875}, A_{1655}/A_{3450}, A_{1320}/A_{3450}, A_{1655}/A_{1070}, A_{1655}/A_{1030}, A_{1560}/A_{1160}, A_{1560}/A_{897} and A_{1320}/A_{1420}, have been used to determine DA by FTIR spectroscopy [131]. The different calibration curves proposed in the literature have different baselines and different characteristic bands for measuring the *N*-acetyl content. Moreover, the validity data of these calibrations depend on the absolute technique used to measure DA, and the conditions under which IR spectra are measured may also have a strong influence on the accuracy of DA values. The IR spectra (shown as absorbance) of α-chitin and β-chitin recorded using different sampling techniques are compared in Figure 9[129].

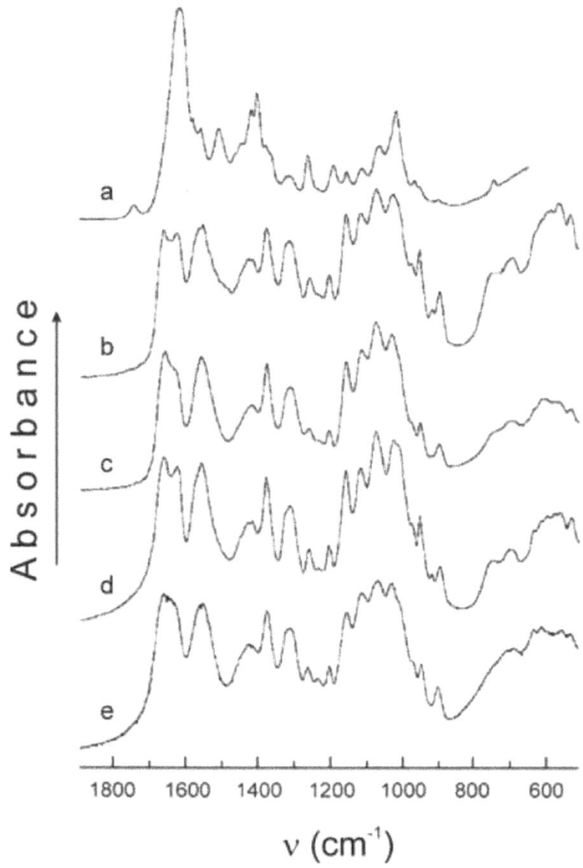

Figure 9 Comparison of IR spectra (shown as absorbance) of α- and β-chitin recorded using different sampling techniques. For α-chitin: (a) ATR on film, (b) DRIFT on powder, (c) Standard transmission on film, (d) Standard transmission on KBr pellet. For β-chitin: (e) Standard transmission on KBr pellets. Reprinted from Polymer 42, 2001, Brugnerotto, J., Lizardi, J., Goycoolea, F.M., Argüelles-Monal, W., Desbrières, J., Rinaudo, M., An infrared investigation in relation with chitin and chitosan characterization, 3569–3580, Copyright (2010), with permission from Elsevier [129].

The ATR spectrum recorded on the film is shown in Figure 9a. This revealed a very low resolution following the collection of 64 scans recorded in typical fashion. The amide I bands (doublet at 1655 and 1625 cm^{-1}) could not be resolved as they were focused into a single band when compared to the standard transmission spectrum of α-chitin obtained in a KBr pellet (Figure 9d). Also, the characteristic band at 1320 cm^{-1} was of very low intensity. By contrast, DRIFT analysis of α-chitin (mixed with HBr) on powder produced a spectrum (Figure 9b) of much better resolution; the amide I band was split into two components. In the transmission spectrum of α-chitin film (Figure 9c) the amide band showed a well-defined peak at 1650 $cm^{-1}$with a minor shoulder at 1625 cm^{-1}. Knowledge of the source of chitin/chitosan, its nature, water content and level of impurities is also important for the accurate calculation of DA values.

Near infra-red (NIR) spectroscopy has also been applied to determine DA [138,139]. The spectra were recorded from 9090–4000 cm^{-1} and second-derivative spectra were used to determine DA; d-glucosamine and d-glucosamine hydrochloride were chosen as model compounds. A reference curve was constructed by plotting the predicted DA value (using NIR data) *versus* DA determined by ^{1}H NMR spectroscopy.

As shown here, choosing an appropriate method among the various IR procedures is a difficult task for researchers.

IR techniques are also employed for the quantitative evaluation of DA, for example, during the *N*-deacetylation of chitin by hot alkali [140], the formation of *N*-acetyl chitosan gel [141], the γ-irradiation of chitosan powder [142] or the synthesis of organic-soluble acetylated chitosan [143].

FTIR Analysis of Chitin and Chitosan Derivatives

The growing interest in the chemical modification of chitin and chitosan to improve their solubility and applications [23–25] has meant that the most important application of IR spectroscopy in this respect is the structural analysis of chemically modified forms of chitin or chitosan. Typical structural analysis of chitin/chitosan derivatives by FTIR spectroscopy involves: (1) FTIR analysis of chitin/chitosan, (2) FTIR analysis of the chemical reagent(s) used in the reaction, (3) FTIR analysis of the chitin/chitosan derivative obtained, (4) identification of differences between the spectra, (5) interpretation of results.

An example of the application of FTIR spectroscopy in the structural analysis of chitin and chitosan derivatives is the FTIR analysis of chitosan-l-glutamic acid aerogel derivative, which is soluble over a wide pH range [144]. The FTIR spectra obtained during these investigations are presented in Figure 10.

A

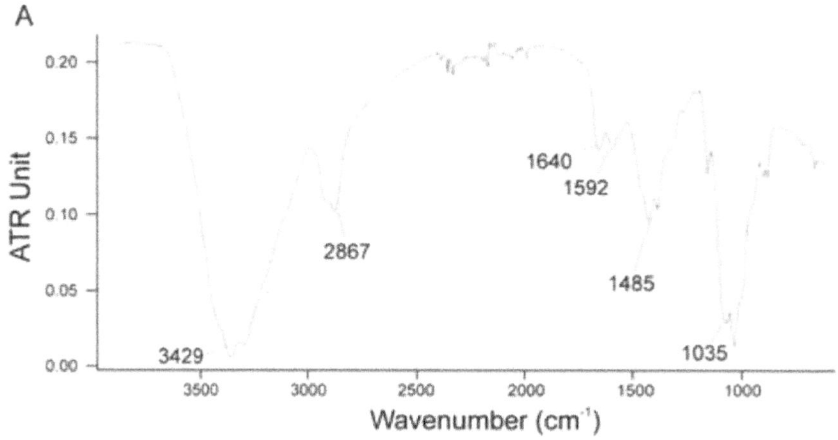

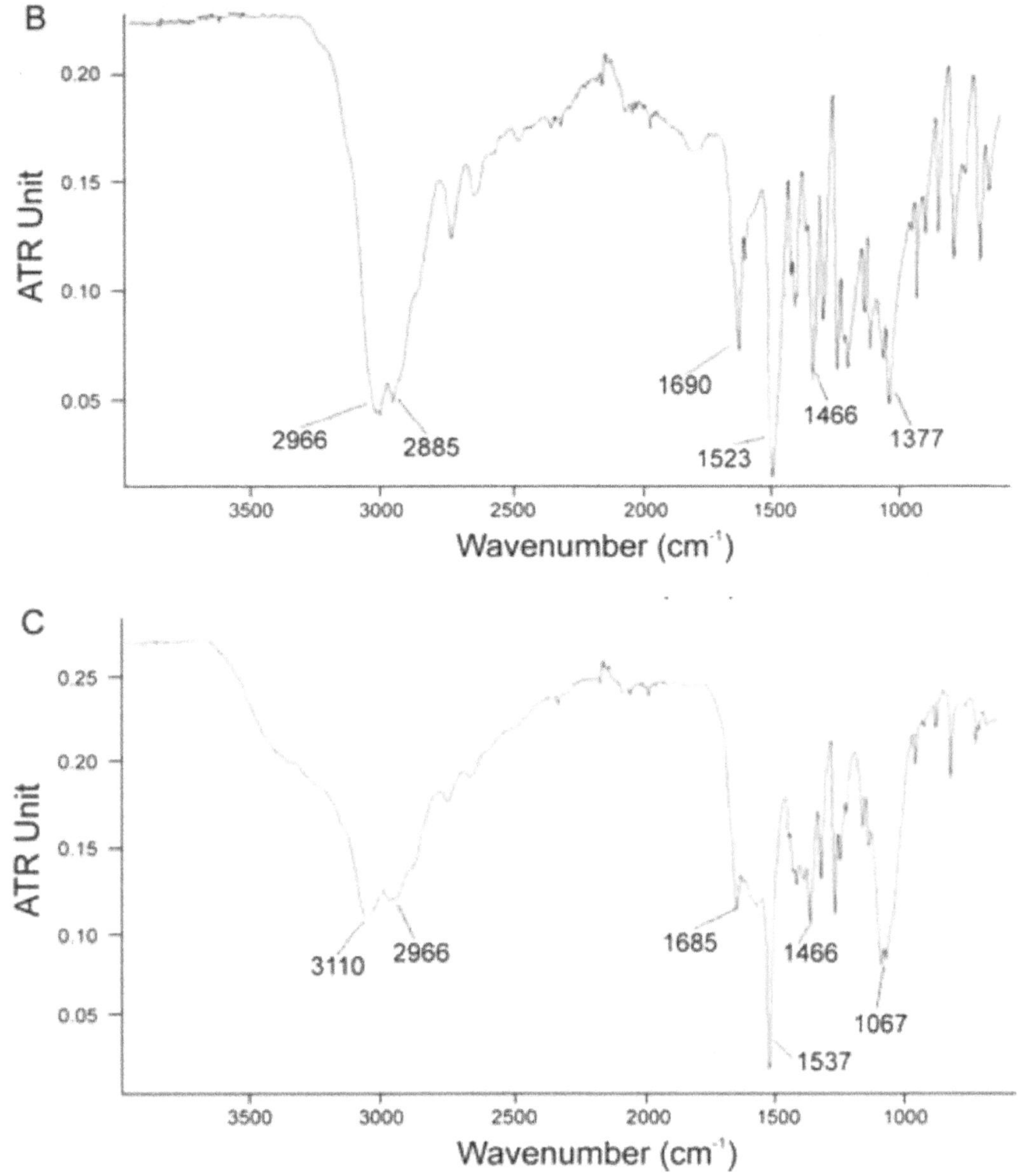

Figure 10 FTIR spectra of (A) chitosan (C), (B) l-glutamic acid (l-GA), and(C) chitosan- l-glutamic acid derivative (Cl-GA). Reprinted from Carbohydrate Polymers 76, 2009, Singh, J., Dutta, P.K., Dutta, J., Hunt, A.J., Macquarrie, D.J., Clark, J.H., Preparation and properties of highly soluble chitosan-L-glutamic acid aerogel derivative, 188–195, Copyright (2010), with permission from Elsevier [144].

The following different characteristic bands were assigned from the FTIR spectra of chitosan, l-glutamic acid (l-GA) and chitosan-l-glutamic acid derivative (Cl-GA) (Figure 10):

- in the IR spectra of chitosan (Figure 10A): 3429 cm^{-1} (O-H stretching overlapping the N-H stretching), 2921 and 2867 cm^{-1} (C-H stretching), 1640 cm^{-1} (amide II band, C-O stretching of the acetyl group), 1592 cm^{-1} (amide II band, N-H stretching) 1485–1380 cm^{-1} (asymmetrical C-H bending of the CH_2 group) and 1035 cm^{-1} (O bridge stretching) of the glucosamine residue.

- in the IR spectra of l-GA (Figure 10B): 2966 cm^{-1} (O-H stretching), 2855 cm^{-1} for (C-H stretching), 1690 cm^{-1} (C=O group) and 1523 cm^{-1} (N-H stretching of the amino group),

- in the IR spectra of Cl-GA derivative (Figure 10C): 3110 and 2966 cm^{-1} (axial OH group of chitosan and glutamic acid), 1685 cm^{-1} (amide linkage), 1556 cm^{-1} (N-H bending and stretching) and 1067 cm^{-1} (C-O-C bridge stretching) of the chitosan residue, 1466 cm^{-1} (the asymmetrical deformation of CH_2).

The following changes were observed in the IR spectra of the chitosan derivative: the C-O adsorption peak of the secondary hydroxyl group enlarged and moved to 1067 cm^{-1}, the intensity of primary alcohol 1035 cm^{-1} (C-O stretching vibration) became much smaller than in chitosan, and a new peak appeared at 2966 cm^{-1}, indicating the incorporation of the l-glutamic acid moieties. These FTIR results provided evidence for the formation of an amide linkage between the COOH group of l-GA and the NH_2 group of the main chitosan chain.

FTIR spectroscopy has been applied to confirm the chemical structure of (1) water-soluble chitosan derivatives such as ethylamine hydroxyethyl chitosan (EHCs) [145], *N*-propyl-*N*-methylene phosphonic chitosan [93], *O*-succinyl-chitosan [146], *N*-alkylated chitosan [147] or methoxy poly(ethylene glycol)-grafted chitosan [148]; (2) organic-soluble derivatives, e.g., acetylated chitosans [149,143]; (3) nanoparticle-chitosan composites [150–152]; 4) hydrogels [153,154] or chitosan-graft copolymers [155–157].

Furthermore, FTIR analyses enabled the interactions occurring between chitin/chitosan and the analysed compounds to be studied and explained [107,158–165]. This technique was applied, for example, in an attempt to obtain an explanation for the sorption mechanism of acid dyes [166–168], phenol and *o*-chlorophenol [169] and fluoride [170,171]. FTIR spectroscopy has also been used to establish differences in structure and the degree of substitution of chitin, chitosan and dibutyrylchitin [172,173].

UV-Vis Spectroscopy

Ultraviolet/visible (UV-Vis) spectroscopy is useful as an analytical technique for two reasons. Firstly, it can be used to identify certain functional groups in molecules, and secondly, it can be used for assaying. Unlike IR spectroscopy, UV-Vis spectroscopy involves the absorption of electromagnetic radiation from the 200–800 nm range and the subsequent excitation of electrons to higher energy states. The absorption of ultraviolet/visible light by organic molecules is restricted to certain functional groups (chromophores) that contain valence electrons of low excitation energy. The UV-Vis spectrum is complex and appears as a continuous absorption band because the superimposition of rotational and vibrational transitions on the electronic transitions gives a combination of overlapping lines. Nowadays, the individual detection of electron transfers without superimposition by neighboring vibrational bands can also be recorded [174]. With UV-Vis spectroscopy it is possible to investigate electron transfers between orbitals or bands of atoms, ions and molecules existing in the gaseous, liquid and solid phase. Analysis of solutions and crystals usually takes place in transmission, whereas powdered samples are often measured in diffuse reflection mode (Diffuse Reflectance Spectroscopy–DRS). Unlike IR spectroscopy, where Fourier transform techniques predominate, dispersive spectrometers are almost exclusively used in UV-Vis spectroscopy because of the large band widths. More details about UV-Vis spectroscopy and its application are presented in many papers and books [174–177].

Typical Conditions of UV-Vis Measurement

UV-Vis spectra of chitin/chitosan derivatives are usually recorded in aqueous acid (acetic acid, phosphoric acid, perchloric acid, hydrochloric acid) solutions in a 1.0 cm quartz cell at ambient temperature [160,178–180]. Sometimes water [181], an aqueous base [182] or DMSO [144] solutions are applied. The Diffuse Reflectance UV-Visible (DRUV) spectra of powdered or film samples are measured [107,112,183–185]. Analysis in the vacuum ultraviolet through the near-infrared range has also been applied [185,186].

UV-Vis Spectra of Chitin and Chitosan

Chitin and chitosan include various ratios of two far-UV chromophoric groups, *N*-acetyl-glucosamine (GlcNAc) and glucosamine (GlcN); as a result, their extinction coefficients for wavelengths shorter than approximately 225 nm is non-zero. Because GlcNAc and GlcN residues show no evidence of interacting within the chitin/chitosan chain, the monomer units contribute in a simple, additive way to the total absorbance of these polymers at a particular wavelength [180]. The UV spectra of mixtures of *N*-acetyl-glucosamine and glucosamine hydrochloride are quite similar to the spectra of chitosan, and the λ_{max} is 201 nm in 0.1 M HCl solution (Figure 11) [180].

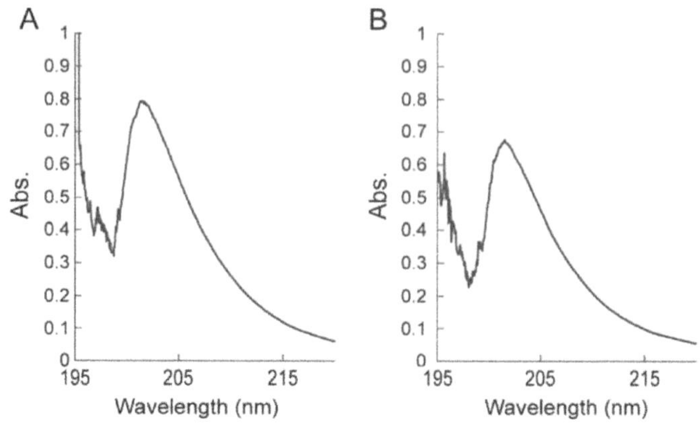

Figure 11 UV spectra: (a) a mixture of *N*-acetyl-glucosamine and glucosamine hydrochloride in 0.1 M hydrochloric acid solution; (b) chitosan in 0.1 M hydrochloric acid solution. Reprinted from Carbohydrate Research 341, 2006, Liu, D., Wei, Y., Yao, P., Jiang, L., Determination of the degree of acetylation of chitosan by UV spectrometry using dual standards, 782–785, Copyright (2010), with permission from Elsevier [180].

UV-Vis spectroscopy can be used for the optical characterization of the source of chitin [112]. The optical spectra for two black coral species, *Antipathes caribbeana* and *A. pennacea*, showed a wide band from 300 nm to 500 nm, with small differences depending on the species. Additionally, the spectra confirmed the presence of Fe^{+3} in *A. caribbeana* and Mn^{+3} in both species.

Determination of the Degree of *N*-Acetylation of Chitin and Chitosan Using Uv-Vis Spectroscopy

The first derivative UV method for DA determination was proposed by Muzzarelli and Rocchetti in 1985 [187]. The principle of this method was based on the absorbance of the intensity of the *N*-acetyl group in chitin or chitosan. The absorbance of *N*-acetyl glucosamine (at maximum wavelength) was linearly dependent on its concentration in the 0.50–5.0 $mg{\cdot}L^{-1}$ range. The limit of detection of GlcNAc in 0.01 M acetic acid was found to be 0.5 $mg{\cdot}L^{-1}$. The DA values were determined in acetic acid solutions at 199 nm [187]. The simplicity and convenience of this method were the best among the available methods, because it required very small amounts of sample and did not require expensive equipment. In addition, the first derivative of the spectra was less affected by background noise and impurities. Several modified first-derivative UV methods have been proposed to improve the convenience and accuracy of measurement [178,180,188–191]. Tan *et al.* [178] compared the results of four methods (^{1}H NMR, UV, ninhydrin assay and potentiometric titration) and suggested the first-derivative UV method as a standard method for the routine determination of DA of chitosan. Pedroni *et al.* [188] analysed chitosan solutions in different solvents; 0.1 M HCl solutions turned out to be the best. Hsiao *et al.* [189] suggested concentrated phosphoric acid as the UV-transparent solvent system. In this way, the UV determination could be validated across the whole DA range. Liu *et al.* [180] employed two standards (*N*-acetylglucosamine and glucosamine hydrochloride) to determine the degree of *N*-acetylation of chitosan by UV spectrophotometry. Wu and Zivanovic [190] plotted the first derivative UV values at those wavelengths against the concentrations of GlcNAc and found that the best linear regression was obtained at 203 nm (the range of 0–50 μg/mL, $R^2 = 0.996$). Da Silva *et al.*[191] put forward a new mathematical expression. This was derived in such a way that DA could be determined directly from the mass concentration of a chitosan solution and the first derivative of its UV spectrum at 202 nm, thus eliminating the need for the empirical correction of curves for highly deacetylated samples. A procedure was proposed for the accurate mass determination of the hygroscopic chitosan.

In summary, UV techniques for DA determination are more sensitive than IR, ^{13}C NMR and ^{15}N NMR spectroscopy, and they give a better accuracy than NIR, IR and NMR methods [122,192]. The first-derivative UV method is the most sensitive of the proposed techniques. The first derivatives of the spectra are less affected by impurities such as protein, and more accurate data can be obtained. In contrast to other techniques, it is possible to analyse the DA values of a chitin/chitosan sample with a high water content. The first-derivative UV methods are easy to carry out, and the equipment is available in most monitoring and research laboratories. These techniques were used to verify the validity of several methods of DA determination in different laboratories.

Application of UV-Vis Spectroscopy to the Analysis of Chitin/Chitosan Based Compounds

The steps in the structural analysis of chitin/chitosan derivatives by UV-Vis spectroscopy are very similar to those in the FTIR methodology (Section 2.2.4). The main difference is connected with the aim of these analyses. IR spectroscopy is used mostly for determining the molecular structure of chitin and chitosan samples, whereas UV-Vis spectroscopy is more often applied to the study of covalent and non-covalent interactions. Since certain functional groups present in organic molecules absorb light at characteristic wavelengths in the UV-Vis region, this technique is applied qualitatively to identify the presence of these groups in samples, supporting structural information obtained from other spectroscopic methods, especially IR.

As mentioned, one of the most important applications of UV-Vis techniques is the characterization of interactions between chitin or chitosan and the target compounds. For example, the analysis of the interaction between chitosan and cellulose, described by Urreaga and de la Orden [107], will be presented in greater detail. The UV-Vis spectra obtained during that study are shown in Figure 12.

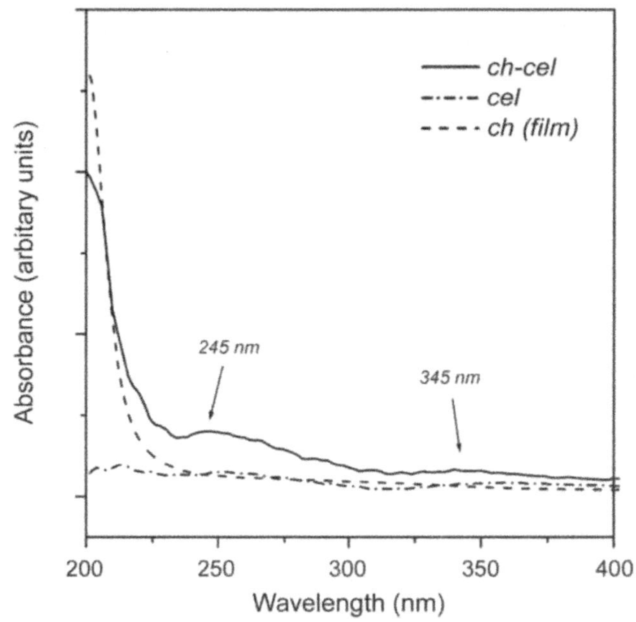

Figure 12 UV–Vis spectrum of a pure chitosan (ch) film measured in transmission mode, and diffuse reflectance UV–Vis spectra of untreated cellulose (cel), and cellulose treated with 1% chitosan (ch–cel). Reprinted from European Polymer Journal 42, 2006, Urreaga, J.M., de la Orden, M.U., Chemical interactions and yellowing in chitosan-treated cellulose, 2606–2616, Copyright (2010), with permission from Elsevier [107].

The spectrum of chitosan-treated cellulose (Figure 12) showed the characteristic chitosan absorption below 220 nm revealing the presence of chitosan. Additionally, this spectrum showed very weak absorptions at $\lambda > 230$ nm, including weak but characteristic bands centred at *circa* 245 nm and 345 nm, which did not appear in the spectra of the starting materials (cellulose and chitin). This observation suggested the existence of chemical interactions between cellulose and chitosan. In order to verify the existence of the above chitosan–cellulose interactions, samples of pure chitosan, pure cellulose and chitosan-coated cellulose were heated at different temperatures and characterized by UV-Vis spectroscopy. The spectrum of the chitosan-treated cellulose was more complex than those of chitin and cellulose and exhibited two new absorptions centred at*circa* 272 and 340 nm. The presence of the band at 272 nm was explained by the independent oxidations of chitosan and cellulose, but the appearance of the 340 nm band again indicated the existence of a chemical reaction in chitosan-treated celluloses, demonstrating that chemical interaction between cellulose and chitosan took place even at moderate temperatures. The new bands, which were observed in materials processed at room temperature, were assigned (also using FTIR and fluorescence spectroscopy) to conjugated imines, produced in the condensation reaction between chitosan amino and carbonyl groups. UV-Vis spectroscopy has also been employed, e.g., to characterize the interactions between chitosan and uranyl ions [160], chitosan and synthetic phospholipid membranes [193], β-cyclodextrin-linked chitosan and p-nitrophenolate [194] and between chitosan and gelatin [195].

UV-Vis techniques were useful for determining the structure of many chitin/chitosan derivatives, for example, chitosan hydrogel membranes obtained by UV- and γ-radiation [196], substituted polyaniline/chitosan composites [163], chitosan-l-glutamic acid aerogel derivatives [144], fluorescent chitosan derivatives containing substituted naphthalimides [197] and dendritic polyaniline nanoparticles synthesized by carboxymethyl chitin templating [103]. They were also employed to determine the actual degree of substitution of fluorescent chitosan (*DS* parameter) and the concentration of the 9-anthrylmethylated unit in modified chitosan [198].

It should also be mentioned that UV-Vis spectroscopy is used to monitor chemical reactions [185,199,200] and the manufacture of complex multilayers, nanoparticles, nanocomposite or multilayer films [183,184,201–203].

UV-Vis spectroscopy is employed to determine the physicochemical properties of chitin/chitosan-based compounds such as the turbidity (optical dispersion, OD) of their solutions [179], the capacity adsorption of chitosan membranes [204], the lower critical solution temperature (LCST) of a chitosan derivative copolymer solution in PBS [205] or the optical properties of nanoparticles [206].

Mass Spectrometry

Mass spectrometry (MS) is a destructive analytical technique used for measuring the characteristics of individual molecules. The basic information obtained from mass spectrometric analysis is the molecular mass of a compound, which is determined by measuring the mass to charge ratio (m/z) of its ion. With the ionization method, full particulars about a molecule's chemical structure can be found. MS can analyse chemicals with a wide mass range–from small molecules to complicated biomolecules such as carbohydrates, proteins, peptides or nucleic acids. The mass spectrometer is composed of three fundamental parts: an ion source, an analyser and a detector. First, the sample must be introduced to the heart of each mass spectrometer–the ion source. There are two possible ways of doing this: through a direct inlet (septum inlet, syringe, probe, *etc.*), or an indirect inlet (gas chromatograph, liquid chromatograph, or capillary electrophoresis column). In the ion source the sample molecules are ionized by one of a variety of methods (electron impact–EI, chemical ionization–CI, electrospray ionization–ESI, matrix assisted laser desorption ionization–MALDI, *etc.*). Next, the stream of ions is transferred to the analyser, where they are sorted and separated according to m/z. The most commonly used analyzers are quadrupoles, ion trap and time of flight. In tandem mass analysis (MS/MS, MS_n) a combination of two or more analysers is used. The MS can be operated in three modes: full scan (scanning of a selected mass range), more sensitive SIM (selected ion monitoring = SIR - single ion recording), and the most sensitive MRM (multiple reaction monitoring). Finally, the ions reach the detector when their energy is converted to electrical signals that the computer can read. In the end, mass spectra are obtained, and identification of compounds is possible [207–209].

Typical Conditions of Mass Spectrometric Analysis of Chitin/Chitosan and Their Derivatives

Different types of mass spectrometer can be chosen according to the information required. Spectrometers with various combinations of inlet systems, ion sources and analysers are used for investigations into chitin, chitosan and their derivatives. In most cases the positive ion mode is preferred.

The first method used was pyrolysis-mass spectrometry [210]. That was back in 1982, but combined with gas chromatography it is still a routine technique [211–215]. In GC/MS, the first sample is pyrolyzed (450–615 °C), after which the volatile pyrolysis products are separated on a capillary column (the temperature program from 35–50 °C with different rates to 220–300 °C). In GC/MS analysis, the full scan mode (m/z from 15–35 to 300–650) and the electron impact mode (70 eV) are commonly used. Products are identified on the basis of their GC retention parameters and their mass spectra [210–215].

GLC-MS is used for analyzing chitosan or its impurities. But first, all the target compounds have to be converted to volatile derivatives (hydrolysis, and reduction of chitin [216,217]; this is then followed by the extraction of impurities [218] and trimethylsililation or acetylation). The products are separated using a GLC nonpolar or medium-polar capillary column with a gradient temperature programme (usually from ~60 °C to ~280 °C with different heating rates). For detection, MS with EI ionization is employed [216–218].

Most of the papers dealing with the mass spectrometric analysis of chitin and chitosan report that MALDI was the ionization method selected. This ion source is most often combined with a time of flight analyser (MALDI-TOF) [219–226]. For this type of analysis chitooligosaccharides were dissolved in water [226], MeOH-H_2O (1:1) [224] or 1% acetic acid solution [225]. The matrix most often used is 2,5-dihydroxybenzoic acid (DHB) [219,221–226]; sometimes it is 2-(4-hydroxyphenylazo)benzoic acid (HABA) dissolved in dioxin [224]. To obtain monosodium and monopotassium carbohydrate adducts in the mass spectrum, NaCl and KCl can be added as a cationating agent [224]. The MS ion source is equipped with a nitrogen laser emitting at 337 nm [225,226]. For these analyses an accelerating potential of 20–30 kV is used, and mass spectra are measured in positive ion mode [221,222]. A modified version of MALDI-TOF was introduced for chitin/chitosan investigations–MALDI-TOF PSD MS (postsource decay mass spectrometry) [227]. Another MS technique with a MALDI source is tandem mass spectrometry, for which an ion trap analyser is very useful. One analyzer allows MS^2 and also MS^3 spectra to be obtained using a collision energy previously established experimentally [228]. MALDI-TOF/TOF equipment can also be applied in chitin and chitosan analysis using tandem mass spectrometry. Precursor ions are accelerated at 8 kV, while their fragment ions are accelerated to 19 kV [229].

Time-of-flight secondary ion mass spectrometry (TOF-SIMS) is another method used during chitosan investigations. This type of spectrometer is equipped with a gallium primary ion beam operated at 15 kV. The secondary ions appearing are extracted (10 kV) and fly through the analyzer (TOF type) to the detector [230,231].

After MALDI, ESI (electrospray ionization) is the second most important ionization technique for chitin/chitosan analysis [219,232–238]. The ESI ion source is usually coupled to an ion trap analyser [235], but a quadrupole type of analyser can also be used [237]. For sample measurements several parameters should be optimized, for example, the source voltage (3.5–4 kV), nitrogen sheath gas rate and capillary temperature. A syringe pump is used (3–4 μL/min) for direct sample injection [235,238]. Tandem mass spectrometry with ESI ionization is preferred for acquiring more structural information. If the triple quad system (QqQ) is used, the first quadrupole selects the precursor ions, the second quadrupole acts as collision chamber, and the third quadrupole analyses the ion products; argon is the neutral collision gas [239,240]. Sometimes, a combination of quadrupole and time-of-flight (Q-TOF) is found useful [241]. Connecting a liquid chromatograph to the MS system is an additional advantage. Separating solutions containing chitosan products on a chromatographic column makes the analysis easier. LC-MS analysis is carried out using the same mass spectrometric modes (positive and negative ion detection, electrospray ionization, scanning and SIM mode, *etc.*). Ion source conditions usually are the same as in classic MS analysis that yields protonated products $[M+H]^+$, sodium adducts $[M+Na]^+$, *etc.* For HPLC separation a different type of column is used, and the mobile phase often consists of acetonitrile or water with added formic acid [242–245]. LC-MS is highly suitable method for investigating chitosan derivatives.

Fast atom bombardment (FAB) is another soft ionization method useful in chitin and chitosan investigations. For this purpose the following compounds can act as the matrix: thioglycerol [246,247], m-nitrobenzyl alcohol [247], glycerol [248–250], dithiothreitol [251], 2,4-dinitrobenzyl alcohol, or dithiothreitol-thioglycerol (1:1, v/v) [223]. The mass spectrometer usually operates in positive-ion mode, most often producing $[M+H]^+$ ions [246,248,249,252] but also $[M+Na]^+$ [250,251] or $[M+K]^+$ [251]. The atom gun is operated at 6–8 kV with xenon as the bombarding gas [246,247,250,251]. This type of ion source cooperates with collision-induced dissociation tandem mass spectrometry (FAB-CID) measurements [246,247,252]. To obtain FAB-CID-MS/MS spectra, the collision is caused by helium, and a 10 kV accelerating voltage is used [246,247].

Sometimes less popular ionization methods are used for chitin/chitosan analysis. One of them is chemical ionization (CI) with the use of ammonia [253] or acetic acid [120] as the reactive gas. In both cases MS works in positive ion mode, generating $[M + NH_4]^+$ and $[M + H]^+$, respectively [120,253]. Another ionization method is plasma desorption by the fission products of californium-252 in a time/flow biochemical mass spectrometer [254].

TC-EA/IRMS (high-temperature conversion-elemental analyser/isotope ratio mass spectrometer) is very rarely used in chitin/chitosan investigations. In this method the first step is the quantitative conversion of oxygen to CO and hydrogen to H_2 at a high temperature. The gaseous products are separated by GC, transferred to IRMS where the ratios of the stable oxygen and hydrogen isotopes in chitin are determined [255,256].

Chitosan TG-MS (thermogravimetry-mass spectrometry) analysis begins with the very rapid heating of the sample from room temperature to 600 °C, after which the gaseous products pass into the mass spectrometer. Ions with m/z in the 40–130 range are analysed by MS with a quadrupole analyzer and Channeltron detector [257].

Mass Spectrometric Determination of the Degree of Polymerization of Chitin and Chitosan

Being a natural co-polymer of glucosamine and *N*-acetylglucosamine, chitosan possesses properties like antimicrobial and antitumor activity, and stimulates plant growth [222,238,258]. This activity is closely related to the molecular weight. This fact must be taken into consideration if one is planning to use chitosan in biological or pharmaceutical applications. Chitooligosaccharides can be obtained by depolymerization. This is done physically [226,238] chemically [245] or enzymatically [244,250,259–262]. In each case, MS can be used to determine the degree of depolymerization of chitosan and to track the kinetics.

One of the physical methods for depolymerizing chitosan is sonolysis. It was found that after ultrasound treatment random depolymerization of chitosan occurred, producing many different chitooligosaccharides (COS). ESI-MS [238] and MALDI-TOF MS (Figure 13) [226] were used to detect the products. The primary chemical structure was found to have remained unchanged and that sonolysis was a good enzymatic or alternative chemical method for obtaining oligosaccharides for commercial use.

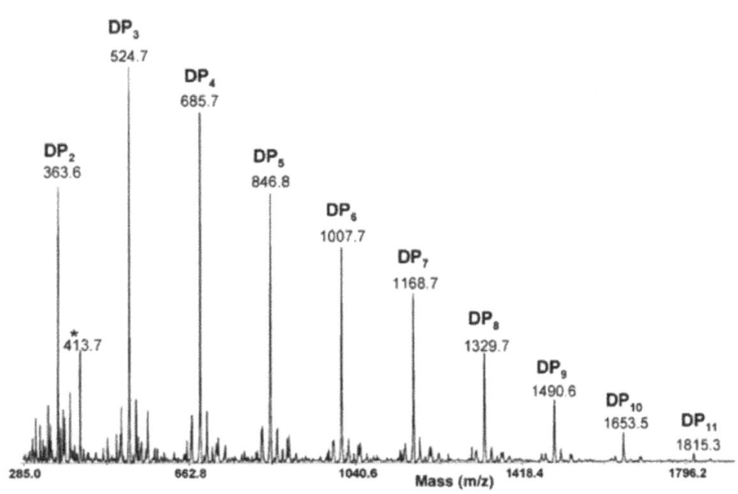

Figure 13 MALDI-TOF MS of deacetylated chitooligosaccharides obtained from ultrasonically treated chitosan. Reprinted with permission from*Biomacromolecules*. 2009, *10*, 1203–1211. Copyright 2010, American Chemical Society [226].

Depolymerization of chitin/chitosan can be achieved by classical chemical methods. The results of LC-MS analysis proved that hydrolysis of chitin from fungi with 6 M HCl yielded glucosamine monomers [245]. Besides providing qualitative chitin composition, the method also makes it possible to track the kinetics of the reaction by ^{15}N labeling of the glucosamine from the fungal chitin. For this purpose, acid-hydrolyzed chitin from the fungus *P. chrysogenum* growing on a minimal medium and on a medium containing added $(^{15}NH_4)_2SO_4$ (rich medium–Blakeslee's formula) was analyzed by LC-MS. Then the analysis on the basis of the peak ratios in mass spectrum was performed (Figure 14) [245].

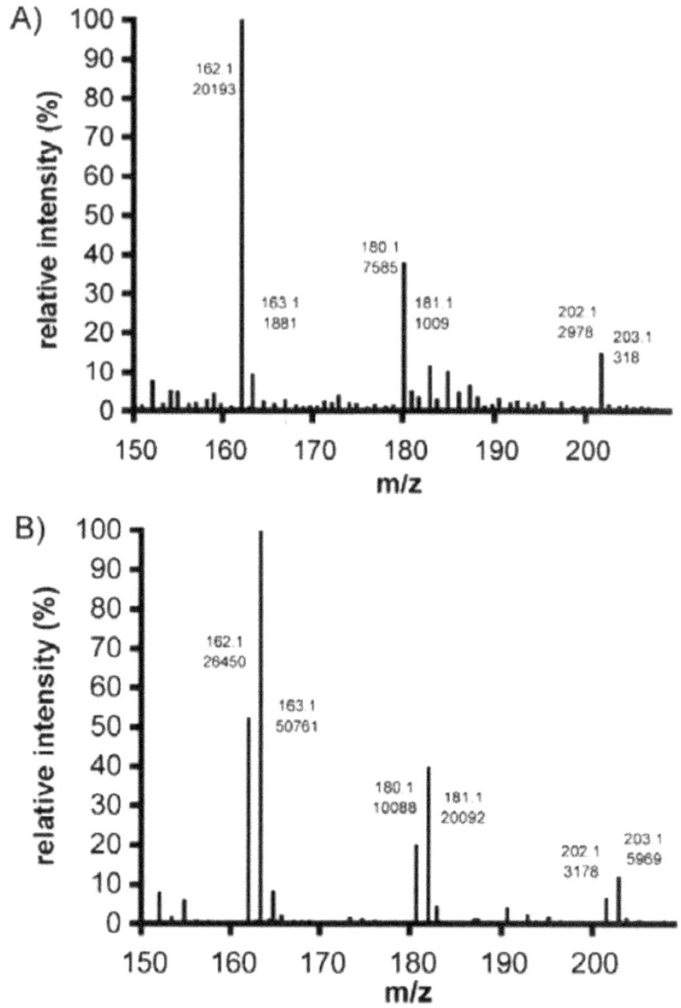

Figure 14 Mass spectra of acid hydrolyzed chitin from *P. chrysogenum* growing on (A) minimal medium, (B) medium with the addition of $(^{15}NH_4)_2SO_4$ (rich medium–Blakeslee's formula). Reprinted with permission from *Biomacromolecules*. 2009, *10*, 793–797. Copyright 2010, American Chemical Society [245].

The enzymatic method of chitin/chitosan depolymerization is the most popular one, and mass spectrometry is often used to detect the products and the efficiency of this process. From MS studies it was found that immobilization of enzymes (papain [259], neutral protease [243]) enhanced depolymerization in the chitosan chain comparing with free enzymes [243,259]. The activities of enzymes isolated from different sources (commercial enzymes [261], isolated from living organisms: *Vibrio harveyi* [244], *Serratia marcensces* [251], *Bacillus circulans* [237], *Amycolatopsis orientalis*, *Streptomyces* sp. [241]) were tested by qualitative MS studies of obtained chitooligosaccharides. It is possible to study enzyme selectivity, and the mechanism of COS bonding to the enzyme. Knowledge of the influence of polymerization and the degree of acetylation is very important in the large-scale production of chitooligosaccharides, and in their use in enzymology or drug design. Table 1 shows a number of examples of the analysis of the degree of polymerization of chitooligosaccharides using mass spectrometry.

Table 1 MS-analyzed chitooligosaccharides obtained after depolymerization using different enzymes

Enzyme	DP analysis MS method	Chitin/chitosan polymerization degree DP	Chitooligosaccharide applicability	Ref.
Papain	MALDI-TOF	DP 2–DP 8	Antibacterial activity against *Bacillus cereus* and *Escherichia coli*	[222]
Immobilized Papain	MALDI-TOF	DP 3–DP 7	Comparing the depolimerization efficiency between free and immobilized papain	[259]
Pronaze	MALDI-TOF	DP 2–DP 9	Antibacterial activity against *Bacillus cereus* and *Escherichia coli*	[222]
Cellulast (Novozymes)	MALDI-TOF	DP 2–DP 8	Induction of defence response of Oryza sativa L. against Pyricularia grisea (Cooke) Sacc.	[258]
Isozyme of pectinase	MALDI-TOF FAB-MS	DP 2–DP 6	Antimicrobial activity *Bacillus cereus* and *Escherichia coli*	[250]
Chitinase	MALDI-TOF	Q1: DP 3–DP 8 Q2: DP 2–DP 12 Q3: DP 2–DP 10	Affect on fungal (alternaria alternate, *Rhisopus stolnifer, Botrytis cinera, Penicillinum expansum*) growth rate	[262]
Chitosanase	MALDI-TOF	DP 2–DP 8	Investigations of prebiotic effect on the *Bifidobacterium bifidum* and Lactobaccillus sp.	[260]
Chitinase	LC-ESI-MS	DP 2–DP 6	Studies of mechanism of bonding COS to enzyme helpful in the drug-screening program (for drugs in allergic asthma)	[244]

To prepare oligosaccharides, chitosan can be hydrolyzed by a mixture of different enzymes (cellulose, alpha amylase, proteinase). The products are analyzed by MALDI-TOF [220].

Mass spectrometry is a good method for the detection of chitooligosaccharides obtained not only by depolymerization, but also as a result of synthesis. FAB-MS and FAB-MS/MS were used to analyse chitinpentaose, chitintetraose, chitintriose and two modified chitin oligosaccharides produced by Rhizobium NodC protein in *Escherichia coli* [246].

Determination of the Degree of N-Acetylation of Chitin and Chitosan Using Mass Spectrometry

The degree of *N*-acetylation is a factor that allows chitin and chitosan to be distinguished. Besides UV, IR and NMR spectroscopy, mass spectrometry is one of the spectral techniques used to establish the degree of*N*-acetylation of chitin, chitosan and their derivatives [122].

For this purpose, the pyrolysis-gas chromatography-mass spectrometry (EI ionization) combination is used. From the EI mass spectra it is possible to identify volatile compounds obtained from chitin and chitosan. The peaks at m/z 60 and m/z 42 are derived from *N*-acetyl-glucosamine, while the ions m/z 80 and m/z 67 are from glucosamine. DA is established on the basis of the peak ratios 80:60, 67:60, 80:42, 80:125, 94:125, 80:110. These ratios increase with decrease in the degree of acetylation [122,211,212].

As in the case of investigations into the degree of polymerization of chitin/chitosan, the usual MS method is MALDI-TOF. The mass spectra of the series of oligosaccharides with DA from 25 to 90% obtained with this method are shown in Figure 15 [240].

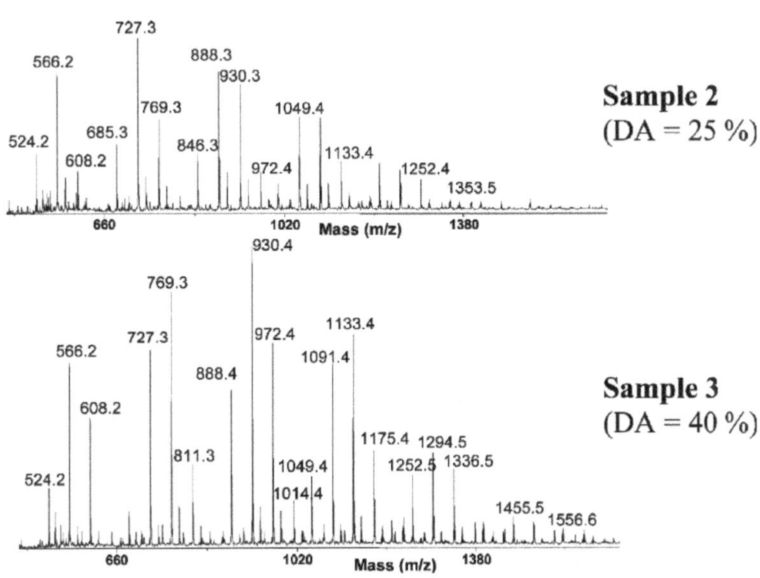

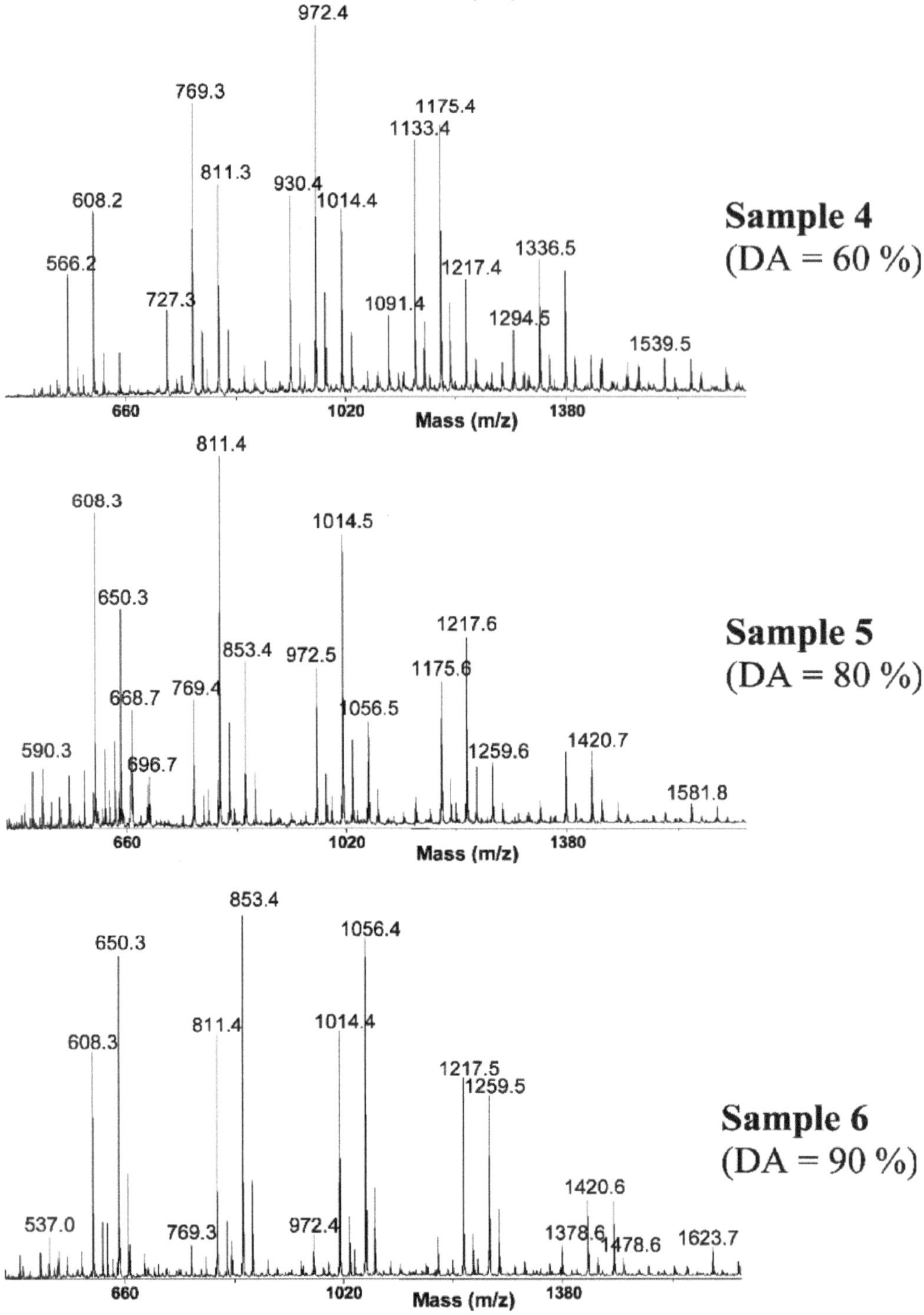

Figure 15 nMALDI-TOF MS of chitooligosaccharides with different degrees of acetylation. Reprinted with permission from *Biomacromolecules*. 2008, *9*, 1731–1738. Copyright 2010, American Chemical Society [240].

As mentioned above, chitin and chitosan play a significant part in inducing defense responses in plants. Some properties are very selective: for example, chitin oligosaccharides can induce lignifications in wounded wheat leaves, while chitosan COS cannot. Mass spectrometry was used to establish DA in the stereocontrolled synthesis of the chitosan dodecamer (FAB-MS) [263], the synthesis of the chitosan tetramer derivative β-d-GlcNAc-(1→4)- β-d-GlcNAc-(1→4)- β-d-GlcNAc-(1→4)-d-GlcN (FAB-MS) [249], the synthesis of a series of oligosaccharides (MALDI-TOF) [264], and the synthesis of partially *N*-acetylated chitin oligomers using *N*-acylated chitotrioses as substrates in a lysozyme-catalysed transglycosylation reaction (MALDI-TOF, ESI-MS) [219]. All of them are used in biological investigations. The degree of acetylation of enzymatic degradation products of chitin and chitosan is measured. For these investigations chitin deacetylase from *Colletotrichum lindemuthianum* was chosen (FAB-MS, MALDI-TOF) [221,249].

Analysis of Chitin and Chitosan Derivatives Using Mass Spectrometry

Some analytical methods based on mass spectrometry force the conversion of chitin and chitosan into the gaseous state. There are two cases when this is required: during analysis of thermal degradation products, and when GLC-MS is used.

The first example is a study of the thermal decomposition products of chitosan, which is done by TG-MS analysis. The chitosan is heated to different temperatures to be degraded, and the gaseous products are detected using a mass spectrometer. From this experiment it is possible to monitor the thermal behavior of chitosan [257].

In comparison with the previous method, using a combination of pyrolysis-gas chromatographymass spectrometry, it is possible to separate a mixture of chitosan or chitosan pyrolysis products. For qualitative analysis the deuterated analogue of *N*-acetyl-glucosamine is used as an internal standard. The main components of the pyrolysates (chitin markers) are acetamide, acetylpyridone and 3-acetamidofuran [213–215,217]. This method is used to confirm the presence of chitin in biological samples [215,255].

For the GLC-MS analysis of chitin and chitosan, these need to be converted to volatile derivatives. For this purpose the chitosan or chitin methanolysis/hydrolysis products are trimethylsilated or acetylated. The results allow the sugar composition of chitin and chitosan to be determined. For qualitative analysis, an internal standard is needed [216,217].

TC-EA/IRMS is a very interesting, but still rarely used method. It is useful in the palaeolimnological study of fossil chironomid head capsules. All organic compounds are converted to CO and H_2. With the isotopic ratio mass spectrometer the ratios of stable oxygen and hydrogen isotopes from chironomid larvae can be determined. Water and diet were also investigated in this experiment to test how the $\delta^{18}O$ and $\delta^{2}H$ in water and the diet influence the $\delta^{18}O$ and $\delta^{2}H$ of chironomid larvae [255,256].

Most of all, however, mass spectrometry, especially LC-MS and LC-MS/MS, is used to monitor the success of chitin and chitosan derivatization reactions. The chitosan is derivatized for different reasons, for example, to sequence mixtures of chitooligosaccharides. For this purpose the reducing end of hetero-chitooligosaccharides was treated with 2-aminoacridone (AMAC) by reductive amination. The products were analyzed using MALDI-TOF-MS (Figure 16), MALDI-TOF/TOF-MS/MS (Figure 17) [229], PSD (postsource decay) MALDI TOF MS [227]. Chitosan sequence determination was a part of an investigation into the inhibition of a family of 18 chitinases. Besides AMAC, 3-(acetylamino)-6-aminoacridine is also used for the reductive amination of chitooligosaccharides. The sequencing information was taken from vMALDI (vacuum MALDI) ion trap MS^n measurements [228].

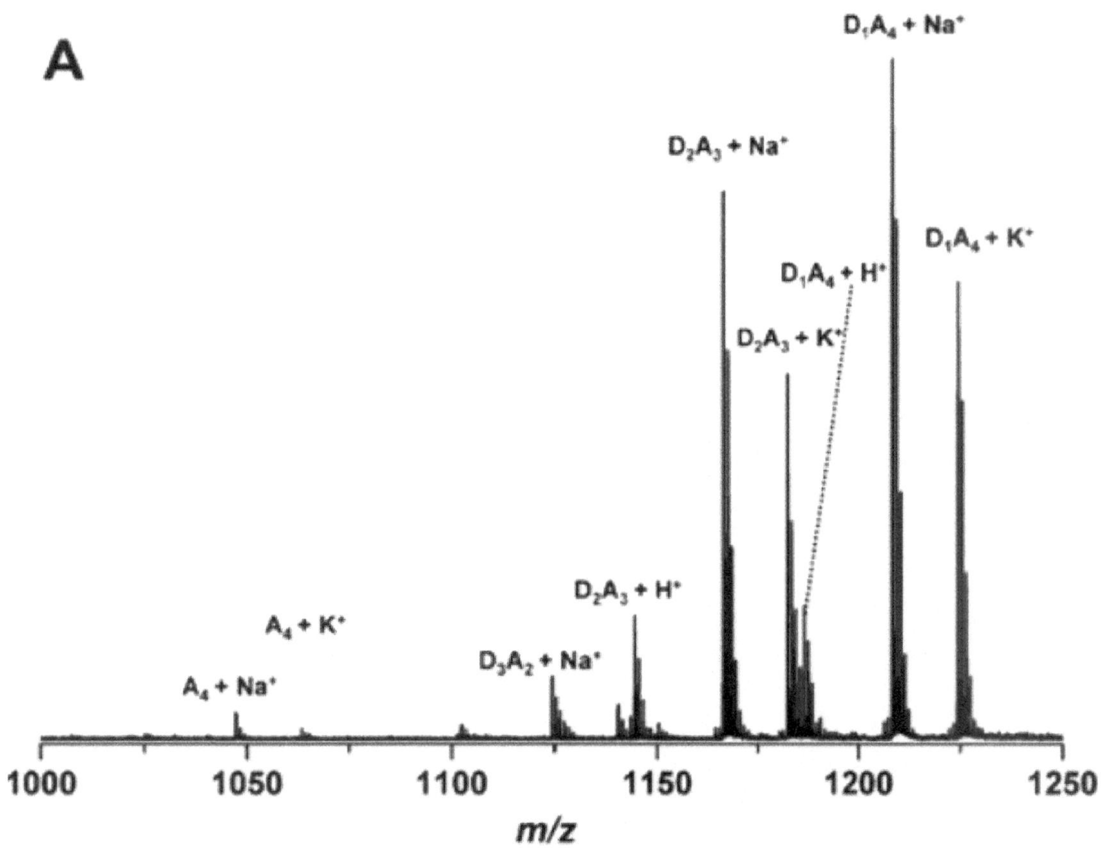

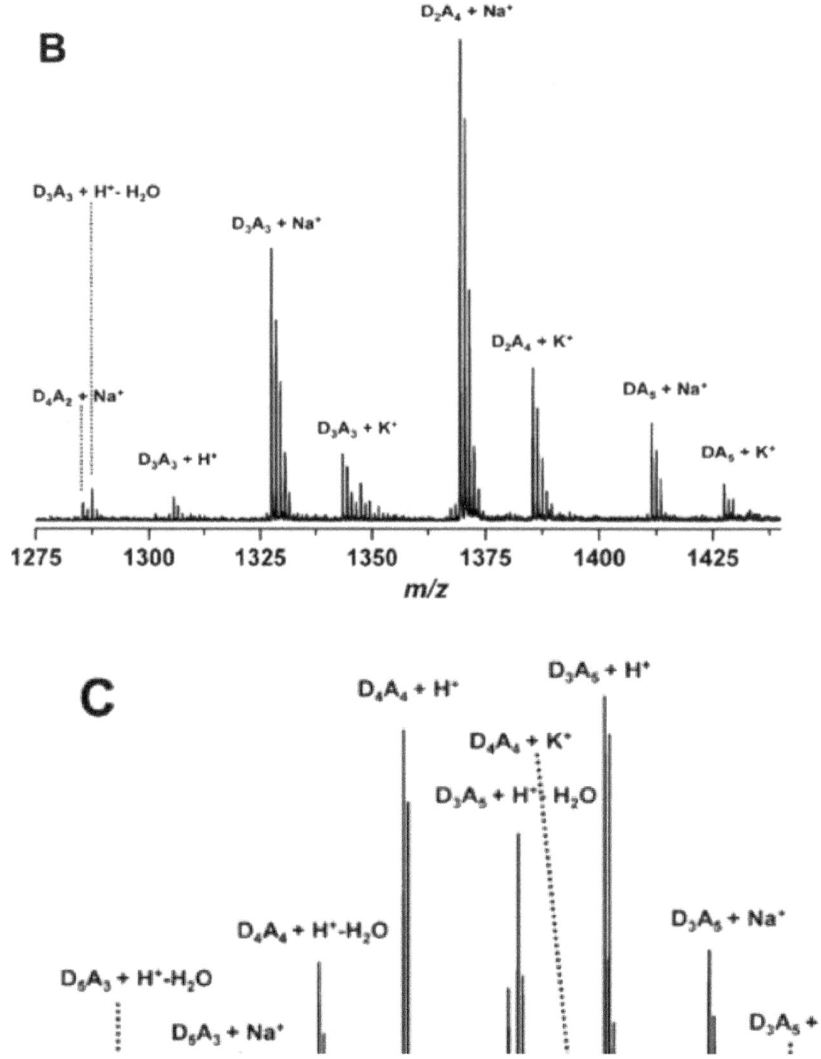

Figure 16 MALDI-TOF-MS of AMAC derivatized chitooligosaccharides: (A) DP 5, (B) DP 6, (C) DP 8. Reprinted from Carbohydrate Polymers 74, 2008, Cederkvist, F.H., Parmer, M.P., Vårum, K.M., Eijsink, V.G.H., Sørlie, M., Inhibition of a family 18 chitinase by chitooligosacharides, 41–49, Copyright (2010), with permission from Elsevier [229].

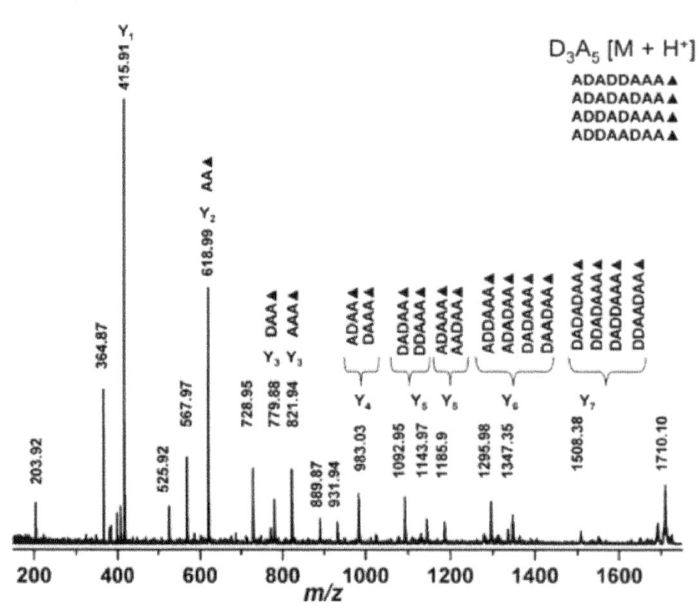

Figure 17 MALDI-TOF/TOF-MS/MS of AMAC derivatized D_3A_5 sugar 1711.80 m/z (▲ - AMAC). Reprinted from Carbohydrate Polymers 74, 2008, Cederkvist, F.H., Parmer, M.P., Vårum, K.M., Eijsink, V.G.H., Sørlie, M., Inhibition of a family 18 chitinase by chitooligosacharides, 41–49, Copyright (2010), with permission from Elsevier [229].

Examples of other chitin/chitosan derivatives analysed by MS are listed in Table 2.

Table 2 MS applications in investigations of chitin/chitosan derivatives

Chitin and chitosan derivative	**MS method**	**Chitooligosaccharide derivative applicability**	**Ref.**
Copper-chitooligosaccharides complexes	ESI-MS, ESI-MS/MS (triple quadrupole, CAD –colision activated dissociation)	Metal-ligand associations studies	[232]
Lipo-chitin oligosaccharides	ESI-MS (quadrupole), FAB-MS, CID-MS/MS (QTof)	Structural studies of lipo-chitin oligosaccharides isolated from bacteria and their role as signal molecules in symbiosis	[247,252]
Products of electrochemical reaction between caffeic acid and glucosamine	ESI-MS	Studies of chitosan–coated electrodes for bimodal sensing	[236]
Methacrylated chitoligosaccharides	MALDI-TOF	Production of biodegradable biopolymers	[225]
Chitosan/tripolyphosphate nanoparticles	ToF-SIMS	Studies of nanoparticles as drug delivery system	[231]
Catechin-modified chitosan	ESI-MS	Creating polymers for technical applications	[235]
Benzenesulfony chitosan, Dinitrobenzenesulfonyl chitosan	MALDI-TOF	Implications for drug detoxification	[265]
Chitosan-g-PEG=X (where X-Man, cholesterol, coumarin, biotin)	MALDI-TOF	Producing copolymers used in active targeting and antiadhesive therapy	[224]
Multilayers consisting of: chitosan, hyaluronan, and poyethyleneimine	ToF-SIMS	Bioactive coating of endovascular stent	[230]
Dodecyl galate (DDG)-chitosan	FAB-MS, ESI-MS	Peroxidaze catalyzed production of biopolymers	[266]

Application of Chitosan in Mass Spectrometric Analysis

Chitin and chitosan may be involved in an analytical process using mass spectrometry not only as an analyte. These polymers can play an indirect role in MS analysis; given the properties of chitin and chitosan, they can be used in pre-MS analysis treatment.

One example is the coating of capillaries for capillary electrophoresis with carboxymethyl chitosan. In contrast to uncoated capillaries, the results are more stable and repeatable. With this type of capillary for CE-MS, a very good separation efficiency, sensitivity and less interference can be achieved [267,268].

Inductively-coupled plasma-mass spectrometry (ICP-MS) is very commonly applied to determine the trace metals in different matrices like water samples. Chitosan-based chelating resins are the basis of the sample pretreatment method. These were tested for the collection and preconcentration of analytes (metals), for matrix elimination, and to find the best adsorption capacity. These novel resins could be:

- *N,N′,N″*-triacetate-type chitosan (EDTriA-type chitosan) [269];
- Cross-linked chitosan with *N*-2-hydroxypropyl iminodiacetic acid groups (CCS-HP/IDA), and cross-linked chitosan with *N,N*- iminodiacetic acid groups (CCS-IDA) [270];
- Cross-linked chitosan modified with catechol and salicylic acid [271];
- Chitosan resin derivatized with 3,4-dihydroxybenzoic acid (CCTS-DHBA) [272];
- Chitosan resin with amino acids [273].

For the same purpose chitosan resin functionalized with histidine was used in a column in combination with an inductively-coupled plasma-atomic emission spectrometer (ICP-AES) [274]. ICP-MS has been used for the meta-analysis of chitosan using samples based on chitosan Chitlacsilver nanoparticles but only to calculate the amount of silver [275].

NMR Spectroscopy

NMR spectroscopy is one of the most powerful techniques for the structural and physicochemical study of organic compounds, both small molecules and polymers. It seems to be highly suitable for studying chitin and chitosan. However, the solubility of these compounds presents a serious problem. As stated earlier, chitin with a higher DA is practically insoluble in all solvents and mixtures of different ones, whereas chitosan is soluble in aqueous acidic solution. Its solubility depends on the degree of *N*-acetylation, the degree of polymerization, the degree of crystallinity, the distribution of GlcNAc and GlcN along the polymer chain, the ionic strength of the solvent and the pH and concentration of chitosan in the solution [276].

Different NMR techniques have been used to study chitin, chitosan and their derivatives, including ^{13}C [277,278] and ^{15}N [279,280] solid-state NMR, and ^{1}H [281–283], ^{31}P [284], and ^{13}C [285] liquid-state NMR. However, only ^{13}C and ^{15}N solid-state NMR can be used to study chitins with a high DA, as this technique does not require the solubilization of the polymer.

Description of NMR Techniques

^{1}H NMR Spectroscopy

The most important factor is to find a proper solvent, which should have good solubility properties towards the target material. When the sample is only partially soluble, no quantitative analysis is accurate and reproducible [280,286]. Moreover, the residual signal of the solvent should not overlap the signals of the sample. The most common solvents for liquid-state NMR spectroscopy are D_2O/DCl, D_2O/CD_3COOD, D_2O/DCOOD.

Before ^{1}H NMR measurement, a sample should be freeze-dried 2–3 times from 99.99% D_2O in order to remove the residual signal of HOD visible in the ^{1}H spectrum. ^{1}H NMR spectra are usually recorded in aqueous acidic solution (pD ~ 5) at a temperature of 30–80 °C; 8–128 scans are done, depending on the concentration of samples. A typical ^{1}H NMR spectrum of chitosan is shown in Figure 18.

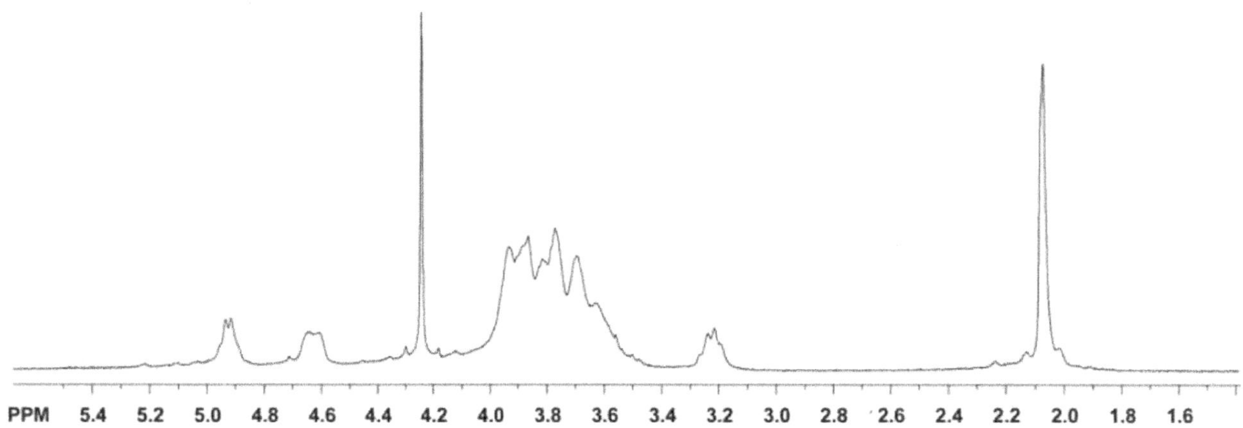

Figure 18 ^{1}H NMR spectrum of chitosan; the DA = 48. Spectrum was recorded at 400 MHz and 80 °C in D_2O (pD ~ 5) relative to internal acetone (δ_H2.225).

The spectrum shows two signals in the anomeric namely H-1 of GlcN (D) at δ ~ 4.9 and H-1 of GlcNAc (~ 4.6. They are shifted to higher values due to the neigh glycosidic and oxygen atoms of the ring. The resonan H-3–H-6 ring protons of GlcN and H-2–H-6 of GlcN present in the middle of the spectrum (δ ~ 3.5–4.0), f a group of broad, overlapping signals. The remainin ring protons of GlcN residues are shifted to characteri... lower values (δ~ 3.2) because of the adjacent amino group. The characteristic signal of the protons in the *N*-acetyl group (GlcNAc) is at δ ~ 2.1.

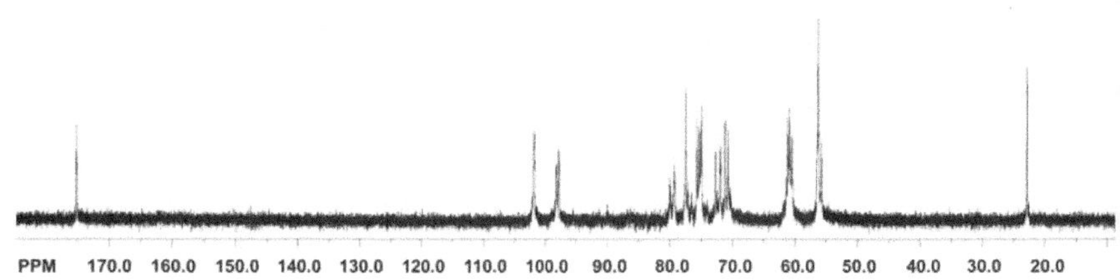

Figure 19 Liquid-state ^{13}C NMR spectrum of chitosan; DA = 48. Spectrum was recorded at 100 MHz and 70 °C in D_2O (pD ~ 5) relative to internal acetone (δ_C 31.45).

^{13}C NMR Spectroscopy

There are two kinds of ^{13}C NMR spectroscopy: liquid-state and solid-state. The first one has the same limitations as ^{1}H NMR spectroscopy, namely, the insufficient solubility of most analyzed materials. Moreover, ^{13}C NMR spectroscopy is much less sensitive than ^{1}H NMR spectroscopy owing to the properties of the carbon nucleus and the only 1% abundance of the ^{13}C isotope in nature. Liquid-state ^{13}C NMR spectra are usually recorded from the same kinds of solutions as are used for producing ^{1}H NMR spectra, but the number of scans has to be much higher–this can vary from several dozen to hundreds of thousands. Solid-state ^{13}C NMR spectroscopy is a much more powerful technique. Chitin can be analysed without any special solubilization of the polymer. This means that quite a large amount of sample can be used, which solves the problem of the low sensitivity of ^{13}C NMR spectroscopy. The ^{13}C NMR spectra of solid samples are generally recorded with magic-angle spinning (MAS) and cross-polarization (CP). MAS averages out dipolar interactions and chemical shift anisotropy, producing highly resolved spectra. CP considerably increases the sensitivity of the technique by reducing the relaxation delay due to the magnetization transfer from the ^{1}H to the ^{13}C spins. The intensities of the ^{13}C NMR signals are influenced by the kinetics of the CP process: different contact times affect the intensities of ^{13}C NMR resonances. Hence, it is very important to use the proper contact and relaxation delay times [287–289]. Duarte *et al.*optimized the relaxation delay (5 s) and the contact time (1 ms) during series assignments of the degree of *N*-acetylation of chitin and deacetylated chitin samples. Care should be taken to ensure that the CP build-up is similar for all the carbon atoms used in the quantitative calculation. Moreover, different chitin/chitosan materials often contain paramagnetic centres, which may distort the intensity of some ^{13}C NMR signals in the spectrum [287].

A typical ^{13}C NMR spectrum (Figure 19) of chitin/chitosan contains anomeric (C-1) signals of both monosaccharide residues (GlcN and GlcNAc) at δ ~ 102–105, C-3 and C-5 signals at δ ~ 73–75, C-6 signals at δ ~ 60, C-4 signals shifted to 81–85 due to glycosylation at position 4, and C-2 resonances shifted to δ ~ 55–57 due to the influence of the attached amino group. The remaining signals of the *N*-Ac group of GlcNAc are at δ ~ 23 and δ ~ 174 (methyl group and carboxylic carbon atoms respectively).

The sets of ^{13}C chemical shifts for chitin samples isolated from different sources were found elsewhere [53].

^{15}N NMR Spectroscopy

Solid-state ^{15}N NMR spectroscopy, like solid-state ^{13}C NMR spectroscopy, is a very useful technique, especially for studying insoluble chitin, but it is more time-consuming owing to the low natural abundance of the ^{15}N nuclide (<0.4%). Even using high-field NMR spectrometers with large amounts of sample, measurement times are usually of the order of 12–24 h, which makes the technique very expensive. Watson*et al.* developed the ^{15}N labeling of fungal chitin, which makes ^{15}N NMR spectroscopy much more effective [245]. The solid-state ^{15}N NMR spectra for the natural abundance of ^{15}N and for ^{15}N-labeled chitin samples needed 60,000 and only 2,000 repetitions respectively. As with solid-state ^{13}C NMR spectroscopy, care should be taken to ensure that CP build-up is similar for both *N*-acetylated and *N*-deacetylated nitrogen centres.

Two different nitrogen atom signals are present in the chitin/chitosan spectrum: one corresponds to the amine group (NH_2 in GlcN; δ ~ 10), the other to the acetamide group (NH-CO-CH_3 in GlcN; δ ~ 110).

^{31}P NMR Spectroscopy

^{31}P NMR spectroscopy is one of the more routine NMR techniques because ^{31}P has an isotopic abundance of 100% and a relatively high magnetogyric ratio. Neither native chitins nor synthesized chitosans contain phosphorus atoms, so they cannot be analysed directly using solid-state ^{31}P NMR spectroscopy. However, it is possible to characterize numerous soluble chitosan derivatives containing phosphorus atoms using both liquid-state and solid-state ^{31}P NMR spectroscopy. Particular types of phosphorus compounds have characteristic chemical shift ranges and are very easily identified in the spectrum.

Two-dimensional (2D) NMR Spectroscopy

2D NMR spectroscopy provides much more information about the molecule than 1D NMR and is very useful for studying complex molecules. Of the many 2D NMR spectroscopic techniques, COSY (DQF-COSY), NOESY and HSQC are the most common.

COSY (COrrelation SpectroscopY) and DQF-COSY (Double Quantum Filtered COrrelation SpectroscopY) allow one to determine which protons are spin-spin coupled. The signals of protons that are two or three bonds apart are visible.

NOESY (Nuclear Overhauser Effect SpectroscopY) shows correlations of all protons which are close enough for dipolar interaction by coupling through space (<5 Å). ^{1}H-^{13}C HSQC (Heteronuclear Single Quantum Coherence) allows one to determine which protons are directly bonded with particular carbon atoms.

Determination of the Degree of Acetylation (DA)

As already stated, determination of DA of chitin and chitosan is very important from the applicative point of view. Apart from UV/Vis and FTIR, NMR spectroscopy is the most powerful technique for such a study. Review articles summarizing the literature to date on the DA determination of chitin and chitosan using NMR methods have been published by Kasaai [122,131,276]. The DA values of chitin and chitosan can be determined on the basis of integrals from different NMR spectra, *i.e.*, liquid state ^{1}H, and ^{13}C (spectra decoupled from ^{1}H during the acquisition time, without the NOE effect) [290], and solid-state ^{13}C and ^{15}N. For instance, for ^{1}H and ^{13}C NMR spectra, the DA can be calculated by comparing the integral of the methyl carbon/protons of the acetyl group to the integrals of other carbons/protons from the main chains. Many equations for the DA calculation have been put forward for different types of NMR spectroscopy. DA of chitosan can be calculated on the basis of proton integrals from ^{1}H NMR spectra using several equations:

$$\%DA = \frac{1/3 * I_{CH_3(A)}}{1/6 * I_{H2-H6(A+D)}} * 100 \text{ [178,291]} \tag{4}$$

$$\%DA = \frac{7(I_{H1(A)} + I_{CH_3(A)})}{4(I_{H1(D)} + I_{H2-H6(A+D)}) + I_{H1(A)} + I_{CH_3(A)}} * 100 \text{ [282]} \tag{5}$$

$$\%DA = \frac{I_{CH_3(A)}}{3 * I_{H1(A+D)}} * 100 \text{ [290]} \tag{6}$$

$$\%DA = \frac{I_{H1(A)} + I_{CH_3(A)}/3}{I_{H1(D)} + I_{H2(D)} + I_{H1(A)} + I_{CH_3(A)}/3} * 100 \text{ [212]} \tag{7}$$

$$\%DA = \frac{1/3 * I_{CH_3(A)}}{I_{H2(D)}} * 100 \text{ [292]} \tag{8}$$

$$\%DA = 100 - \frac{I_{H1(D)}}{I_{H1(D)} + I_{CH_3(A)}/3} * 100 \text{ [293]} \tag{9}$$

$$\%DA = 100 - \frac{I_{H1(D)}}{I_{H1(D)} + I_{H1(A)}} * 100 \text{ [293]} \tag{10}$$

and the averaged value of three equations [128]:

$$\%DA = 100 - \frac{I_{H2(A)} + I_{H3-H6(A+D)}}{I_{H2(D)}} * 100 \tag{11}$$

$$\%DA = 100 - \frac{I_{H2(A)} + I_{H3-H6(A+D)}}{I_{CH_3(A)}} * 100 \tag{12}$$

$$\%DA = 100 - \frac{I_{H2(D)}}{I_{CH_3(A)}} * 100 \tag{13}$$

On the basis of carbon atom integrals from ^{13}C NMR spectra, DA of chitosan is calculated mainly according to equation [130,278,288,289,294]:

$$\%DA = \frac{I_{CH_3(A)}}{I_{C1-C6(A+D)}/6} * 100 \tag{14}$$

Since solid-state ^{15}N NMR spectra of chitin and chitosan show two very well resolved nitrogen atom signals, DA can be calculated using a simplified equation [279,280]:

$$\%DA = \frac{I_{N(A)}}{I_{N(A)} + I_{N(D)}} * 100 \tag{15}$$

There are many reports on the use of different NMR spectroscopy methods for determining DA of chitin and chitosan. In 1990, Hirai *et al.* suggested using ^{1}H NMR to study the degree of acetylation of chitosan with different DA (3–40%) [291]. The proposed method was more effective, precise and simple compared to the conventional colloid titration and elemental analysis methods, which were also applied during the study. One of the first assignments of the ^{1}H and ^{13}C chemical shifts of chitosan using ^{1}H, ^{13}C, and COSY NMR spectra was also reported [291]. In 1990, Pelletier *et al.* used solid-state ^{13}C NMR to determine DA of chitin and chitosan products [295]. The percentage of deacetylation was calculated by comparing the area of CH_3resonance to the resonances of the all the remaining carbons. Raymond *et al.* used the same approach [278] to compare DA of chitosan obtained by ^{13}C CP-MAS NMR spectroscopy and conductometric titration. Although both methods yielded similarly good results, the authors concluded that NMR was somewhat limited when DA was low, and likewise conductometric titration when DA values were high [278]. Ottøy *et al.* investigated chitosans fractionated into acid-soluble and acid-insoluble fractions after heterogeneous alkaline deacetylation [294] and obtained the DA values for both groups of fractions using ^{13}C CP-MAS NMR spectroscopy. Comparison of these results with those from ^{1}H NMR spectroscopy for the acid-soluble fractions showed them to be consistent.

One of the first applications of solid-state ^{15}N NMR spectroscopy to the study of chitin and chitosan was described by Yu *et al.* [279]. The authors used this technique to determine the degree of *N*-acetylation of chitin and chitosan, finding that solid state ^{15}N NMR was the more reliable method, especially for the determination of the extent to which poly(3-hydroxybutyrate) was grafted onto chitosan; the drawback of this method is that it is more time-consuming owing to the low natural abundance of ^{15}N. These authors noted that ^{15}N NMR spectra were easier to analyse than ^{13}C CP-MAS spectra, in which all the oxygen-bearing carbons of both chitin and chitosan overlap; this makes calculating DA more difficult [279].

In 2000 Heux *et al.* compared DA calculated from ^{1}H liquid-state NMR, and ^{13}C and ^{15}N solid-state NMR in the whole range of acetyl content from 0 to 100% [280]. The authors found that all three methods were in good agreement; nevertheless, the limitation of solid-state NMR was reliable detection at DA < 5% due to a small distortion of the spectrum baseline and signal broadening. They also found that ^{15}N CP-MAS NMR spectroscopy was particularly powerful for calculating the acetyl content in complex associations of chitin and other polysaccharides [280]. Sets of ^{13}C and ^{15}N CP-MAS NMR spectra are illustrated in Figures 20 and 21, respectively. The spectra were recorded using the following parameters: for the ^{13}C spectra contact time = 1 ms, recycle delay = 4 s, and 10,000 scans were acquired; for the ^{15}N spectra the corresponding values were 2 ms, 1 s and 100,000 scans [280].

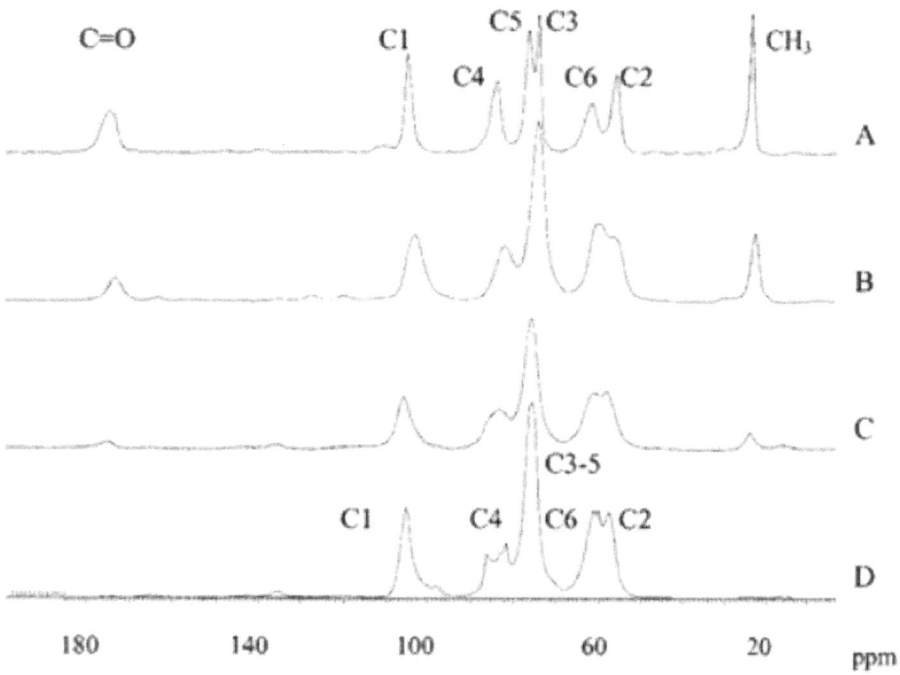

Figure 20 ^{13}C CP-MAS NMR spectra of samples A-D (with decreasing DA; ~100%, ~60%, ~20%, ~0%). Reprinted with permission from*Biomacromolecules*. 2008, *1*, 746–751. Copyright 2010, American Chemical Society [280].

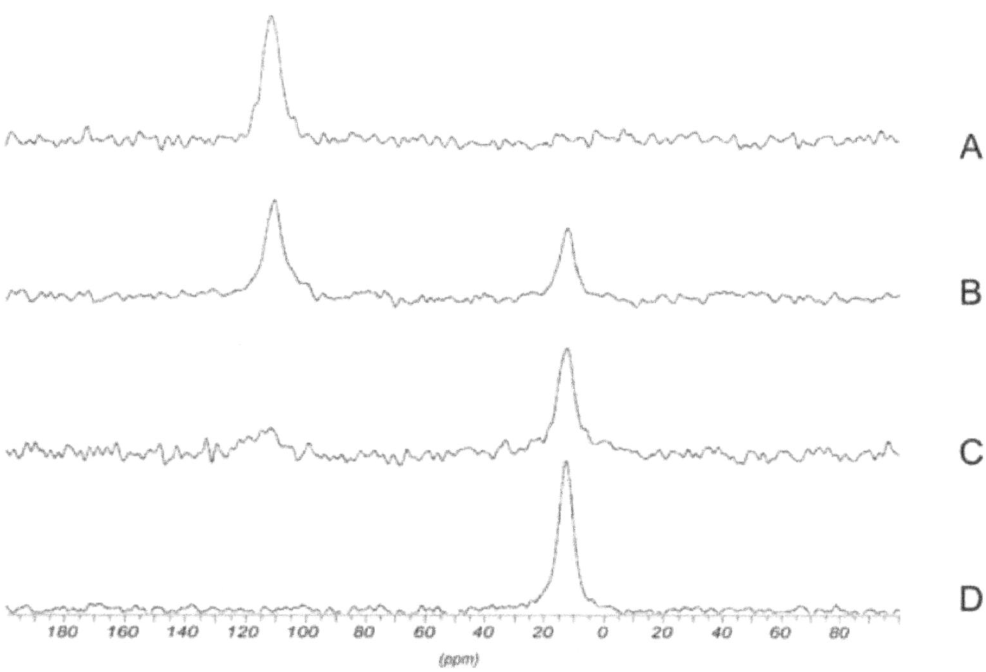

Figure 21 ^{15}N CP-MAS NMR spectra of samples A-D (with decreasing DA; ~100%, ~60%, ~20%, ~0%). Reprinted with permission from Biomacromol. 2000, 4, 746–751. Copyright 2010, ACS Publications [280].

Detailed descriptions of the determination of DA of chitin or chitosan using ^{1}H NMR spectroscopy [128,178,212,282,290,292,293,296], and ^{13}C CP-MAS NMR spectroscopy [130, 287–289,297] can be found elsewhere.

It should be mentioned that ^{1}H NMR spectroscopy is the best choice for determining DA of chitosan. ^{1}H NMR does not require previous calibrations and enables the accurate determination of even low DA values. Moreover, the importance of this method in the study of chitosan is also demonstrated by the fact that ^{1}H NMR data are usually used as standards for calibrating alternative methods [296]. Lavertu *et al.* validated the ^{1}H NMR spectroscopic method for the determination of DA using equations 1, 6a, and 6b. DA values calculated using three different combinations of peak intensities were very close to each other, demonstrating that the technique is also internally consistent. Moreover, the authors discussed the precision, ruggedness, robustness, specificity, stability and accuracy of the technique [293]. It was found to be simple, rapid and more precise than other known techniques like IR or titration.

Determination of the Pattern of N-acetylation

Most properties of chitin and chitosan and further applications depend strongly not only on DA, but probably also on the pattern of acetylation (PA) [2]. PA is a parameter describing the distribution of GlcNAc and GlcN residues along the polysaccharide chain and can be determined by ^{1}H and ^{13}C NMR spectroscopy using appropriate integrals. In the ^{1}H NMR spectrum, the nearest neighbor effect on resonances is visible for the anomeric proton of GlcNAc (sequences AD and AA) and proton H-1 of GlcN (sequences DD and DA) (Figure 22). The rest of the signals in the ^{1}H NMR spectrum are unsuitable for PA assignment as they overlap [282].

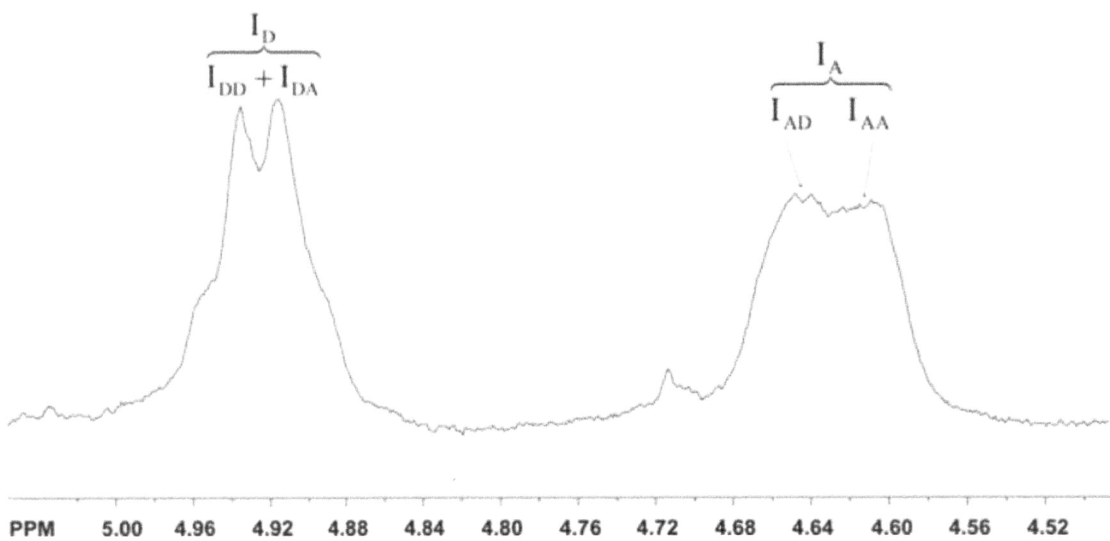

Figure 22 Section of the anomeric region of the ^{1}H NMR spectrum of chitosan; DA = 48. Spectrum was recorded at 400 MHz and 80 °C in D_2O (pD ~ 5) relative to internal acetone (δ_H 2.225). Appropriate intensities of sequences are labeled.

In general, PA in the chitosan samples can be calculated from the C-3, C-5 and C-6 carbon signals in the ^{13}C NMR [297], although detailed inspection of ^{13}C NMR spectra showed that the highest resolution with only low overlapping signals for the four diad frequencies (F_{AA}, F_{AD}, F_{DA} and F_{DD}) were achieved for the carbon C-5 signals [285] (Figure 23). For the other carbon resonances diad, triad and tetrad peaks were superimposed on neighboring peaks, which precludes a quantitative extraction of signal areas.

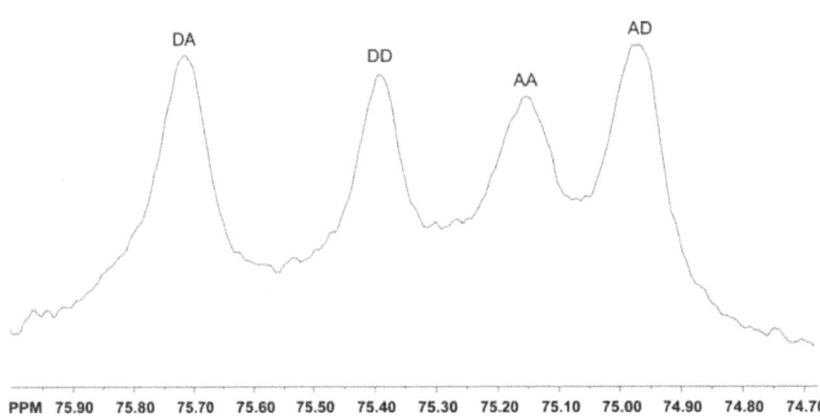

Figure 23 Sections of the ^{13}C NMR spectra of chitosan; DA = 48, showing the C-5 signals region. Spectrum was recorded at 100 MHz and 70 °C in D_2O (pD ~ 5) relative to internal acetone (δ_C 31.45). Appropriate sequences are labeled.

The relative experimental intensities of ^{1}H or ^{13}C resonances in chitosan samples should be normalized according to Bernoullian statistics [298,299] and can be presented as:

$$F_{AD} = \frac{I_{AD} + I_{DA}}{I_{AD} + I_{DA} + I_{DD} + I_{AA}} \quad (16)$$

$$F_{AA} = \frac{I_{AA}}{I_{AD} + I_{DA} + I_{DD} + I_{AA}} \quad (17)$$

$$F_{DD} = \frac{I_{DD}}{I_{AD} + I_{DA} + I_{DD} + I_{AA}} \quad (18)$$

where I_{AD}, I_{DD}, I_{AA} and I_{DA} are the respective intensities of signals AA, DD, AA and DA in the samples. In the above equation F_{AA} (F_{DD}) is the probability that two A(D) groups are adjacent to each other and F_{AD} the probability that one group A has D as a neighbor, and *vice versa*. The data should be transformed into one parameter PA, which describes the acetylation pattern as described before [298,299]:

$$PA = \frac{F_{AD}}{(2F_{AA}) + F_{AD}} + \frac{F_{AD}}{(2F_{DD}) + F_{AD}} \quad (19)$$

If the statistics are consistent with the Bernoullian model for polymers, the values PA = 0, 1 and 2 respectively indicate a perfect block, random distribution, and the alternating distribution of *N*-acetyl groups along the chitin/chitosan chain (Figure 24).

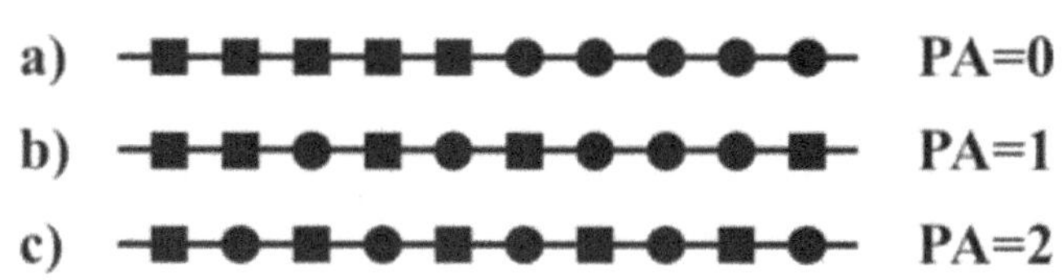

Figure 24 The distribution of GlcNAc (▪) and GlcN (●) residues along the polysaccharide chain at DA = 0.5; (a)–perfect block (PA = 0), (b)–random distribution (PA = 1), and (c)–alternating distribution (PA = 2).

Determination of PA in a chitosan sample using NMR data requires spectra with sufficient resolution of the diad frequencies. In order to get well resolved spectra, the raw chitosan should be depolymerized by nitrous acid. Additionally, pD ~ 4 and higher measurement temperatures (60–80 °C) respectively improve the solubility of chitosan and the reduce viscosity of the solution.

Vårum *et al.* have reported methods for determining the nearest-neighbor (diad) frequencies F_{AA}, F_{AD}, F_{DA}, and F_{DD} of chitosans using ^{1}H and ^{13}C NMR spectroscopy [282,297]. In the first paper, DA and PA of different chitosans obtained by the depolymerization of chitin under homogeneous and heterogeneous conditions were determined by ^{1}H NMR spectroscopy. Chitosan prepared by *N*-deacetylation under homogeneous conditions gave values for the diad frequencies that were roughly consistent with a random distribution of the *N*-acetyl groups. Samples prepared under heterogeneous conditions had a frequency of the AA diad slightly higher than that for a random (Bernoullian) distribution. This means that the GlcNAc units of chitosans prepared by *N*-deacetylation under heterogeneous conditions had a slightly more blockwise distribution than those prepared under homogeneous ones [282]. ^{13}C NMR spectroscopy results revealed that both groups of chitosans (*N*-deacetylated under homogeneous and heterogeneous conditions) gave values for the diad and triad frequencies that were consistent with a random arrangement of GlcN and GlcNAc residues along the chitosan chain [297]. Kumirska *et al.* compared the most frequently used methods for the PA determination of different chitosan samples [300]. These authors proved that ^{13}C NMR spectroscopy based on the analysis of carbon C-5 signals was clearly the most suitable method for determining PA. A number of validation parameters of that method were presented (specificity, sensitivity, repeatability and reproducibility) in that paper [300].

Weinhold *et al.* improved the chitosan pattern determination proposed by Vårum *et al.* [297] through the implementation of a line-fitting procedure for ^{13}C NMR spectra [285]. The authors assigned PA (0.5–1.5) to 32 chitosans and did not find any evidence for the existence of a clear blockwise or clear alternating chitosan preparation, although the various samples were produced under different conditions. Moreover, it was shown for the first time that PA was exponentially correlated with DA:

$$PA = 1.11 - 0.58 * e^{\left(\frac{-DA}{0.13}\right)} \quad (20)$$

Kumirska *et al.* applied a similar approach for the determination of PA of low-molecular-weight chitosans used in biomedical applications in order to assess the need for the initial depolymerization of the samples [301]. PA values of low-molecular-weight chitosan preparations (<41 kg/mol) were estimated without initial degradation of the native samples, whereas for higher-molecular-weight samples (>41 kg/mol) limited degradation was recommended.

Martinou *et al.* implemented ^{1}H and ^{13}C NMR spectroscopy to monitor the mode of action of chitin deacetylase from *Mucor rouxii* on fully water-soluble, partially *N*-acetylated chitosans [302]. The chitosan was enzymatically deacetylated and the nearest-neighbor frequencies (F_{AA}, F_{AD}, F_{DA} and F_{DD}) were determined by ^{13}C NMR spectroscopy, showing that the transition frequencies F_{AD} and F_{DA} were lower than expected from a random (Bernoullian) distribution in the further enzymatically deacetylated chitosans, while F_{AA} and F_{DD} were higher compared to the random distribution. Such a change in the frequencies associated with the enzymatic deacetylation of chitosans was qualitatively in agreement with an enzyme operating according to a multiple-attack mechanism (chitin deacetylase did not preferentially attack any sequence in the chitosan molecule) [302].

Study of Chitin and Chitosan Derivatives

Chitin and chitosan derivatives can be characterized by all of the above-mentioned NMR techniques in both the liquid and solid states. Very often ^{1}H and ^{13}C NMR spectroscopy are used together to study soluble compounds, and when phosphorus derivatives are involved ^{31}P NMR spectroscopy is additionally invoked. For instance, ^{1}H NMR spectroscopy can be used to assign the degree of substitution (DS) of chitosan by modifying groups. Tommerass *et al.* studied the *N*-alkylated trimer 2- acetamido-2-deoxy-d-glucopyranosyl-β-(1→4)-2-acetamido-2-deoxy-d-glucopyranosyl-β-(1→4)-2,5- anhydro-d-mannofuranose (A–A–M) on a fully *N*-deacetylated chitosan using different NMR techniques. The authors used ¹H NMR spectroscopy to assign the degree of substitution of chitosan (the DA<0.001) by the A-A-M group [303]. The DS values (0.07, 0.23, 0.40) were calculated from the methyl resonance of the *N*-acetyl group at δ 2.1 compared to H-2 of substituted and unsubstituted chitosan residues at δ 3.0–3.4 (Figure 25).

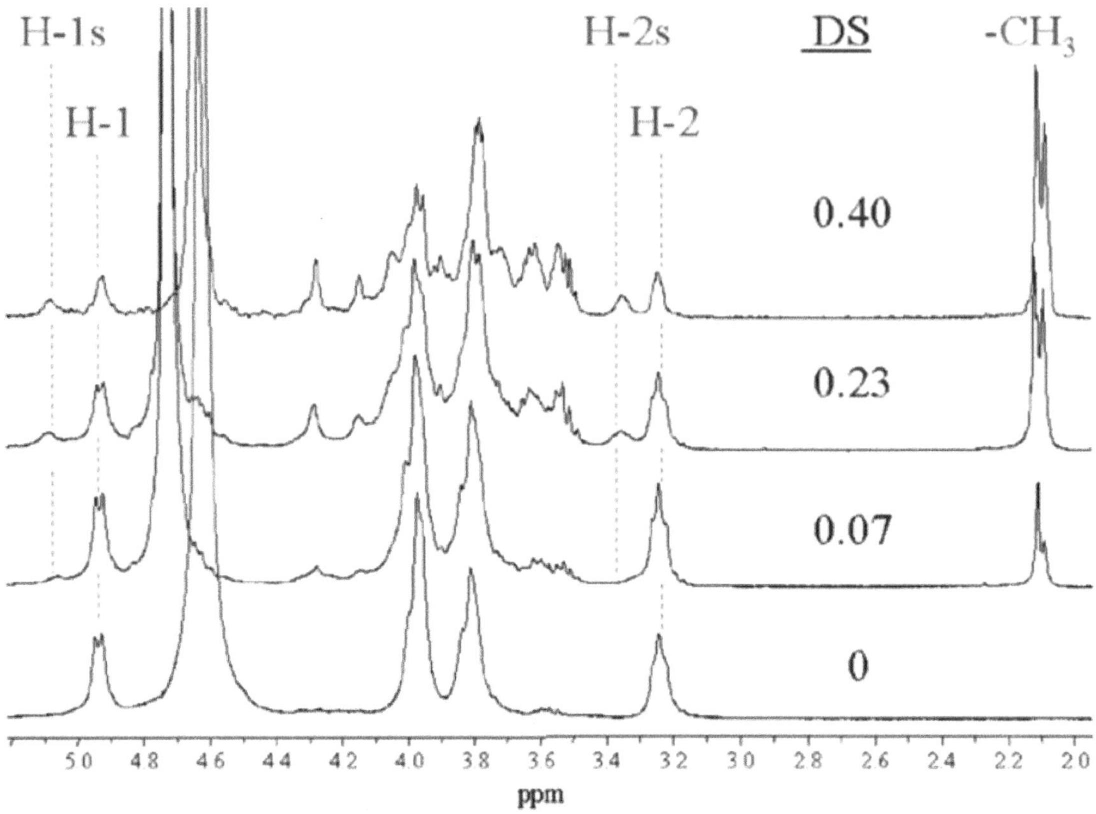

Figure 25 ¹H NMR spectra of chitosan (DP = 25, DA < 0.001) *N*-alkylated with the *N*-acetylated trimer A–A–M. Reprinted from Carbohydrate Research 337, 2002, Tømmeraas, K., Köping-Höggård, M., Vårum, K.M., Christensen B.E., Artursson, P., Smidsrød, O., Preparation and characterization of chitosans with oligosacharide branches, 2455–2462, Copyright (2010), with permission from Elsevier [303].

¹H NMR spectroscopy was used to characterize a number of chitin and chitosan derivatives, *inter alia*, glycol chitosan [304], hydroxypropyl chitosan [155], *N*-(4-bromonaphthalimide)-chitosan [197], acylated chitosan [143,173], *N*-alkylated chitosan [147], and various *N*-alkyl and *N*-benzyl chitosans [305]. ¹³C NMR spectroscopy was applied to study many soluble derivatives of chitin and chitosan: 6-carboxychitin [306], 6-*O*-carboxymethyl-chitin and 3,6-*O*-carboxymethyl-chitin [307], and C-6 and C-3 sulphated chitosan [308].¹H and ¹³C NMR spectroscopy were utilised to characterize *N,N,N*-trimethyl chitosan chloride [309], 2-*N*-(2-ethoxycarbonylethyl)chitin [310], *O*-succinyl-chitosan [146], chitin-graft-poly(ε-caprolactone) [311], *N*-carboxybutylchitosan and 5-methylpyrrolidinone chitosan [312], and chitosan derivatives with galactose groups [313]. ¹H and ¹³C NMR spectroscopy were supplemented by DQF-COSY NMR spectroscopy for the study of the *N*-alkylated trimer 2-acetamido-2-deoxy-d-glucopyranosyl-β-(1→4)-2-acetamido-2-deoxy-d-glucopyranosyl- β-(1→4)-2,5-anhydro-d-mannofuranose on a fully *N*-deacetylated chitosan [303], and 2D¹H,¹³C HSQC NMR spectroscopy for chitins sulphated at C-3 and C-6 [314], as well as COSY and ¹H,¹³C HETCOR (hetero-correlation spectroscopy) for the study of *O,N*-carboxymethylchitosans [281].

Solid-state ¹³C CP-MAS NMR spectroscopy was used to characterize *O*-hydroxypropyl chitin [315], chitin grafted with poly(acrylic acid) [316], derivatives of chitosan and crown ethers [317], dibutyrylchitin [172], and gels obtained by reacting chitosan with diethyl squarate and 1,1,3,3-tetramethoxypropane [318,319].

³¹P NMR spectroscopy in combination with other NMR techniques was used in the study of numerous chitin and chitosan derivatives containing phosphorus atoms: *N*-(diisopropylphosphono-thiooxomethyl) chitosan, *N*-(2-diethylphosphono-ethyl)-chitosan, and *N*-(2,2-diethylbisphosphono-ethyl)- chitosan [320], phosphorylated chitosan and chitin [321–323], phosphorylcholine chitosan [324], *N*-propyl-*N*-methylene phosphonic chitosan [93], 6-deoxy(diethyl phosphate)-chitin, [325], *O*-ethyl phosphonate chitosan [326], and diethyl phosphate chitosan [327].

Physicochemical Characterization of Chitin and Chitosan

Tanner *et al.* [328] were among the first to attempt to analyze α- and β-chitin polymorphs by NMR spectroscopy. Spectra of different α- and β-chitin were identified, but broad lines and distinct peak asymmetry made it difficult to interpret the spectra, to measure accurate values of chemical shifts, and to make useful comparisons between the spectra of those two polymorphs. In 2004 Jang *et al.* recorded solid-state ^{13}C CP–MAS NMR spectra for α-chitin, β-chitin, and γ-chitin [52]. The NMR spectra of α-chitin and β-chitin were clearly distinguished, but the γ-chitin spectrum was almost the same as that of α-chitin. In the β-chitin spectrum, a signal assigned to C3 and C5 appeared at δ 74, whereas in the α-chitin one, the C3 and C5 appeared as two sharply resolved signals at δ 73 and δ 75, respectively. This was attributed to the different configurations of C3 and C5 resulting from the hydrogen bonds formed. α-Chitin, having an antiparallel structure, has intersheet and intrasheet hydrogen bonding, but β-chitin, having a parallel structure, forms only intrasheet hydrogen bonding. A detailed explanation of the two kinds of hydrogen bonding in α-chitin using ^{13}C CP-MAS NMR spectroscopy was published by Kameda *et al.* [329]. The peaks corresponding to the C3 and C5 carbon atoms of γ-chitin were resolved at δ 73 and δ 75 because γ-chitin formed intersheet hydrogen bonding [52].

In a similar way, Cárdenas *et al.* characterized different polymorphs of chitins isolated from squid, shrimp, prawn, lobsters and king crabs [53].

Using ^{13}C CP-MAS NMR spectroscopy Cortizo *et al.* identified β-chitin isolated from the squid pen *Illex argentinus* [330], and Manni *et al.* isolated α-chitin from shrimp waste [331].

Other Applications of NMR Techniques

NMR techniques can also be used to examine different chemical and physiochemical processes by the analysis of final or intermediate products. Toffey *et al.* [332] used solid-state ^{13}C-NMR spectroscopy to investigate the thermally-induced conversion into chitin of a water-soluble solid consisting of an ionic complex of chitosan, acetic acid and chitosonium acetate. They recorded the ratio of the signal at δ 179 (the carboxylic atom of the acetate group) and the signal at δ 173 (the carboxylic atom of the *N*-acetyl group) during the conversion.

Holme *et al.* studied the thermal depolymerization of chitosan chloride in the solid state by analyzing the products using ^{1}H and ^{13}NMR spectroscopy [333]. The rate of acid hydrolysis of the glycosidic bond in chitosan solutions was found to be in the order A–A ≈ A–D >> D–A ≈ D–D.

Einbu and Vårum reported studies of the depolymerization and de-*N*-acetylation reaction using the chitin dimer (GlcNAc-GlcNAc) and the monomer GlcNAc as respective model substances. The rates of the reactions were determined as a function of acid concentration and temperature using ^{1}H NMR data [334]. In a subsequent paper, the authors described their study of chitin hydrolysis by concentrated and deuterated hydrochloric acid, which was a very suitable solvent for the direct characterization of the chemical composition of chitin by ^{1}H NMR spectroscopy. The rate of hydrolysis of a glycosidic linkage following an*N*-acetylated unit was found to be 54-times higher than the rate of *N*-deacetylation and 115-times higher than the rate of hydrolysis of a glycosidic linkage following a *N*-deacetylated unit [335].

NMR spectroscopy can be utilized to study the enzymatic degradation of chitin. Martinou *et al.* used ^{1}H and ^{13}C NMR spectroscopy to monitor the mode of action of chitin deacetylase from *Mucor rouxii* [302], while Sorbotten *et al.* studied the degradation products of chitosans using chitinase β from *Serratia marcescens* by ^{1}H NMR [336].

NMR spectroscopy, molecular modeling (MM) and molecular dynamics (MD) are excellent tools for studying interactions between chitin/chitosan and proteins or peptides. Colombo *et al.* used molecular dynamics (MD) simulations based on NOE NMR to study the complex between hevein and *N,N'*-diacetylchitobiose and *N,N',N''*-triacetylchitotriose. The results of the simulations showed that a carbohydrate oligomer was able to move on the surface of the relatively flat binding pocket of hevein, therefore occupying different binding subpockets [337]. NMR spectroscopy and MD were also used for studying the interaction of a variety of modified hevein domains with chitooligosaccharides [338]. Aboitiz *et al.* utilized NMR and MM to demonstrate that trisaccharides containing GalNAc and ManNAc residues were also recognized by hevein domains. Their observations indicated that the present nature of the modifications of chitotriose at either the non-reducing end (GalNAc instead of GlcNAc) or at the reducing end (ManNAc instead of GlcNAc) did not modify the mode of binding of the trisaccharide to hevein [339].

Other Spectroscopic Methods

Apart from the spectroscopic techniques commonly used in the structural analysis of chitin, chitosan and their derivatives, which have been described above, there are many other spectroscopic methods applied to the analysis of these polymers. They include Scanning Electron Spectroscopy (SEM), Scanning Electron Spectroscopy coupled with Energy Dispersive Spectroscopy (EDAX), Scanning Electron Spectroscopy coupled with Transmission Electron Microscope (SEM-TEM system), Circular Dichroism Spectroscopy (CD) and Inductively Coupled Plasma Spectroscopy (ICP).

SEM is an extremely useful method for the visual confirmation of the morphology and physical state of the surface [44]. SEM has been used, for example, for determining the surface morphology of fungi, and chitin and chitosan in crab shells [340–342], for examining the morphology of deacetylated chitosan powder and chitosan films [140], and for characterizing the surface of chitosan and various types of crosslinked deacetylated chitin before and after metal binding [90]. Nowadays this technique is applied mostly to the determination of new chitin and chitosan derivatives, for example, to study the microstructure of chitosan(chitin)/cellulose biosorbents [49], to characterize the novel nanocomposite scaffold of chitosan and bioactive glass ceramic nanoparticles [343], to characterize the morphology of the surface and the cross-section of chitosan-silica hybrid membranes [344] and to investigate the morphology of vulcanized natural rubber/chitosan blends [91].

Morphological analysis of chitin/chitosan and their derivatives has also been carried out using the SEM-TEM system, for instance, to reveal the structural features of a chitosan/hydroxyapatite nanocomposite [345] or to observe the changes in morphology during the *N*-deacetylation of chitin nanowhiskers to a chitosan nanoscaffold [115].

EDAX is used, among other things, to determine the metal uptake mechanism on chitosan, a complex phenomenon involving nodule formation on the polymer surface, ion adsorption and ion absorption [346]. This technique is also applied to determine the porosity of chitosan beads and membranes, and the diffusion of metal ions through them [44].

Another spectroscopic technique applied to chitin and chitosan investigations is Inductively Coupled Plasma Spectroscopy [ICP] [49]. This has been used to determine the amount of metal ions adsorbed onto chitosan/chitin cellulose biosorbents. The adsorption capacity of these chitosan biosorbents was found to be Cu(II) (0.417 mmol/g) > Zn(II) (0.303 mmol/g) > Cr(VI) (0.251 mmol/g) > Ni(II) (0.225 mmol/g) > Pb(II) (0.127 mmol/g). Inductively-coupled plasma spectrometry was also used for the same reason by Gamage in 2007 [347].

CD spectroscopy measures differences in the absorption of left-handed polarized light *versus* right-handed polarized light that arise due to structural asymmetry [348]. A CD spectrometer records this phenomenon as a function of wavelength. Kittur *et al.* [349] used CD to investigate the degree of *N*-acetylation of low-molecular-weight chitosan (LMWC) prepared from pectolytic hydrolysates of chitosan. Circular dichroism spectra showed a decrease in the segment of *N*-acetylated glucosamine units in LMWC.

In general spectroscopic techniques are very useful and important in the structural analysis of chitin, chitosan and their derivatives, although we need to mention that there are many other non-spectroscopic analytical methods that are routinely used in such studies: viscosity measurements, thermogravimetric analysis (TGA), differential scanning calorimetry (DSC), isothermal titration calorimetry (ITC), cyclic voltamperometry, high-performance liquid chromatography (HPLC) and size exclusion chromatography (SEC).

CONCLUSIONS

Chitin and chitosan are natural aminopolysaccharides with unique structures and interesting properties such as biocompatibility and biodegradability; they are non-toxic, have a wide range of applications, and the raw material sources for their production are unlimited. Chemical modification of these polymers results in improved solubility in water or organic solvents, which enhances their biological activities and favors the continuous development of their applications as new functional biomaterials with excellent potential in various fields. In order to acquire a deeper understanding of the mechanism of these properties, it is necessary for chitin/chitosan and their derivatives to be structurally and physicochemically well characterized. Knowledge of the microstructure of these compounds is essential for an understanding of structure–property–activity relationships. Although spectroscopic methods are of great use to scientists working in the chitin/chitosan field, many researchers lack specialist knowledge of these techniques. This paper focuses on the practical aspects of using the five spectroscopic methods most often applied – X-ray spectroscopy, infrared (IR) and UV-Vis-spectroscopy, mass spectrometry (MS) and nuclear magnetic resonance spectroscopy (NMR-spectroscopy) – in the structural investigation and physicochemical characterization of chitin, chitosan and their derivatives. The application of other spectroscopic methods is also discussed. This paper provides hands-on information about these valuable research tools, emphasizing practical aspects such as sample preparation or typical measurement conditions, their limitations and advantages, and interpretation of results. The usefulness of these methods in establishing and confirming molecular structures, determining physicochemical parameters such as the degree of *N*-acetylation, pattern of *N*-acetylation, degree of polymerization, molecular weight, crystallinity, the sequence or degree of substitution, as well as in monitoring reactions, controlling the purity of these compounds and in characterizing their intra- and intermolecular interactions, is discussed. It should also be mentioned that the structural analysis of chitin/chitosan and their derivatives is based on the application of at least a few different types of spectroscopic methods during the same study. The main reason for this is the possibility of obtaining different but compatible and complete information about the structure and physicochemical properties of these compounds, which is impossible using only one spectroscopic technique. Continuous improvements in sampling techniques, analysis software and instrumentation hardware, mean that spectroscopic methods have revolutionized the chemical and physicochemical characterization of chitin/chitosan samples, and are now routinely used by scientists working in the chitin/chitosan field.

ACKNOWLEDGEMENTS

The authors express their gratitude for the financial support provided by the Polish Ministry of Research and Higher Education under grant DS/8200-4-0085-10.

REFERENCES

1. Muzzarelli RAA, Jeuniaux C, Gooday GW. Chitin in Nature and Technology. Plenum Publishing Corporation; New York, NY, USA: 1986.
2. Rinaudo M. Chitin and chitosan: Properties and application. Prog Polym Sci. 2006;31:603–632.
3. Roberts GAF. Chitin Chemistry. 1st ed. MacMillan; London, UK: 1992.
4. Austin PR. Chitin solutions and purification of chitin. Methods Enzymol. 1988;161:403–407.
5. Kurita K. Controlled functionalization of the polysaccharide chitin. Progr Polym Sci. 2001;26:1921–1971.
6. Muzzarelli RAA, editor. Natural Chelating Polymers. Pergamon Press; New York, NY, USA: 1973. p. 83.
7. Pillai CKS, Paul W, Sharma CP. Chitin and chitosan polymers: Chemistry, solubility and fiber formation. Progr Polym Sci. 2009;34:641–678.
8. Ravi Kumar MNV. A review of chitin and chitosan applications. React Funct Polym. 2000;46:1–27.
9. Roberts GAF. The Road is long... Adv Chitin Sci. 2007;10:3–10.
10. Tajik H, Moradi; M, Rohani SMR, Erfani AM, Jalali FSS. Preparation of Chitosan from Brine Shrimp (*Artemia urmiana*) Cyst Shells and Effects of Different Chemical Processing Sequences on the Physicochemical and Functional Properties of the Product. Molecules. 2008;13:1263–1274.
11. Muzzarelli RAA, Muzzarelli C. Chitosan chemistry:

Relevance to the biomedical sciences. Adv Polym Sci. 2005;186:151–209.

12. Arai K, Kinumaki T, Fujita T. Toxicity of chitosan. Bull Tokai Reg Fish Lab. 1968;43:89–94.

13. Di Martino A, Sittinger M, Risbud MV. Chitosan: A versatile biopolymer for orthopaedic tissue-engineering. Biomaterials. 2005;26:5983–5990.

14. Crini G, Badot P-M. Application of chitosan, a natural aminopolysaccharide, for dye removal from aqueous solutions by adsorption processes using batch studies: A review of recent literature. Prog Polym Sci.2008;33:399–447.

15. Shahidi F, Arachchi JKV, Jeon Y-J. Food applications of chitin and chitosans. Trends Food Sci Technol.1999;10:37–51.

16. Dutta PK, Tripathi S, Mehrotra GK, Dutta J. Perspectives for chitosan based antimicrobial films in food applications. Food Chem. 2009;114:1173–1182.

17. Thacharodi D, Panduranga Rao K. Development and *in vitro* evaluation of chitosan based transdermal drug delivery systems for controlled delivery of propranolol hydrochloride. Biomaterials. 1995;16:145–148.

18. Cho Y-W, Jang J, Park CR, Ko S-W. Preparation and solubility in acid and water of partially deacetylated chitins. Biomacromolecules. 2000;1:609–614.

19. Kim I-Y, Seo S-J, Moon H-S, Yoo M-K, Park I-Y, Kim B-C, Cho C-S. Chitosan and its derivatives for tissue engineering applications. Biotechnol Adv. 2008;26:1–21.

20. Van der Lubben IM, Verhoef JC, Borchard G, Junginger HE. Chitosan and its derivatives in mucosal drug and vaccine delivery. Eur J Pharm Sci. 2001;14:201–207.

21. Alves NM, Mano JF. Chitosan derivatives obtained by chemical modifications for biomedical and environmental applications. Int J Biol Macromol. 2008;43:401–414.

22. Clasen C, Wilhelms T, Kulicke W-M. Formation and Characterization of Chitosan Membranes. Biomacromolecules. 2006;7:3210–3222.

23. Sajomsang W, Tantayanon S, Tangpasuthadol V, Thatte M, Daly WH. Synthesis and characterization of *N*-aryl chitosan derivatives. Int J Biol Macromol. 2008;43:79–87.

24. Silva SS, Menezes SMC, Garcia RB. Synthesis and characterization of polyurethane-g-chitosan. Eur Polym J. 2003;39:1515–1519.

25. Zhao Q, Qian J, An Q, Gao C, Gui Z, Jin H. Synthesis and characterization of soluble chitosan/sodium carboxymethyl cellulose polyelectrolyte complexes and the pervaporation dehydration of their homogeneous membranes. J Membr Sci. 2009;333:68–78.

26. Al Sagheer FA, Al-Sughayer MA, Muslim S, Elsabee MZ. Extraction and characterization of chitin and chitosan from marine sources in Arabian Gulf. Carbohydr Polym. 2009;77:410–419.

27. Campana-Filho SP, De Britto D, Curti E, Abreu FR, Cardoso MB, Battisti MV, Sim PC, Lavall RL. Extraction, structures and properties of α- and β-chitin. Quim Nova. 2007;30:644–650.

28. Noishiki Y, Takami H, Nishiyama Y, Wada M, Okada S, Kuga S. Alkali-induced conversion of β-chitin to α-chitin. Biomacromolecules. 2003;4:896–899.

29. Synowiecki J, Al-Khateeb NA. Production, Properties, and Some New Applications of Chitin and Its Derivatives. Crit Rev Food Sci Nutr. 2003;43:145–171.

30. Hayes M, Carney B, Slater J, Brück W. Mining marine shellfish wastes for bioactive molecules: Chitin and chitosan -Part A: Extraction methods. Biotechnol J. 2008;3:871–877.

31. Struszczyk MH. Chitin and Chitosan: Part I. Properties and production. Polimery. 2002;47:316–325.

32. Aranaz I, Mengíbar M, Harris R, Paños I, Miralles B, Acosta N, Galed G, Heras Á. Functional Characterization of Chitin and Chitosan. Curr Chem Biol. 2009;3:203–230.

33. Shirai K, Palella D, Castro Y, Guerrero-Legarreta I, Saucedo-Castaneda G, Huerta-Ochoa S, Hall GM. Characterization of chitins from lactic acid fermentation of prawn wastes. In: Chen RH, Chen HC, editors. Advances in Chitin Science. III. Elsevier; Taiwan: 1998. pp. 103–110.

34. Kurita K, Tomita K, Ishi S, Nishimura S-I, Shimoda K. β-Chitin as a convenient starting material for acetolysis for efficient preparation of *N*-acetylchitooligosaccharides. J Polym Sci Part A Polym Chem.1993;31:2393–2395.

35. Aiba S. Studies on chitosan: 3. Evidence for the presence of random and block copolymer structures in partially *N*-acetylated chitosans. Int J Biol Macromol. 1991;13:40–44.

36. Weinhold MX, Sauvageau JCM, Keddig N, Matzke M, Tartsch B, Grunwald I, Kübel C, Jastorff B, Thöming J. Strategy to improve the characterization of chitosan for sustainable biomedical applications: SAR guided multi-dimensional analysis. Green Chem. 2009;11:498–509.

37. Struszczyk MH, Struszczyk KJ. Medical Application of Chitin and Its Derivatives. Polish Chitin Society; Lotz, Poland: 2007. pp. 139–147. Monograph XII.

38. Guo T. More power to X-rays: New developments in X-ray spectroscopy. Laser Photon. 2009;6:591–622.

39. Broglie M. Sur une nouveau procédé permettant d'obtenir la photographie des spectres de raies des rayons Röntgen. Comptes Rendus. 1913;157:924–926.

40. Siegbahn K. Electron Spectroscopy for Chemical Analysis (ESCA) Philos Trans R Soc Lond A Math Phys Sci. 1970;268:33–57.

41. Cullity BD. Elements of X-ray Diffraction. Addison-Wesley; Reading, MA, USA: 1978.

42. Clark GL, Smith AF. X-ray diffraction studies of chitin, chitosan and derivatives. J Phys Chem.1937;40:863–879.

43. Ogawa K, Yui T, Miya M. Dependence on the preparation procedure of the polymorphism and crystallinity of chitosan membranes. Biosci Biotech Biochem. 1992;56:858–862.

44. Varma AJ, Deshpande SV, Kennedy JF. Metal complexation by chitosan and its derivatives: a review. Carbohydr Polym. 2004;55:77–93.

45. Minke R, Blackwell J. The structure of α-chitin. J Mol Biol. 1978;120:167.

46. Matienzo LJ, Winnacker SK. Dry Processes for Surface Modification of a Biopolymer: Chitosan.Macromol Mater Eng. 2002;287:871–880.

47. Veleshko AN, Rumyantseva EV, Veleshko IE, Teterin AYu, Maslakov KI, Teterin YuA, Kulyukhin SA, Vikhoreva GA. X-ray Photoelectron Study of Complexation between Uranyl Group and Chitosan. Radiochemistry. 2008;5:523–529.

48. Dambies L, Guimon C, Yiacoumi S, Guibal E. Characterization of metal ion interactions with chitosan by X-ray photoelectron spectroscopy. Colloids Surf A. 2001;177:203–214.

49. Xiaoqi S, Bo P, Yang J, Ji C, Deqian L. Chitosan (Chitin)/ Cellulose Composite Biosorbents Prepared Using Ionic Liquid for Heavy Metal Ions Adsorption. Am Inst Chem Eng. 2009;55:2062–2069.

50. Kurmaev EZ, Shin S, Watanabe M, Eguchi R, Ishiwata Y, Takeuchi T, Moewesd A, Ederere DL, Gaof Y, Iwamig M, Yanagiharah M. Probing oxygen and nitrogen bonding sites in chitosan by X-ray emission. J Electron Spectrosc Relat Phenom. 2002;125:133–138.

51. Klepka M, Lawniczak-Jablonska K, Demchenko IN, Nedelko N, Ślawska-Waniewska A, Rodrigues CA, Bordini C. X-ray Absorption Spectroscopy investigation of Fe in metal-chitosan complexes. J Synchr Rad Nat Sci. 2006;5:220–223.

52. Jang MK, Kong BG, Jeong YI, Lee CH, Nah JW. Physicochemical characterization of α-chitin, β-chitin and γ-chitin separated from natural resources. J Polym Sci Part A Polym Chem. 2004;42:3423–3432.

53. Cárdenas G, Carbrera G, Taboada E, Miranda SP. Chitin characterization by SEM, FTIR, XRD, and 13C cross polarization/mass angle spinning NMR. J Appl Polym Sci. 2004;93:1876–1885.

54. Kim SS, Kim SH, Lee YM. Preparation, characterization, and properties of β-chitin and *N*-acetylated β-chitin. J Polym Sci Part B: Polym Phys. 1996;34:2367–2374.

55. Yen MT, Mau JL. Selected physical properties of chitin prepared from shiitake stipes. LWT– Food Sci Technol. 2007;40:558–563.

56. Muzzarelli C, Francescanheli O, Tosi G, Muzzarelli RAA. Susceptibility of dibutyryl chitin and regenerated chitin fibres to deacylation and depolymerization by lipases. Carbohydr Polym. 2004;56:137–146.

57. Wada M, Saito Y. Lateral thermal expansion of chitin crystals. J Polym Sci Part B Polym Phys.2001;39(1):168–174.

58. Feng F, Liu Y, Hu K. Influence of alkali-freezing treatment on the solid state structure of chitin.Carbohydr Res. 2004;339:2321–2324.

59. Muzzarelli RAA, Morganti P, Morganti G, Palombo P, Palombo M, Biagini G, Mattioli Belmonte M, Giantomassi F, Orlandi F, Muzzarelli C. Chitin nanofibrils/chitosan composites as wound medicaments. Carbohydr Polym. 2007;70:274–284.

60. Abdou ES, Nagy KSA, Elsabee MZ. Extraction and characterization of chitin and chitosan from local sources. Bioresour Technol. 2008;99:1359–1367.

61. Samuels RJ. Solid state characterization of the structure of chitosan films. J Polym Sci Part B: Polym Phys. 1981;19:1081–1105.

62. Sakurai M, Takagi M, Takahashi T. Crystal structure of chitosan. I. Unit cell parameters. Seni Gakkaishi.1984;40:T-246.

63. Ogawa K, Hirano S, Miyanishi T, Yui T, Watanabe T. A new polymorph of chitosan. Macromolecules.1984;17:973–975.

64. Ogawa K, Inukai S. X-Ray diffraction study of sulfuric, nitric, and halogen acid salts of chitosan.Carbohydr Res. 1987;160:425–433.

65. Cairns P, Miles MJ, Morris VJ, Ridout MJ, Brownsey GJ, Winter WT. X-ray fibre diffraction studies of chitosan and chitosan gels. Carbohydr Res. 1992;235:23–28.

66. Saito H, Tabeta R. High-resolution solid-state ^{13}C NMR study of chitosan and its salts with acids: Conformational characterization of polymorphs and helical structures as viewed from the conformation dependent ^{13}C chemical shifts. Macromolecules. 1987;20:2424–2430.

67. Yui T, Imada K, Okuyama K, Obata Y, Suzuki K, Ogawa K. Molecular and Crystal Structure of the Anhydrous Form of Chitosan. Macromolecules. 1994;27:7601–7605.

68. Mazeau K, William WT, Chanzy H. Molecular and Crystal Structure of a High-Temperature Polymorph of Chitosan from Electron Diffraction Data. Macromolecules. 1994;27:7606–7612.

69. Okuyama K, Noguchi K, Miyazawa T, Yui T, Ogawa K. Molecular and Crystal Structure of Hydrated Chitosan. Macromolecules. 1997;30:5849–5855.

70. Ogawa K, Oka K, Miyanishi T, Hirano S. Chitin, Chitosan and Related Enzymes. Academic Press; Orlando, FL, USA: 1984. pp. 327–345.

71. Ogawa K. Effect of Heating an Aqueous Suspension of Chitosan on the Crystallinity and Polymorphs.Agric Biol Chem. 1991;55:2375–2379.

72. Kawada J, Abe Y, Yui T, Okuyama K, Ogawa K. Crystalline transformation of chitosan from hydrated to anhydrous polymorph via chitosan monocarboxylic acid salts. J Carbohydr Chem. 1999;18:559–571.

73. Lima IS, Airoldi C. A thermodynamic investigation on chitosan–divalent cation interactions.Thermochim Acta. 2004;421:133–139.

74. Zhang Y, Xue C, Xue Y, Gao R, Zhang X. Determination of the degree of deacetylation of chitin and chitosan by X-ray powder diffraction. Carbohydr Res. 2005;340:1914–1917.

75. Focher B, Beltranme PL, Naggi A, Torri G. Alkaline Ndeacetylation of chitin enhanced by flash treatments: reaction kinetics and structure modifications. Carbohydr Polym. 1990;12:405–418.

76. Struszczyk H. Microcrystalline Chitosan. I Preparation and Properties of Microcrystalline Chitosan. J Appl Polym Sci. 1987;33:177–189.

77. Schiffman JD, Stulga LA, Schauer CL. Chitin and Chitosan: Transformations Due to the Electrospinning Process. Polym Eng Sci. 2009;49:1918–1928.

78. Zhang Y, Xue C, Li Z, Zhang Y, Fu X. Preparation of half-deacetylated chitosan by forced penetration and its properties. Carbohydr Polym. 2006;65:229–234.

79. Fernandez Cerveraa M, Heinamaki J, Rasanenc M, Maunuc SL, Karjalainenb M, Nieto Acostaa OM, Iraizoz Colartea A, Yliruusib J. Solid-state characterization of chitosans derived from lobster chitin.Carbohydr Polym. 2004;58:401–408.

80. Toffey A, Glasser WG. Chitin Derivatives. II. Time–Temperature–Transformation Cure Diagrams of the Chitosan Amidization Process. J Appl Polym Sci. 1999;73:1879–1889.

81. Ogawa K, Yui T. Crystalllinity of partialy *N*-acetylated chitosan. Biosci Biotech Biochem.1993;57:1466–1469.

82. Seoudi R, Nada AMA. Molecular structure and dielectric properties studies of chitin and its treated by acid, base and hypochlorite. Carbohydr Polym. 2007;68:728–733.

83. Nishino T, Matsui R, Nakamae K. Elastic Modulus of the Crystalline Regions of Chitin and Chitosan. J Polym Sci Part B Polym Phys. 1999;37:1191–1196.

84. Ogawa K. X-ray study of Chitosan L- and D- Ascorbates. Chem Mater. 1996;8:2349–2351.

85. Muzzarelli RAA. Chitin. Pergamon Press; Oxford, UK: 1977.

86. Ogawa K, Oka K. X-ray Study of Chitosan-Transition Metal Complexes. Chem Mater. 1993;5:726–728.

87. Schlick S. Binding Sites of Cu2+ in Chitin and Chitosan. An Electron Spin Resonance Study.Macromolecules. 1986;19:192–195.

88. El-Sherbiny IM. Synthesis, characterization and metal uptake capacity of a new carboxymethyl chitosan derivative. Eur Polym J. 2009;45:199–210.

89. Hai-Bing Li, Yuan-Yin C, Shi-Lan L. Synthesis, Characterization, and Metal Ions Adsorption Properties of Chitosan–Calixarenes (I) J Appl Polym Sci. 2003;89:1139–1144.

90. Trimukhe KD, Varma AJ. A morphological study of heavy metal complexes of chitosan and crosslinked chitosans by SEM and WAXRD. Carbohydr Polym. 2008;71:698–702.

91. Jobish J, Vijayalakshmi R. Thermal stability, morphology, and X-ray diffraction studies of dynamically vulcanized natural rubber/chitosan blends. J Mater Sci. 2009;44:4087–4094.

92. Sahoo S, Sasmal A, Nanda R, Phani AR, Nayak PL. Synthesis of chitosan– polycaprolactone blend for control delivery of ofloxacin drug. Carbohydr Polym. 2010;79:106–113.

93. Zuńiga A, Debbaudt A, Albertengo L, Rodríguez MS. Synthesis and characterization of *N*-propyl-*N*-methylene phosphonic chitosan derivative. Carbohydr Polym. 2010;79:475–480.

94. Mohamed KR, Mostafa AA. Preparation and bioactivity evaluation of hydroxyapatite-titania/chitosangelatin polymeric biocomposites. Mater Sci Eng A. 2008;28:1087–1099.

95. Nagahama H, Maeda H, Kashiki T, Jayakumar R, Furuike T, Tamura H. Preparation and characterization of novel chitosan/gelatin membranes using chitosan hydrogel. Carbohydr Polym. 2009;76:255–260.

96. Tretenichenko EM, Datsun VM, Ignatyuk LN, Nud'ga LA. Preparation and Properties of Chitin and Chitosan from a Hydroid Polyp. Russ J Appl Chem. 2006;79:1341–1346.

97. Aji PM, Laborie MPG, Oksman K. Cross-Linked Chitosan/Chitin Crystal Nanocomposites with Improved Permeation Selectivity and pH Stability. Biomacromolecules. 2009;10:1627–1632.

98. Kurmaev EZ, Shin S, Watanabe M, Eguchi R, Ishiwata Y, Takeuchi T, Moewesd A, Ederere DL, Gaof Y, Iwamig M, Yanagiharah M. Probing oxygen and nitrogen bonding sites in chitosan by X-ray emission. J Electron Spectrosc Relat Phenom. 2002;125:133–138.

99. Stuart BH. Infrared Spectroscopy: Fundamentals and Applications (Analytical Techniques in the Sciences (AnTs) *) John Wiley & Sons Ltd; Chichester, UK: 2004.

100. Smith BC. Fundamentals of Fourier Transform Infrared Spectroscopy. CRC Press; Boca Raton, FL, USA: 1996.

101. Günzler H, Gremlich HU. IR Spectroscopy: An Introduction. Wiley-VCH; Weinherm, Germany: 2002.

102. Griffiths P, De Haseth JA. Fourier Transform Infrared Spectrometry. 2nd ed. Wiley- Interscience; New York, NY, USA: 2007.

103. Thanpitcha T, Sirivat A, Jamieson AM, Rujiravanit R. Dendritic polyaniline nanoparticles synthesized by carboxymethyl chitin templating. Eur Polym J. 2008;44:3423–3429.

104. Bourtoom T, Chinnan MS. Preparation and properties of rice starch-chitosan blend biodegradable film.LWT-Food Sci Technol. 2008;41:1633–1641.

105. Yang J, Yao Z, Tang C, Darvell BW, Zhang H, Pan L, Liu J, Chen Z. Growth of apatite on chitosan-multiwall carbon nanotube composite membranes. Appl Surf Sci. 2009;255:8551–8555.

106. Sajomsang W, Ruktanonchai UR, Gonil P, Nuchuchua O. Mucoadhesive property and biocompatibility of methylated N-aryl chitosan derivatives. Carbohydr Polym. 2009;78:945–952.

107. Urreaga JM, de la Orden MU. Chemical interactions and yellowing in chitosan-treated cellulose. Eur Polym J. 2006;42:2606–2616.

108. Darmon SE, Rudall KM. Infra-red and X-ray studies of chitin. Disc Faraday Soc. 1950;9:251–260.

109. Pearson FG, Marchessault RH, Liang CY. Infrared spectra of crystalline polysaccharides. V. Chitin. J Polym Sci. 1960;13:101–116.

110. Brunner E, Ehrlich H, Schupp P, Hedrich R, Hunoldt S, Kammer M, Machill S, Paasch S, Bazhenov VV, Kurek DV, Arnold T, Brockmann S, Ruhnow M, Born R. Chitin-based scaffolds are an integral part of the skeleton of the marine demosponge *Ianthella basta*. J Struct Biol. 2009;168:539–547. [PMC free article]

111. Focher B, Naggi A, Torri G, Cosani A, Terbojevich M. Structural differences between chitin polymorphs and their precipitates from solutions-evidence from CP-MAS ^{13}CNMR, FT-IR and FT-Raman spectroscopy. Carbohydr Polym. 1992;17:97–102.

112. Juárez-de la Rosa BA, Ardisson P-L, Azamar-Barrios JA, Quintana P, Alvarado-Gil JJ. Optical, thermal, and structural characterization of the sclerotized skeleton of two antipatharian coral species. Mater Sci Eng C. 2007;27:880–885.

113. Furuhashi T, Schwarzinger C, Miksik I, Smrz M, Beran A. Molluscan shell evolution with review of shell calcification hypothesis. Comp Biochem Physiol B Biochem Mol Biol. 2009;154:351–371.

114. Paulino AT, Simionato JI, Garcia JC, Nozaki J. Characterization of chitosan and chitin produced from silkworm chrysalides. Carbohydr Polym. 2006;64:98–103.

115. Lertwattanaseri T, Ichikawa N, Mizoguchi T, Tanaka Y, Chirachanchai S. Microwave technique for efficient deacetylation of chitin nanowhiskers to a chitosan nanoscaffold. Carbohydr Res. 2009;344:331–335.

116. Pranshanth KVH, Kittur FS, Tharanathan RN. Solid state structure of chitosan prepared under different*N*-deacetylating conditions. Carbohydr Polym. 2002;50:27–33.

117. Wu T, Zivanovic S, Draughon FA, Conoway WS, Sams CE. Physicochemical Properties and Bioactivity of Fungal Chitin and Chitosan. J Agric Food Chem. 2005;53:3888–3894.

118. Prashanth KVH, Tharanathan RN. Crosslinked chitosan-preparation and characterization. Carbohydr Res. 2006;341:169–173.

119. Nah J-W, Jang M-K. Spectroscopic Characterization and Preparation of Low Molecular, Water-Soluble Chitosan with Free-Amine Group by Novel Method. J Polym Sci Part A Polym Chem. 2002;40:3796–3803.

120. Dong Y, Wang H, Zheng W, Zhao Y, Bi D, Zhao L, Li X. Liquid crystalline behaviour of chitooligosaccharides. Carbohydr Polym. 2004;57:235–240.

121. Li QD, Dunn ET, Grandmaison EW, Goosen MFA. Applications and Properties of Chitosan. J Bioact Compat Polym. 1992;7:370–397.

122. Kasaai MR. Various Methods for Determination of the Degree of *N*-Acetylation of Chitin and Chitosan: A Review. J Agric Food Chem. 2009;57:1667–1676.

123. Sabnis S, Block LH. Improved infrared spectroscopic method for the analysis of degree of *N*-deacetylation. Polym Bull. 1997;39:67–71.

124. Muzzarelli RAA, Tanfani F, Scarpini G, Laterza G. The degree of acetylation of chitins by gas chromatography and infrared spectroscopy. J Biochem Bioph Methods. 1980;2:299–306.

125. Domszy JG, Roberts GAF. Evaluation of infrared spectroscopic techniques for analysing chitosan. Die Makromolekulare Chemie. 1985;186:1671–1677.

126. Baxter A, Dillon M, Taylor KD, Roberts GAF. Improved method for IR determination of the degree of*N*-acetylation of chitosan. Inter J Biolog Macromol. 1992;14:166–169.

127. Shigemasa Y, Matsuura H, Sashiwa H, Saimato H. An improved IR spectroscopic determination of degree of deacetylation of chitin. In: Domard A, Jeuniaux C, Muzzarelli RAA, Roberts GAF, editors.Advances in Chitin Science. Vol. 1. André Publisher; Lyon, France: 1996. pp. 204–209.

128. Shigemasa Y, Matsuura H, Sashiwa H, Saimato H. Evaluation of different absorbance ratios from infrared spectroscopy for analyzing the degree of deacetylation in chitin. Int J Biol Macromol. 1996;18:237–242.

129. Brugnerotto J, Lizardi J, Goyoolea FM, Argülles-Monal W, Desbrières J, Rinaudo M. An infrared investigation in relation with chitin and chitosan characterization. Polymer. 2001;42:3569–3580.

130. Duarte ML, Ferreira MC, Marvão MR, Rocha J. An optimized method to determine the degree of acetylation of chitin and chitosan by FTIR spectroscopy. Int J Biol Macromol. 2002;31:1–8.

131. Kasaai MR. A review of several reported procedures to determine the degree of *N*-acetylation for chitin and chitosan using infrared spectroscopy. Carbohydr Polym. 2008;71:497–508.

132. Duarte ML, Ferreira MC, Marvão MR, Rocha J. Chitin and chitosan: An optimized methodology by FTIR spectroscopy. In: Uragami T, Kurita K, Fukamizo T, editors. Chitin and Chitosan in Life Science.Kodansha Scientific Ltd; Tokyo, Japan: 2001. pp. 86–90.

133. Moore GK, Roberts GAF. Determination of the degree of *N*-acetylation of chitosan. Int J Biol Macromol. 1980;2:115–116.

134. Dong Y, Xu C, Wang J, Wu Y, Wang M, Ruan Y. Influence of degree of deacetylation on critical concentration of chitosan/dichlorocatic acid liquid crystalline solution. J Appl Polym Sci. 2002;83:1204–1208.

135. Qin C, Li H, Xiao Q, Liu Y, Zhu J, Du Y. Water-solubility of chitosan and its antimicrobial activity.Carbohydr Polym. 2006;63:367–374.

136. Miya M, Iwamoto R, Yoshikawa S, Mima S. I.R spectroscopic determination of CONH content in highly deacetylated chitosan. Int J Biol Macromol. 1980;2:323–324.

137. Qu X, Wirsén A, Albertsson A-C. Effect of lactic/glycolic acid side chains on the thermal degradation kinetics of chitosan derivatives. Polymer. 2000;41:4841–4847.

138. Rathke TD, Hudson SM. Determination of the degree of *N*-acetylation in chitin and chitosan as well as their monomer sugar ratios by near infrared spectroscopy. J Polym Sci A Polym Chem. 1993;31:749–753.

139. Vårum KM, Egelandsdal B, Ellekjaer MR. Characterization of partially *N*-acetylated chitosans by near infrared spectroscopy. Carbohydr Polym. 1995;28:187–193.

140. Baskar D, Sampath Kumar TS. Effect of deacetylation time on the preparation, properties and swelling behavior of chitosan films. Carbohydr Polym. 2009;78:767–772.

141. Vachoud L, Zydowicz N, Domarad A. Formation and

characterisation of a physical chitin gel.Carbohydr Res. 1997;302:169–177.

142. Zainol I, Akil HM, Mastor A. Effect of γ-irradiation on the physical and mechanical properties of chitosan powder. Mater Sci Eng C. 2009;29:292–297.
143. Ma G, Yang D, Kennedy JF, Nie J. Synthesize and characterization of organic-soluble acylated chitosan. Carbohydr Polym. 2009;75:390–394.
144. Singh J, Dutta PK, Dutta J, Hunt AJ, Macquarrie DJ, Clark JH. Preparation and properties of highly soluble chitosan–l-glutamic acid aerogel derivative. Carbohydr Polym. 2009;76:188–195.
145. Xie Y, Liu X, Chen Q. Synthesis and characterization of water-soluble chitosan derivate and its antibacterial activity. Carbohydr Polym. 2007;69:142–147.
146. Zhang C, Ping Q, Zhang H, Shen J. Synthesis and characterization of water-soluble *O*-succinyl- chitosan. Eur Polym J. 2003;39:1629–1634.
147. Ma G, Yang D, Zhou Y, Xiao M, Kennedy JF, Nie J. Preparation and characterization of water-soluble*N*-alkylated chitosan. Carbohydr Polym. 2008;74:121–126.
148. Jeong Y-I, Kim D-G, Jang M-K, Nah J-W. Preparation and spectroscopic characterization of methoxy poly(ethylene glycol)-grafted water-soluble chitosan. Carbohydr Res. 2008;343:282–289.
149. Zong Z, Kimura Y, Takahashi M, Yamane H. Characterization of chemical and solid state structures of acylated chitosans. Polymer. 2000;41:899–906.
150. Kaushik A, Khan R, Solanki PR, Pandey P, Alam J, Ahmad S, Malhotra BD. Iron oxide nanoparticles–chitosan composite based glucose biosensor. Biosens Bioelectron. 2008;24:676–683.
151. Borges O, Borchard G, Verhoef JC, de Sousa A, Junginger HE. Preparation of coated nanoparticles for a new mucosal vaccine delivery system. Int J Pharm. 2005;299:155–166.
152. Yang D, Li J, Jiang Z, Lu L, Chen X. Chitosan/TiO_2 nanocomposite pervaporation membranes for ethanol dehydration. Chem Eng Sci. 2009;64:3130–3137.
153. Tan H, Chu CR, Payne KA, Marra KG. Injectable *in situ* forming biodegradable chitosan–hyaluronic acid based hydrogels for cartilage tissue engineering. Biomaterials. 2009;30:2499–2506. [PMC free article]
154. Madhumathi K, Shalumon KT, Divya Rani VV, Tamura H, Furuike T, Selvamurugan N, Nair SV, Jayakumar R. Wet chemical synthesis of chitosan hydrogel–hydroxyapatite composite membranes for tissue engineering applications. Int J Biol Macromol. 2009;45:12–15.
155. Xie W, Xu P, Wang W, Liu Q. Preparation and antibacterial activity of a water-soluble chitosan derivative. Carbohydr Polym. 2002;50:35–40.
156. Wu C-S. A comparison of the structure, thermal properties, and biodegradability of polycaprolactone/chitosan and acrylic acid graftedpolycaprolactone/chitosan. Polymer. 2005;46:47–155.
157. Munro NH, Hanton LR, Moratti SC, Robinson BH. Synthesis and characterisation of chitosan-graft-poly(OEGMA) copolymers prepared by ATRP. Carbohydr Polym. 2009;77:496–505.
158. Wu Z, Feng W, Feng Y, Liu Q, Xu X, Sekino T, Fujii A, Ozaki M. Preparation and characterization of chitosan-grafted multiwalled carbon nanotubes and their electrochemical properties. Carbon. 2007;45:1212–1218.
159. Winie T, Arof AK. FT-IR studies on interactions among components in hexanoyl chitosan-based polymer electrolytes. Spectrochim Acta Part A. 2006;63:677–684.
160. Piron E, Domarad A. Interaction between chitosan and uranyl ions. Part 2. Mechanism of interaction.Int J Biol Macromol. 1998;22:33–40.
161. Sipos P, Berkesi O, Tombácz E, StPierree TG, Webb J. Formation of spherical iron(III) oxyhydroxide nanoparticles sterically stabilized by chitosan in aqueous solutions. J Inorg Biochem. 2003;95:55–63.
162. Bhattarai SR, Bahadur KCR, Aryal S, Khil MS, Kim HY. *N*-Acylated chitosan stabilized iron oxide nanoparticles as a novel nano-matrix and ceramic modification. Carbohydr Polym. 2007;69:467–477.
163. Yavuz AG, Uygun A, Bhethanabotla VR. Substituted polyaniline/chitosan composites: Synthesis and characterization. Carbohydr Polym. 2009;75:448–453.
164. Baran EJ. Spectroscopic investigation of the $VO2^{+}$/chitosan interaction. Carbohydr Polym. 2008;74:704–706.
165. Phisalaphong M, Jatupaiboon N. Biosynthesis and characterization of bacteria cellulose–chitosan film. Carbohydr Polym. 2008;74:482–488.
166. Kamari A, Wan Ngah WS, Liew LK. Chitosan and chemically modified chitosan beads for acid dyes sorption. J Environ Sci. 2009;21:296–302.
167. Rosa S, Laranjeira MCM, Riela HG, Fávere VT. Cross-linked quaternary chitosan as an adsorbent for the removal of the reactive dye from aqueous solutions. J Hazard Mater. 2008;155:253–260.
168. Sakkayawong N, Thiravetyan P, Nakbanpote W. Adsorption mechanism of synthetic reactive dye wastewater by chitosan. J Colloid Interface Sci. 2005;286:36–42.
169. Nadavala SK, Swayampakula K, Boddu VM, Abburi K. Biosorption of phenol and *o*-chlorophenol from aqueous solutions on to chitosan–calcium alginate blended beads. J Hazard Mater. 2009;162:482–489.
170. Sundaram CS, Viswanathan N, Meenakshi S. Fluoride sorption by nano-hydroxyapatite/ chitin composite. J Hazard Mater. 2009;172:147–151.
171. Ma W, Ya F-Q, Han M, Wang R. Characteristics of equilibrium, kinetics studies for adsorption of fluoride on magnetic-chitosan particie. J Hazard Mater. 2007;143:296–302.
172. Van de Velde K, Kiekens P. Structure analysis and degree of substitution of chitin, chitosan and dibutyrylchitin by FT-IR spectroscopy and solid state ^{13}C NMR. Carbohydr Polym. 2004;58:409–416.
173. Mine S, Izawa H, Kaneko Y, Kadokawa J. Acetylation of α-chitin in ionic liquids. Carbohydr Res.2009;344:2263–2265.

174. Hunger M, Weitkamp J. *In situ* IR, NMR, EPR, and UV/Vis spectroscopy: Tools for new insight into the mechanisms of heterogeneous catalysis. Angew Chem Int Ed. 2001;40:2954–2971.

175. Hollas JM. Modern Spectroscopy. 3rd ed. J. Wiley & Sons; Chichester, UK: 1996.

176. Ojeda CB, Rojas FS. Recent developments in derivative ultraviolet/visible absorption spectrophotometry. Anal Chim Acta. 2004;518:1–24.

177. Förster H. UV/VIS Spectroscopy. Springer; Berlin/Heidelberg, Germany: 2004.

178. Tan SC, Khor E, Tan TK, Wong SM. The degree of deacetylation of chitosan: advocating the first derivative UV-spectrophotometry method of determination. Talanta. 1998;45:713–719.

179. De Souza HKS, Bai G, do Pilar Gonçalves M, Bartos M. Whey protein isolate–chitosan interactions: A calorimetric and spectroscopy study. Thermochim Acta. 2009;495:108–114.

180. Liu D, Wei Y, Yao P, Jiang L. Determination of the degree of acetylation of chitosan by UV spectrophotometry using dual standards. Carbohydr Res. 2006;341:782–785.

181. Fu Z-S, Sun B-B, Chen J, Yuan L. Preparation and photochromism of carboxymethyl chitin derivatives containing spirooxazine moiety. Dyes Pigm. 2008;76:515–518.

182. Sahoo S, Sasmal A, Nanda R, Phani AR, Nayak PL. Synthesis of chitosan–polycaprolactone blend for control delivery of ofloxacin drug. Carbohydr Polym. 2010;79:106–113.

183. Yang M, Yang Y, Yang H, Shen G, Yu R. Layer-by-layer self-assembled multilayer films of carbon nanotubes and platinum nanoparticles with polyelectrolyte for the fabricationofbiosensors.Biomaterials.2006;27:246–255.

184. Lin Q-K, Ren K-F, Ji J. Hyaluronic acid and chitosan-DNA complex multilayered thin film as surface-mediated nonviral gene delivery system. Colloids Surf B. 2009;74:298–303.

185. Ramaprasad AT, Rao V, Sanjeev G, Ramananic SP, Sabharwal S. Grafting of polyaniline onto the radiation crosslinked chitosan. Synth Met. 2009;159:1983–1990.

186. Nosal WH, Thompson DW, Yan L, Sarkar S, Subramanian A, Woollam JA. UV–vis–infrared optical and AFM study of spin-cast chitosan films. Colloids Surf B. 2005;43:131–137.

187. Muzzarelli RAA, Rocchetti R. Determination of the degree of acetylation of chitosans by first derivative ultraviolet spectrophotometry. Carbohydr Polym. 1985;6:461–472.

188. Pedroni VI, Gschaider ME, Schulz PC. UV spectrophotometry: Improvements in the study of the degree of acetylation of chitosan. Macromol Biosci. 2003;3:531–534.

189. Hsiao HY, Tsai CC, Chen S, Hsieh BC, Chen RLC. Spectrophotometric determination of deacetylation degree of chitinous materials dissolved in phosphoric acid. Macromol Biosci. 2004;4:919–921.

190. Wu T, Zivanovic S. Determination of the degree of acetylation (DA) of chitin and chitosan by an improved first derivative UV method. Carbohydr Polym. 2008;73:248–253.

191. Da Silva RMP, Mano JF, Reis RL. Straightforward determination of the degree of *N*-acetylation of chitosan by means of first-derivative UV spectrophotometry. Macromol Chem Phys. 2008;209:1463–1473.

192. Khan TA, Peh KK, Ch'ng HS. Reporting degree of deacetylation values of chitosan: The influence of analytical methods. J Pharm Pharm Sci. 2002;5:205–212.

193. Liu H, Du Y, Wang X, Sun L. Chitosan kills bacteria through cell membrane damage. Int J Food Microbiol. 2004;95:147–155.

194. Tanida F, Tojima T, Han S-M, Nishi N, Tokura S, Sakairi N, Seino H, Hamada K. Novel synthesis of a water-soluble cyclodextrin-polymer having a chitosan skeleton. Polymer. 1998;39:5261–5263.

195. Mi F-L. Synthesis and characterization of a novel chitosan-gelatin bioconjugate with fluorescence emission. Biomacromolecules. 2005;6:975–987.

196. Felinto MCFC, Parra DF, da Silva CC, Angerami J, Oliveira MJA, Lugão AB. The swelling behavior of chitosan hydrogels membranes obtainedby UV- and γ-radiation. Nucl Instrum Methods Phys Res Sect B.2007;265:418–424.

197. Munro NH, Hanton LR, Robinson BH, Simpson J. Synthesis and characterisation of fluorescent chitosan derivatives containing substituted naphthalimides. React Funct Polym. 2008;68:671–678.

198. T⊘mmeraas K, Strand SP, Tian W, Kenne L, Vårum KM. Preparation and characterisation of fluorescent chitosans using 9-anthraldehyde as fluorophore. Carbohydr Res. 2001;336:291–296.

199. Kang B, Dai Y, Zhang H, Chen D. Synergetic degradation of chitosan with gamma radiation and hydrogen peroxide. Polym Degrad Stab. 2007;92:359–362.

200. Sun Ch, Qu R, Chen H, Ji C, Wang C, Sun Y, Wang B. Degradation behavior of chitosan chains in the 'green' synthesis of gold nanoparticles. Carbohydr Res. 2008;343:2595–2599.

201. Huang H, Yang X. Synthesis of chitosan-stabilized gold nanoparticles in the absence/presence of tripolyphosphate. Biomacromolecules. 2004;5:2340–2346.

202. Chen X, Zhang X, Yang W, Evans DG. Biopolymer-manganese oxide nanoflake nanocomposite films fabricated by electrostatic layer-by-layer assembly. Mater Sci Eng C. 2009;29:284–287.

203. Huang H, Yang X. Chitosan mediated assembly of gold nanoparticles multilayer. Colloids Surf A.2003;226:77–86.

204. Larena A, Cáceres DA. Variability between chitosan membrane surface characteristics as function of its composition and environmental conditions. Appl Surf Sci. 2004;238:273–277.

205. Mao Z, Ma L, Yan J, Yan M, Gao C, Shen J. The gene

transfection efficiency of thermoresponsive*N,N,N*-trimethyl chitosan chloride-g-poly(*N*-isopropylacrylamide) copolymer. Biomaterials. 2007;28:4488–4500.

206. Zhang Q, Zhang L, Li J. Fabrication and electrochemical study of monodisperse and size controlled Prussian blue nanoparticles protected by biocompatible polymer. Electrochim Acta. 2008;53:3050–3055.

207. Johnstone RAW, Rose ME. Mass Spectrometry for Chemists and Biochemists. 2nd ed. Cambridge University Press; Cambridge, UK: 1996.

208. Siuzdak G. The Expanding Role of Mass Spectrometry in Biotechnology. MCC Press; San Diego, CA, USA: 2006.

209. De Hoffmann E, Stroobant V. Mass Spectrometry: Principles and Applications. 3rd ed. J. Wiley & Sons; Chichester, UK: 2007.

210. Mattai J, Hayes ER. Characterization of chitosan by pyrolysis—mass spectrometry. J Anal Appl Pyrolysis. 1982;3:327–334.

211. Nieto JM, Peniche-Covas C, Padrón G. Characterization of chitosan by pyrolysis-mass spectrometry, thermal analysis and differential scanning calorimetry. Thermochim Acta. 1991;176:63–68.

212. Sato H, Mitzutani S, Tsuge S. Determination of the degree of acetylation of chitin/chitosan by pyrolysis-gas chromatography in the presence of oxalic acid. Anal Chem. 1998;70:7–12.

213. Bierstedt A, Stankiewicz BA, Briggs DEG, Evershed RP. Quanitative and qualitative analysis of chitin in fossil arthropods using a combination of colorimetric assay and pyrolysis-gas chromatography-mass spectrometry. Analyst. 1998;123:19–145.

214. Steinbrecht RA, Stankiewicz BA. Molecular composition of the wall of insects alfactory sensilla-the chitin question. J Insect Physiol. 1999;45:785–790.

215. Furuhashi T, Beran A, Blazso M, Czegeny Z, Schwarzinger C, Steiner G. Pyrolysis GC/MS and IR spectroscopy in chitin Analysis of Molluscan shells. Biosci Biotechnol Biochem. 2009;73:93–103.

216. Lee JW, Deng F, Yeomans WG, Allen AL, Gross RA, Kaplan DL. Direct incorporation of glucosamine and *N*-acetylglucosamie into exopolymers by *Gluconacetobacter xylinus* (=*Acetobacter xylinum*) ATCC 10245: production of chitosan-cellulose and chitin-cellulose exopolymers. Appl Environ Microbiol.2001;67:3970–3975. [PMC free article]

217. Vesentini D, Steward D, Singh AP, Ball R, Daniel G, Franish R. Chitosan-mediated changes in cell wall composition, morphology and ultrastructure in two wood-inhabiting fungi. Mycol Res. 2007;111:875–890.

218. Cunha AG, Fernandes SCM, Freire CSR, Silvestre AJD, Neto CP, Gandini A. what is the real value of chitosan's surface energy? Biomacromolecules. 2008;9:610–614.

219. Akiyama K, Kawazu K, Kobayashi A. A novel method for chemo-enzymatic synthesis of elicitor-active chitosan oligomers and partially *N*-deacetylated chitin oligomers using *N*-acylated chitotrioses as substrates in a lysozyme-catalyzed transglycosylation reaction system. Carbohydr Res. 1995;279:151–160.

220. Zhang H, Du Y, Yu X, Mitsutomi M, Aiba S. Preparation of chitooligosaccharides from chitosan by complex enzyme. Carbohydr Res. 1999;320:257–260.

221. Tokuyasu K, Mitsutomi M, Yamaguchi I, Hayashi K, Mori Y. Recognition of chitooligosaccharides and their *N*-acetyl groups by putative subsites of chitin deacetylase from a Deuteromycete, *Colletotrichum lindemuthianum*. Biochemistry. 2000;39:8837–8843.

222. Vishu Kumar AB, Varadaraj MC, Gowda LR, Tharanathan RN. Characterization of chito-oligosaccharides prepared by chitosanolysis with the aid of papain and Pronase, and their bactericidal action against *Bacillus cereus* and *Escherichia coli*. Biochem J. 2005;391:167–175. [PMC free article]

223. Makino A, Kurosaki K, Ohmae M, Kobayashi S. Chitinase-catalyzed synthesis of alternatingly *N*-deacetylated chitin: A chitin-chitosan hybrid polysaccharide. Biomacromolecules. 2006;7:950–957.

224. Fernandez-Megia E, Novoa-Carballal R, Quiñoá E, Riguera R. Conjugation of bioactive ligands to PEC-grafted chitosan at the distal end of PEG. Biomacromolecules. 2007;8:833–842.

225. Boesel LF, Reis RL, San Roman J. Innovative Approach for producing injectable, biodegradable materials using chitooligosaccharides and Green Chemistry. Biomacromolecules. 2009;10:465–470.

226. Popa-Nita S, Lucas J-M, Ladavière C, David L, Domard A. Mechanisms involved during the ultrasonically induced depolymerization of chitosan: characterization and control. Biomacromolecules.2009;10:1203–1211.

227. Bahrke S, Einarsson JM, Gislason J, Haebel S, Letzel MC, Peter-Katalinić J, Peter MG. Sequence analysis of chitooligosaccharides by matrix-assisted laser desorption ionization postsource decay mass spectrometry. Biomacromolecules. 2002;3:696–704.

228. Haebel S, Bahrke S, Peter MG. Quantitative sequencing of complex mixtures of heterochitooligosaccharides by vMALDI-linear ion trap mass spectrometry. Anal Chem. 2007;79:5557–5566.

229. Cederkvist FH, Parmer MP, Vårum KM, Eijsink VGH, SØrlie M. Inhibbition of a family 18 chitinase by chitooligosaccharides. Carbohydr Polym. 2008;74:41–49.

230. Thierry B, Winnik FM, Merhi Y, Silver J, Tabrizian M. Bioactive coatings of endovascular stents based on polyelectrolyte multilayers. Biomacromolecules. 2003;4:1564–1571.

231. Grenha A, Seijo B, Serra C, Remuñán-López C. Chitosan nanoparticle-loaded mannitol macrospheres: structure and surface characterization. Biomacromolecules. 2007;8:2072–2079.

232. Shahgholi M, Callahan JH, Rappoli BJ, Rowley DA. Investigation of copper-saccharide complexation reactions using potentiometry and electrospray mass spectrometry. J Mass Spectrom. 1997;32:1080–1093.

233. Wu L-Q, Gadre AP, Yi H, Kastantin MJ, Rubloff GW,

Bentley WE, Payne GF, Ghodssi R. Voltage-dependent assembly of the polysaccharide chitosan onto an electrode surface. Langmuir. 2002;18:8620–8625.

234. Liew S-T, Wei A. Preparation of orthogonally protected chitosan oligosaccharides: observation of an anomalous remote substituent effect. Carbohydr Res. 2002;337:1319–1324.

235. Wu L-Q, Embree HD, Balgley BM, Smith PJ, Payne GF. Utilizing renewable resources to create functional polymers: chitosan-based associative thickener. Environ Sci Technol. 2002;36:3446–3454.

236. Liu Y, Gaskell KJ, Cheng Z, Yu L, Payne GF. Chitosan-coated electrodes for bimodal sensing: selective post-electrode film reaction for spectroelectrochemmical analysis. Langmuir. 2008;24:7223–7231.

237. Yao Y-Y, Shrestha KL, Wu Y-J, Tasi H-J, Chen C-Y, Yang J-M, Ando A, Cheng C-Y, Li Y-K. Structural simulation and protein engineering to convert an endo-chitosanase to an exo-chitosanase. Protein Eng Des Sel. 2008;21:561–566.

238. Li J, Cai J, Fan L. Effect of sonolysis on kinetics and physicochemical properties of treated chitosan. J Appl Polym Sci. 2008;109:2417–2425.

239. Kerwin JL, Whitney DL, Sheikh A. Mass spectrometric profiling of glucosamine, glucosamine polymers and their catecholamine adducts. Model reactions and cuticular hydrolysates of *Toxorhynchites amboinensis* (Culicidae) pupae. Insect Biochem Mol Biol. 1999;29:599–607.

240. Trombotto S, Ladavière C, Delolme F, Domard A. Chemical preparation and structural characterization of a homogenous series of chitin/chitosan oligomers. Biomacromolecules. 2008;9:1731–1738.

241. Dennhart N, Fukamizo T, Brzezinski R, Lacombe-Harvey M-E, Letzel T. Oligosaccharide hydrolysis by chitosanase enzymes monitored by real-time electrospray ionization-mass spectrometry. J Biotechnol.2008;134:253–260.

242. Ishimizu T, Mitsukami Y, Shinkawa T, Natsuka S, Hase S, Miyagi M, Sakiyama F, Norioka S. Presence of asparagine-linked *N*-acetyloglucosamine and chitobiose in *Pyrus pyrifolia* S-RNases associated with gametophytic self-incompatibility. Eur J Biochem. 1999;263:624–634.

243. Li J, Du YM, Liang HB, Yao PJ, Wei YA. Effect of immobilized neutral protease on the preparation and physicochemical properties of low molecular weight chitosan and chito-oligomers. J Appl Polym Sci.2006;102:4185–4193.

244. Suginata W, Pantoom S, Prinz H. Substrate binding modes and anomer selectivity of chitinase A from *Vibrio harveyi*. J Chem Biol. 2009;2:191–202. [PMC free article]

245. Watson HR, Apperley DC, Dixon DP, Edwards R, Hodgson DR. An efficient method for ^{15}N-labeling of chitin in fungi. Biomacromolecules. 2009;10:793–797.

246. Kamst E, van der Drift KMGM, Thomas-Oates JE, Lugtenberg BJJ, Spaink HP. Mass spectrometric analysis of chitin oligosaccharides produced by *Rhizobium* NodC Protein in *Escherichia coli*. J Bacteriol.1995;177:6282–6285. [PMC free article]

247. Van der Drift KMGM, Spaink HP, Bloemberg GV, van Brussel AAN, Lugtenberg BJJ, Haverkamp J, Thomas-Oates JE. *Rhizobium leguminarosum* bv. *trifolii* produces lipo-chitin oligosaccharides with *nodE*-dependent highly unsaturated fatty acyl moieties. J Biol Chem. 1996;271:22563–22569.

248. Tokuyasu K, Ono H, Ohnishi-Kameyama M, Hayashi K, Mori Y. Deacetylation of chitin oligosaccharides of dp 2–4 by chitin deacetylase from *Colletotrichum lindemuthianum*. Carbohydr Res.1997;303:353–358.

249. Tokuyasu K, Ono H, Mitsutomi M, Hayashi K, Mori Y. Synhesis of a chitosan tetramer derivative, β-D-GlcNAc-(1→4)-β-D-GlcNAc-(1→4)-β-D-GlcNAc-(1→4)-D-GlcN through a partial N-acetylation reaction by chitin deacetylase. Carbohydr Res. 2000;325:211–215.

250. Kittur FS, Vishu Kumar AB, Varadaraj MC, Tharanathan RN. Chitoologosaccharides-preparation with the aid of pectinase isozyme from *Aspergillus niger* and their antibacterial activity. Carbohydr Res.2005;340:1239–1245.

251. Aronson NN, Jr, Halloran BA, Alexyev MF, Amable L, Madura JD, Paspulati L, Worth C, Van Roey P. Family 18 chitinase-oligosaccharides substrate interaction: subsite preference and anomer selectivity of *S. Marcescens* chitinase A. Biochem J. 2003;376:87–95. [PMC free article]

252. Van der Drift KMGM, Olsthoorn MMA, Brüll LP, Blok-Tip L, Thomas-Oates JE. Mass spectrometric analysis of lipo-chitin oligosaccharides-signal molecules mediating the host-specific legume-rhizobium symbiosis. Mass Spectrom Rev. 1998;17:75–95.

253. Kollár R, Petráková E, Ashwell G, Robbins PW, Cabib E. Architecture of the yeast cell wall. J Biol Chem. 1995;270:1170–1178.

254. Lopatin SA, Ilyn MM, Pustobaev VN, Bezchetnikova ZA, Varlamov VP, Davankov VA. Mass-spectrometric analysis of *N*-acetylchitooligosaccharides prepared through enzymatic hydrolysis of chitosan.Anal Biochem. 1995;227:285–288.

255. Verbruggen F, Heiri O, Reichart G-J, De Leeuw JW, nierop KGJ, Lotter AF. Effects of chemical pretreatment on δ^{18}O measurements, chemical composition, and morphology of chironomid head capsules. J Paleolimnol. 2009 doi: 10.1007/s10933-009-9374-z. [Cross Ref]

256. Wang YV, O'Brien DM, Jenson J, Francis D, Wooler MJ. The influence of diet and water on the stable oxygen and hydrogen isotope composition of Chironomidae (Diptere) with paleoecological implications.Oecologia. 2009;160:225–233.

257. López FA, Mercê ALR, Alguacil FJ, López-Delgado A. A kinetic study on the thermal behavior of chitosan. J Therm Anal Calorim. 2008;91:633–639.

258. Rodríguez AT, Ramírez MA, Cárdenas RM, Hernández AN, Velázquez MG, Bautista S. Induction of defence response of *Oryza sativa* L. against *Pyricularia grisea* (Cooke) Sacc. by treating seeds with chitosan and hydrolyzed chitosan. Pestic Biochem Physiol. 2007;89:206–215.

259. Lin H, Wang H, Xue C, Ye M. Preparation of chitosan oligomers by immobilized papain. Enzyme Microb Technol. 2002;31:588–592.

260. Lee H-W, Park Y-S, Jung J-S, Shin W-S. Chitosan oligosaccharides, dp 2–8, have prebiotic effect on the *Bifidobacterium bifidium* and *Lactobacillus* sp. Anaerobe. 2002;8:319–324.

261. Li J, Du Y, Yang J, Feng T, Li A, Chen P. Preparation and characterisation of low molecular weight chitosan and chito-oligomers by a commercial enzyme. Polym Degrad Stab. 2005;87:441–448.

262. Oliveira EN, Jr, El Gueddari NE, Moerschbacher BM, Peter MG, Franco TT. Growth of phytopathogenic fungi in the presence of partially acetylated chitooligosaccharides. Mycopathologia.2008;166:163–174.

263. Kuyama H, Nakahara Y, Nucada T, Ito J, Nakahara Yo, Ogawa T. Stereocontrolled synthesis of chitosan dodecamer. Carbohydr Res. 1993;243:C1–C7.

264. Cabrera JC, Messiaen J, Cambier P, Van Cutsem P. Size, acetylation and concentration of chitooligosaccharide elicitors determine the switch from defence involving PAL activation to cell death and water peroxide production in Arabidopsis cell suspensions. Physiol Plant. 2006;127:44–56.

265. Lee D-W, Baney RH. Oligochitosan derivatives bearing electron-deficient aromatic rings for adsorption of Amitriptyline: Implications for drug detoxification. Biomacromolecules. 2004;5:1310–1315.

266. Vachoud L, Chen T, Payne GF, Vazquez-Duhalt R. Peroxidase catalyzed grafting of gallate esters onto the polysaccharide chitosan. Enzyme Microb Technol. 2001;29:380–385.

267. Fu X, Huang L, Zhai M, Li W, Liu H. Analysis of natural carbohydrate biopolimer-high molecular chitosan and carboxymethyl chitosan by capillary zone electrophoresis. Carbohydr Polym. 2007;68:511–516.

268. Fu X, Liu Y, Li W, Pang N, Nie H, Liu H, Cai Z. Analysis of aristolochic acids by CE-MS with carboxymethyl chitosan-coated capillary. Electrophoresis. 2009;30:1783–1789.

269. Katarina RK, Takayanagi T, Oshita K, Oshima M, Motomizu S. Sample pretreatment Rusing chitosan-based chelating resin for the determination of trace metals in seawater Samales by inductively coupled plasma-mass spektrometry. Anal Sci. 2008;24:1537–1544.

270. Gao Y, Oshita K, Lee K-H, Oshima M, Motomizu S. Development of kolumn-pretreatment chelating resins for matrix elimination/multi-element determination by inductively coupled plasma-mass spectrometry.Analyst. 2002;127:1713–1719.

271. Oshita K, Takayanagi T, Oshima M, Motomizu S. Adsorption properties of ionic species on cross-linked chitosans modified with catechol and salicylic acid moieties. Anal Sci. 2008;24:665–668.

272. Sabarudin A, Oshima M, Takayanagi T, Hakim L, Oshita K, Yun Hua Gao, Motomizu S. Functionalization of chitosan with 3,4-dihydroxybenzoic acid for the adsorption/collection of uranium in water samples and its determination by inductively coupled plasma-mass spectrometry. Anal Chim Acta.2007;581:214–220.

273. Oshita K, Takayanagi T, Oshima M, Motomizu S. Adsorption behaviour of cationic and anionic species on chitosan resins possessing amino acids moieties. Anal Sci. 2007;23:1431–1434.

274. Hosoba M, Oshita K, Katarina RK, Takayanagi T, Oshima M, Motomizu S. Synthesis of novel chitosan resin possessing histidine moiety and its application to the determination of trace silver by ICP-AES coupled with triplet automated-pretreatment system. Anal Chim Acta. 2009;639:51–56.

275. Travan A, Pelillo C, Donati I, Marsich E, Benincasa M, Scarpa T, Semeraro S, Turco G, Gennaro R, Paoletti S. Non-cytotoxic silver nanoparticle-polysaccharide nanocomposites with antimicrobial activity. Biomacromolecules. 2009;10:1429–1435.

276. Kasaai MR. Determination of the degree of *N*-acetylation for chitin and chitosan by various NMR spectroscopy techniques: A review. Carbohydr Polym. 2010;79:801–810.

277. Pelletier A, Lemire I, Sygusch J, Chornet E, Overend RP. Chitin/chitosan transformation by thermo-mechano-chemical treatment including characterization by enzymatic depolymerization. Biotechnol Bioeng.1990;36:310.

278. Raymond L, Morin FG, Marchessault RH. Degree of deacetylation of chitosan using conductometric titration and solid-state NMR. Carbohydr Res. 1993;246:331.

279. Yu G, Morin FG, Nobes GAR, Marchessault RH. Degree of Acetylation of Chitin and Extent of Grafting PHB on Chitosan Determined by Solid State 15N NMR. Macromolecules. 1999;32:518–520.

280. Heux L, Brugnerotto J, Desbrières J, Versali M-F, Rinaudo M. Solid State NMR for Determination of Degree of Acetylation of Chitin and Chitosan. Biomacromolecules. 2000;1:746–751.

281. Rinaudo M, Le Dung P, Gey C, Milas M. Substituent distribution on *O,N*-carboxymethylchitosans by^{1}H and ^{13}C NMR. Int J Biol Macromol. 1992;14:122–128.

282. Vårum KM, Anthonsen MW, Grasdalen H, Smisr⊘d O. Determination of the degree of *N*-acetylation and the distribution of *N*-acetyl groups in partially *N*-deacetylated chitins (chitosans) by high-field n.m.r. spectroscopy. Carbohydr Res. 1991;211:17–23.

283. Desbrières J, Martinez C, Rinaudo M. Hydrophobic derivatives of chitosan: Characterization and rheological behavior. Int J Biol Macromol. 1996;19:21–28.

284. Lebouc F, Dez I, Madec P-J. NMR study of the phosphonomethylation reaction on chitosan. Polym.2005;46:319–325.

285. Weinhold MX, Sauvageau JCM, Kumirska J, Thöming J. Studies on acetylation patterns of different chitosan preparations. Carbohydr Polym. 2009;78:678–684.

286. Yang BY, Montgomery R. Degree of acetylation of heteropolysaccharides. Carbohydr Res.2000;323:156–162.

287. Duarte ML, Ferreira MC, Marvao MR, Rocha J.

Determination of the degree of acetylation of chitin materials by 13C CP/MAS NMR spectroscopy. Int J Biol Macromol. 2001;28:359–363.

288. Tolaimate A, Desbričres J, Rhazi M, Alagui A, Vincendon M, Vottero P. On the influence of deacetylation process on the physicochemical characteristics of chitosan from squid chitin. Polymer.2000;41:2463–2469.

289. Guinesi LS, Cavalheiro ETG. The use of DSC curves to determine the acetylation degree of chitin/chitosan samples. Thermochim Acta. 2006;444:128–133.

290. Rinaudo M, Milas M, Dung PL. Characterization of chitosan. Influence of ionic strength and degree of acetylation on chain expansion. Int J Biol Macromol. 1993;15:281–285.

291. Hirai A, Odani H, Nakajima A. Determination of degree of deacetylation of chitosan by ^{1}H NMR spectroscopy. Polym Bullet. 1991;26:87–94.

292. Kasaai MR, Arul J, Chin SL, Charlet G. The use of intense femtosecond laser pulses for the fragmentation of chitosan. J Photochem Photobiol A. 1999;120:201–205.

293. Lavertu M, Xia Z, Serreqi AN, Berrada M, Rodrrigues A, Wang D, Buschmann MD, Gupta A. A validated H NMR method for the determination of the degree of deacetylation of chitosan. J Pharm Biomed Anal. 2003;32:1149–1158.

294. Ott⊘y MH, Vårum KM, Smidsr⊘d O. Compositional heterogeneity of heterogeneously deacetylated chitosans. Carbohydr Polym. 1996;29:17–24.

295. Pelletier A, Lemire I, Sygusch J, Chornet E, Overend RP. Chitin/chitosan transformation by thermo-mechano-chemical treatment including characterization by enzymatic depolymerization. Biotechnol Bioeng.1990;36:310–315.

296. Fernandez-Megia E, Novoa-Carballal R, Quiñoá E, Riguera R. Optimal routine conditions for the determination of the degree of acetylation of chitosan by ^{1}H-NMR. Carbohydr Polym. 2005;61:155–161.

297. Vårum KM, Anthonsen MW, Grasdalen H, Smidsr⊘d O. ^{13}C NMR studies of the acetylation sequences in partially *N*-acetylated chitins (chitosans) Carbohydr Res. 1991;217:19–27.

298. Bovey P, Mireau A. NMR of Polymers. Academic Press; San Diego, CA, USA: 1996.

299. Mirau A. A Practical Guide to Understanding the NMR of Polymers. Academic Press; Hoboken, NJ, USA: 2005.

300. Kumirska J, Weinhold MX, Steudte S, Thöming J, Brzozowski K, Stepnowski P. Determination of the pattern of acetylation of chitosan samples: Comparison of evaluation methods and some validation parameters. Int J Biol Macromol. 2009;45:56–60.

301. Kumirska J, Weinhold MX, Sauvageau JCM, Thöming J, Kaczyński Z, Stepnowski P. Determination of the pattern of acetylation of low molecular weight chitosan used in biomedical applications. J Pharm Biochem Anal. 2009;50:587–590.

302. Martinou A, Bouriotis V, Stokke BT, Vårum KM. Mode of action of chitin deacetylase from *Mucor rouxii* on partially *N*-acetylated chitosans. Carbohydr Res. 1998;311:71–78.

303. Tommeraas K, Koping-Hoggard M, Vårum KM, Christensen BE, Artursson P, Smidsrod O. Preparation and characterisation of chitosans with oligosaccharide branches. Carbohydr Res.2002;337:2455–2462.

304. Knight DK, Shapka SN, Amsden BG. Structure, depolymerization, and cytocompatibility evaluation of glycol chitosan. J Biomed Mater Res Part A. 2007;83:787–798.

305. Rabea EI, Badawy ME, Rogge TM, Stevens CV, Höfte M, Steurbaut W, Smagghe G. Insecticidal and fungicidal activity of new synthesized chitosan derivatives. Pest Manag Sci. 2005;61:951–960.

306. Sun L, Du Y, Yang Y, Shi X, Li J, Wang X, Kennedy JF. Conversion of crystal structure of the chitin to facilitate preparation of a 6-carboxychitin with moisture absorption–retention abilities. Carbohydr Polym.2006;66:168–175.

307. Chen L, Du Y, Wu H, Xiao L. Relationship between Molecular Structure and Moisture- Retention Ability of Carboxymethyl Chitin and Chitosan. J Appl Polym Sci. 2002;83:1233–1241.

308. Vikhoreva G, Bannikova G, Stolbushkina P, Panov A, Drozd N, Makarov V, Varlamov V, Gal'braikh L. Preparation and anticoagulant activity of a low-molecular-weight sulfated chitosan. Carbohydr Polym.2005;62:327–332.

309. Sieval AB, Thanou M, Kotzé AF, Verhoef JC, Brussee J, Junginger HE. Preparation and NMR characterization of highly substituted *N*-trimethyl chitosan chloride. Carbohydr Polym. 1998;36:157–165.

310. Aoi K, Seki T, Okada M, Sato H, Mizutani S, Ohtani H, Tsuge S, Shiogai Y. Synthesis of a novel *N*-selective ester functionalized chitin derivative and water-soluble carboxyethylchitin. Macromol Chem Phys.2000;201:1701–1708.

311. Detchprohm S, Aoi K, Okada M. Synthesis of a Novel Chitin Derivative Having Oligo(ecaprolactone) Side Chains in Aqueous Reaction Media. Macromol Chem Phys. 2001;202:3560–3570.

312. Rinaudo M, Desbrieres J, Le Dung P, Thuy Binh P, Dong NT. NMR investigation of chitosan derivatives formed by the reaction of chitosan with levulinic acid. Carbohydr Polym. 2001;46:339–348.

313. Zhang C, Ping Q, Ding Y, Cheng Y, Shen J. Synthesis, Characterization, and Microsphere Formation of Galactosylated Chitosan. J Appl Polym Sci. 2004;91:659–665.

314. Zou Y, Khor E. Preparation of sulfated-chitins under homogeneous conditions. Carbohydr Polym.2009;77:516–525.

315. Park IK, Park YH. Preparation and Structural Characterization of Water-Soluble *O*-Hydroxypropyl Chitin Derivatives. J Appl Polym Sci. 2001;80:2624–2632.

316. Tanodekaewa S, Prasitsilpa M, Swasdisonb S, Thavornyutikarna B, Pothsreea T, Pateepasen R.

Preparation of acrylic grafted chitin for wound dressing application. Biomaterials. 2004;25:1453–1460.

317. Changhong P, Weijun Y, Motang T. Chemical Modification of Chitosan: Synthesis and Characterization of Chitosan-Crown Ethers. J Appl Polym Sci. 2003;87:2221–2225.

318. De Angelis AA, Capitani D, Crescenzi V. Synthesis and ^{13}C CP-MAS NMR Characterization of a New Chitosan-Based Polymeric Network. Macromolecules. 1998;31:1595–1601.

319. Capitani D, De Angelis AA, Crescenzi V, Masci G, Segre AL. NMR study of a novel chitosan-based hydrogel. Carbohydr Polym. 2001;45:245–252.

320. Lebouc F, Dez I, Gulea M, Madec P-J, Jaffres P-A. Synthesis of Phosphorus-Containing Chitosan Derivatives. Phosphorus Sulfur Silicon Relat Elem. 2009;184:872–889.

321. Jayakumar R, Nagahama H, Furuike T, Tamura H. Synthesis of phosphorylated chitosan by novel method and its characterization. Int J Biol Macromol. 2008;42:335–339.

322. Jayakumar R, Egawa T, Furuike T, Nair SV, Tamura H. Synthesis, Characterization, and Thermal Properties of Phosphorylated Chitin for Biomedical Applications. Polym Eng Sci. 2009;49:844–849.

323. Wang X, Ma J, Wang Y, He B. Structural characterization of phosphorylated chitosan and their applications as effective additives of calcium phosphate cements. Biomaterials. 2001;22:2247–2255.

324. Meng S, Liu Z, Zhong W, Wang Q, Du Q. Phosphorylcholine modified chitosan: Appetent and safe material for cells. Carbohydr Polym. 2007;70:82–88.

325. Zou Y, Khor E. Preparation of C-6 Substituted Chitin Derivatives under Homogeneous Conditions. Biomacromolecules. 2005;6:80–87.

326. Palma G, Casals P, Cárdenas G. Synthesis and Characterization of New Chitosan-O-Ethyl Phosphonate. J Chil Chem Soc. 2005;50:597–602.

327. Cárdenas G, Cabrera G, Taboada E, Rinaudo M. Synthesis and Characterization of Chitosan Alkyl phosphates. J Chil Chem Soc. 2006;51:815–820.

328. Tanner SF, Chanzy H, Vincendon M, Roux JC, Gaill F. High resolution solid-state carbon-13 nuclear magnetic resonance study of chin. Macromolecules. 1990;23:3576–3583.

329. Kameda T, Miyazawa M, Ono H, Yoshida M. Hydrogen Bonding Structure and Stability of α-Chitin Studied by ^{13}C Solid-State NMR. Macromol Biosci. 2005;5:103–106.

330. Cortizo MS, Berghoff CF, Alessandrini JL. Characterization of chitin from *Illex argentinus* squid pen.Carbohydr Polym. 2008;74:10–15.

331. Manni L, Ghorbel-Bellaaj O, Jellouli K, Younes I. Nasri, Extraction and Characterization of Chitin, Chitosan, and Protein Hydrolysates Prepared from Shrimp Waste by Treatment with Crude Protease from*Bacillus cereus* SV1. Appl Biochem Biotechnol. 2010 doi: 10.1007/s12010-009-8846-y. [Cross Ref]

332. Toffey A, Samaranayake G, Frazier CE, Glasser WG. Chitin derivatives. I. Kinetics of the heat-induced conversion of chitosan to chitin. J Appl Polym Sci. 1996;60:75–85.

333. Holme HK, Foros H, Pettersen H, Dornish M, Smidsrod O. Thermal depolymerization of chitosan chloride. Carbohydr Polym. 2001;46:287–294.

334. Einbu A, Vårum KM. Depolymerization and de-*N*-acetylation of Chitin Oligomers in Hydrochloric Acid. Biomacromolecules. 2007;8:309–314.

335. Einbu A, Vårum KM. Characterization of Chitin and Its Hydrolysis to GlcNAc and GlcN.Biomacromolecules. 2008;9:1870–1875.

336. S∅rbotten A, Horn SJ, Eijsink VGH, Vårum KM. Degradation of chitosans with chitinase B from*Serratia marcescens*. Production of chito-oligosaccharides and insight into enzyme processivity. FEBS J.2005;272:538–549.

337. Colombo G, Meli M, Cañada J, Asensio JL, Jiménez-Barberob J. Toward the understanding of the structure and dynamics of protein–carbohydrate interactions: molecular dynamics studies of the complexes between hevein and oligosaccharidic ligands. Carbohydr Res. 2004;339:985–994.

338. Chávez MI, Andreu C, Vidal P, Aboitiz N, Freire F, Groves P, Asensio JL, Asensio G, Muraki M, Cañada FJ, Jiménez-Barbero J. On the Importance of Carbohydrate-Aromatic Interactionsfor the Molecular Recognition of Oligosaccharides by Proteins: NMR Studies of the Structure and Binding Affinity of AcAMP2-like Peptideswith Non-Natural Naphthyl and Fluoroaromatic Residues. Chem Eur J.2005;11:7060–7074.

339. Aboitiz N, Cañada FJ, Hušáková L, Kuzma M, Křen V, Jiménez-Barbero J. Enzymatic synthesis of complex glycosaminotrioses and study of their molecular recognition by hevein domains. Org Biomol Chem.2004;2:1987–1994.

340. Chan H-Y, Chen M-H, Yuan G-F. Fungal chitosan. Fungal Sci. 2007;16:39–52.

341. Yen M-T, Mau J-L. Physico-chemical characterization of fungal chitosan from shiitake stipes. LWT–Food Sci Technol. 2007;40:472–479.

342. Yen M-T, Yang J-H, Mau J-L. Physicochemical characterization of chitin and chitosan from crab shells. Carbohydr Polym. 2009;75:15–21.

343. Peter M, Binulal NS, Soumya S, Nair SV, Furuike T, Tamura H, Jayakumar R. Nanocomposite scaffolds of bioactive glass ceramic nanoparticles disseminated chitosan matrix for tissue engineering applications. Carbohydr Polym. 2010;79:284–289.

344. Xi F, Wu J, Lin X. Novel nylon-supported organic–inorganic hybrid membrane with hierarchical pores as a potential immobilized metal affinity adsorbent. J Chromatogr A. 2006;1125:38–51.

345. Rusua VM, Nga C-H, Wilkec M, Tierscha B, Fratzld P, Petera MG. Size-controlled hydroxyapatite nanoparticles as self-organized organic–inorganic composite materials. Biomaterials. 2005;26:5414–5426.

346. Maruca R, Suder BJ, Wightman JP. Interaction of heavy metals with chitin and chitosan. III. Chromium.J Appl Polym Sci. 1982;27:4827–4837.

347. Gamage A, Shahidi F. Use of chitosan for the removal of metal ion contaminants and proteins from water. Food Chem. 2007;104:989–996.

348. Atkins P, Paula J. Elements of Physical Chemistry. 4th ed. Oxford University Press; New York, NY, USA: 2005.

349. Kittur FS, Vishu Kumar AB, Tharanathan RN. Low molecular weight chitosans-preparation by depolymerization with Aspergillus niger pectinase, and characterization. Carbohydr Res. 2003;338:1283–1290.

CHAPTER 12

INTERFERENCE OF LIGHT IN A MICHELSON-MORLEY INTERFEROMETER: A QUANTUM OPTICAL APPROACH

Ø. Langangen[1,2] A. Vaskinn[3] and B.-S. Skagerstam[3, 4]

[1]Institute of Theoretical Astrophysics, University of Oslo, Blindern, 0315 Oslo, Norway

[2]Department of Biology, Centre for Ecological and Evolutionary Synthesis (CEES), University of Oslo, Blindern, 0316 Oslo, Norway

[3]Department of Physics, The Norwegian University of Science and Technology, 7491 Trondheim, Norway

[4]Centre for Advanced Study (CAS), Drammensveien 78, 0271 Oslo, Norway

ABSTRACT

The temporal coherence interference properties of light as revealed by single detector intensity measurements in a Michelson-Morley interferometer (MMI) is often described in terms of classical optics. We show, in a pedagogical manner, how such features of light also can be understood in terms of a more general quantum-optics framework. If a thermal reference source is used in the MMI local oscillator port in combination with a thermal source in the signal port, the interference pattern revealed by single detector intensity measurements shows a distinctive dependence on the differences in the temperature of the two sources. A related method has actually been used to perform high-precision measurements of the cosmic microwave background radiation. The general quantum-optics framework allows us to consider any initial quantum state. As an example, we consider the interference of single photons as a tool to determine the peak angular-frequency of a single-photon pulse interfering with a single-photon reference pulse. A similar consideration for laser pulses, in terms of coherent states, leads to a different response in the detector. The MMI experimental setup is therefore an example of an optical device where one, in terms of intensity measurements, can exhibit the difference between classical and quantum-mechanical light.

INTRODUCTION

In 2006, Smoot and Mather shared the Nobel Prize in physics "for their discovery of the black-body form and anisotropy of the cosmic microwave background radiation (CMB)" [1]. These exciting discoveries were a breakthrough in modern cosmology by the CMB anisotropy and the strong validation of the black-body spectrum as predicted by the Big Bang theory. The discovery of the black-body form of the CMB spectrum and the high-precision measurement of the CMB temperature (see e.g., [2]) relied heavily on the so-called Far-Infrared Absolute Spectrophotometer (FIRAS) [3] on board the Cosmic Background Explorer (COBE) [4, 5]. In short, the FIRAS is a Michelson-Morley interferometer enabling a comparison of the interference patterns between an observed source and a reference black-body source on board the COBE satellite.

Citation: Ø. Langangen, A. Vaskinn, and B.-S. Skagerstam, "Interference of Light in a Michelson-Morley Interferometer: A Quantum Optical Approach," International Journal of Optics, vol. 2012, Article ID 408067, 10 pages, 2012. doi:10.1155/2012/408067

In this paper, we will make use of Glauber's theory for photon detection [6, 7] (for a guide to the early literature see e.g., [8–10] and for textbook accounts see e.g., [11–14]) together with elementary quantum mechanics to show how the principles of the FIRAS can be understood in a straight-forward manner using a quantum-optics frame-work. Interference phenomena in classical optics are described in terms of classical electromagnetic fields. Such classical fields can be obtained in terms of expectation values of the observable electromagnetic field operators using very special quantum states, that is, coherent states (see e.g., [8–10]). Quantum states with a fixed number of photons, like single-photon states, have, however, no average electromagnetic field, and the interference effects of such states therefore requires an extension of classical considerations. The use of quantum optics enables us to actually consider arbitrary quantum states of the source and reference source, thereby extending the realm of classical optics.

In Section 2, we recapitulate the principles of Glaubers photon detection theory [6, 7] and the transformation properties of a quantum field in a beamsplitter (see e.g., [15–17]). The Glauber theory of optical coherence is by now well established and plays a central role in fundamental studies of quantum interference effects of photon quantum states (see e.g., [18–20]). In Section 3, we consider temporal interference effects in the Michelson-Morley interferometer for pure quantum states like single-photon states as well as classical states corresponding to coherent states (see e.g., [8–10]). In Section 4, we explain the principles of interference of thermal light in the Michelson-Morley interferometer by using only vacuum as the reference source and reproduce known expressions. Here, we also show that the detection intensity obtained with a vacuum reference source does not depend on the details of signal quantum state considered, thermal or not, but only on the average number of photons present. In Section 5, we consider the full system with an observed thermal source combined with a thermal reference source. With the results obtained, we then recover the basic principle of FIRAS and how it may be used as a high-precision thermometer. In Section 6, we, finally, give some concluding remarks.

Our presentation extends a recent presentation by Donges [21] and illustrates, for example, that a quantum-mechanical treatment directly leads to the concept of a thermal coherence length without explicitly making use of results from classical physics like the Wiener-Khintchine theorem as in [21] and that any quantum statecan be considered.

THE MICHELSON-MORLEY INTERFEROMETER

We consider the classical Michelson-Morley interferometer (MMI) as illustrated in Figure 1, where the so-called temporal coherence properties of the radiation field are probed (for an early account see e.g., [22]). Figure 1 is a simplified and schematic version of the FIRAS experimental setup. In order to understand the appearance of interference effects in the MMI, we first discuss the separate parts of the MMI before we consider the full setup with the presence of a reference beam.

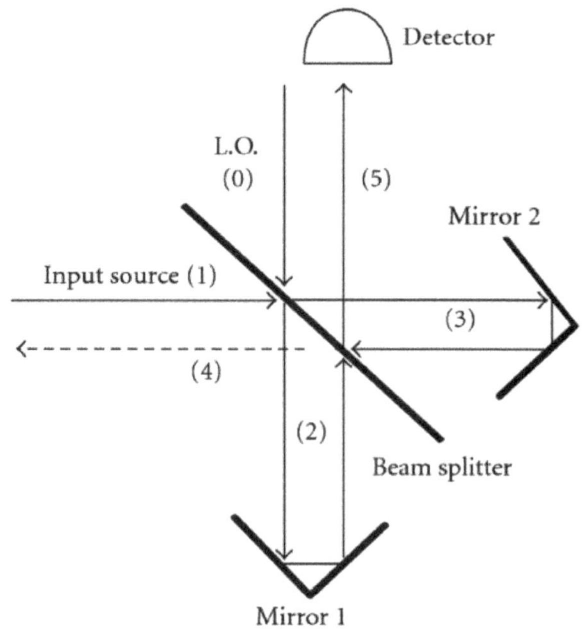

Figure 1: Schematic drawing of the MMI setup considered in this paper. The thick and diagonal line is representing a beam splitter with a transmission coefficient T. The movable mirrors 1 and 2 are identical. The numbers in the parenthesis are referring to the indices used to label the mode operators of the various light beams as discussed in the main text. The signal port (1) and the local oscillator port LO (0) are prepared with various quantum states, like Fock states, coherent states, or states with random phases like thermal states. The output mode (4) will not be considered in the present paper.

Glauber's Theory of Photon Detection

Let us outline a simple, but not unrealistic, model of a photon detector situated at the space-time point $(\vec{x}, t)$. In this simplified model of a photon detection process [6, 7], the detection of a photon is described by an annihilation of a photon at the detector which modifies the initial state $|\text{in}\rangle$ as follows:

$$|\text{in}\rangle \rightarrow \vec{E}^{(+)}(\vec{x}, t)|\text{in}\rangle. \tag{1}$$

Here the time-evolution is implemented in the Heisenberg picture instead of a Schrödinger picture with space-and time-dependent "wave-functions" for photons. The Heisenberg picture seems actually to be mandatory since position-dependent wave functions appear to be ill defined for massless particles with a nonzero spin (see e.g., [23]). The observable electric field operator in Heisenberg picture, $\vec{E}(\vec{x}, t) = \sum_m \vec{E}_m(\vec{x}, t)$, is then described in terms of a suitable normal mode expansion, indexed by mode the number m, as a superposition of positive and negative angular-frequency contributions as follows:

$$\vec{E}_m(\vec{x}, t) = \vec{E}_m^{(+)}(\vec{x}, t) + \vec{E}_m^{(-)}(\vec{x}, t), \tag{2}$$

where $\vec{E}_m^{(+)}(\vec{x}, t)$ $(\vec{E}_m^{(-)}(\vec{x}, t))$ contains an annihilation (creation) operator for a photon with mode number m. According to the basic Born rule in quantum mechanics, the probability to detect the system in a final state $|\text{out}\rangle$, after the single-photon absorption process, is then proportional to $|\langle \text{out}|\vec{E}^{(+)}(\vec{x},t)|\text{in}\rangle|^2$. Since the exact details of the final states are, in general, unknown we sum over all possible final states $|\text{out}\rangle$, that is, we consider $\sum_{\text{out}} |\langle \text{out}|\vec{E}^{(+)}(\vec{x},t)|\text{in}\rangle|^2 = \langle \text{in}|\vec{E}^{(-)}(\vec{x},t)\vec{E}^{(+)}(\vec{x},t)|\text{in}\rangle$, where use has been made of the completeness of all possible final states, that is, $\sum|\text{out}\rangle\langle\text{out}|=1$. In general, we also have to consider not only a pure initial quantum state but also a quantum state as described by a density matrix. This leads to a description of the observed intensity I which we write in the form

$$I = \text{Tr}\left[\rho_{\text{in}}\vec{E}^{(-)}(\vec{x},t)\vec{E}^{(+)}(\vec{x},t)\right], \tag{3}$$

where ρ_{in} is the density matrix describing the initial state. Since we will not be interested in the absolute normalization of the observed intensity I, we can neglect normalization constants that may enter into I. It is a remarkable fact that an analysis of single-photon interference in a Young interferometer using such a quantum-mechanical description of the photon detection process [24], as well as a proper experimental investigation of single-photon interference [25, 26], is fairly recent in the history physics.

For reasons of clarity, we will now consider a typical plane wave normal mode expansion of the electromagnetic field observable $\vec{E}(\vec{x},t)$ in terms of a mode sum over wave-vectors $\vec{k}$ and polarization-vector labels λ, that is, we have for its positive angular-frequency part that

$$\vec{E}^{(+)}(\vec{x},t) = i\sum_{\vec{k}\lambda}\sqrt{\frac{\hbar\omega_k}{2V\varepsilon_0}}\vec{\varepsilon}_{\vec{k}\lambda}a_{\vec{k}\lambda}e^{i\vec{k}\cdot\vec{x}-i\omega_k(t-t_0)}, \tag{4}$$

with $\omega_k = c|\vec{k}|$. The annihilation operator $a_{\vec{k}\lambda}$, and the corresponding unit polarization vector $\vec{\varepsilon}_{\vec{k}\lambda}$, denotes the normal mode considered. t_0 is a suitable initial-time parameter which will be used, in a manner to be clarified below, to express the dependence on various propagation times for the optical paths as given in the MMI setup, a procedure as is well known in classical optics (see e.g., [12, 22]). If the direction of the light beams considered is well defined, the dependence of the detector position $\rightarrow x$ can be neglected in the expression for I. In general, spatial modulations of the measured intensity are expected [27, 28]. A theoretical analysis of such effects along lines as discussed in the literature (see e.g., [29, 30]) will, however, not enlighten the issues we are addressing in the present paper. V is a quantization volume that will be allowed to be arbitrarily large at an appropriate late stage of our calculations.

Since the dependence of normalization constants will be irrelevant, and since we will only consider polarization-independent optical devices, we make use of a scalar notation. We therefore suppress the wave vector and polarization labels and with $\omega \equiv \omega_k$ we write $a(\omega) \equiv a_{\vec{k}\lambda}$ such that $[a(\omega),a^\dagger(\omega)] = \delta_{\omega\omega'}$ in terms of a discrete Kronecker delta $\delta_{\omega\omega'}$. We also make use of the following convenient notation [12] for the positive angular-frequency part of the electric field at the position of the detector at time t:

$$E^{(+)}(t) = i\sum_{\omega}\sqrt{\delta\omega}\omega^{1/2}a(\omega)e^{i\phi(t)}, \tag{5}$$

with the prescription $\sum_\omega \delta\omega \to \int d\omega\omega^{d-1}$ in the infinite volume limit $V\rightarrow\infty$ and where d is the number of space-dimensions. Here $(t)\equiv-\omega(t-\tau)$ now denotes a mode-dependent optical phase which explicitly takes the source-detection propagation time into account in terms of the time-delay τ, which will be evaluated in detail for the MMI-setup below. In (5), the mode operator (ω) ($\equiv a_5(\omega)$ in Figure 1) will later be expressed in terms of a superposition of the input mode operators ($\equiv a_0(\omega)$ and $\equiv a_1(\omega)$ in Figure 1). If the detector time t enters explicitly into the detection intensity equation (3), we will perform a time average which corresponds to a finite detector time-resolution window.

The corresponding time average of the observed intensity $I\equiv(t)$ will be denoted by $\langle I\rangle$, that is,

$$\langle I\rangle = \frac{1}{T_{\text{int}}}\int_{-T_{\text{int}}/2}^{T_{\text{int}}/2} dtI(t), \tag{6}$$

where the time T_{int} of integration, as for example, the time during which an actual measurement proceeds, is chosen to be sufficiently large in comparison with typical time scales of intensity fluctuations. The time-averaged observed intensity $\langle I\rangle$ will in general, as we will see explicitly below, be a function of a time-delay τdepending on the actual experimental set up.

For a finite quantization volume V, ω can be regarded to be discrete and it may be assumes that d=1. When appropriate we will, however, also consider d=3 in order to compare with related results in the literature [21,22]. Our main results will, however, not be very sensitive to the choice of d. Using the same notation as above, a single-photon quantum state $|f\rangle$, with an angular-frequency distribution given by

$$|f\rangle = \sum_{\omega}\sqrt{\delta\omega}f(\omega)|1_\omega\rangle \equiv \left(f,a^\dagger\right)|0\rangle, \tag{7}$$

where $|1_\omega\rangle = a^\dagger(\omega)|0\rangle$ denotes a single-photon state with angular-frequency ω, normalized according to $\langle 1_\omega|1_{\omega'}\rangle = \delta_{\omega\omega'}$, and where $|0\rangle$ denotes the vacuum state. We also make use of the notation $(f,a^\dagger) = (f^*,a)^\dagger \equiv \sum_\omega\sqrt{\delta\omega}f(\omega)a^\dagger(\omega)$. The state $|f\rangle$ above is an eigenstate of the number operator $\hat{N} = \sum_\omega a^\dagger(\omega)a(\omega)$, that is, a Fock state, with, of course, an eigenvalue corresponding to one particle present. Normalization of the state $|f\rangle$ for d=1 therefore corresponds to

$$\langle f\mid f\rangle = 1 = \sum_{\omega}\delta\omega f^*(\omega)f(\omega) = \int_0^\infty d\omega|f(\omega)|^2, \tag{8}$$

in the large-volume V limit. In order to make our presentation quantitative, we will, for reasons of simplicity, consider real-valued, single-photon, angular-frequency distributions (ω) such that

$$f(\omega) = \frac{1}{N}\exp\left(-\frac{(\omega-\overline{\omega})^2}{2\sigma^2}\right), \tag{9}$$

with a mean angular-frequency $\overline{\omega}$ and width σ and where the normalization constant N is given by

$$|N|^2 = \frac{\sigma\sqrt{\pi}}{2}\left(1+\frac{2}{\sqrt{\pi}}\int_0^{\overline{\omega}/\sigma}dxe^{-x^2}\right). \tag{10}$$

This choice of frequency distribution makes it possible to actually carry out all relevant expressions analytically. In obtaining the properly normalized expression Equation (9), we keep $\omega \geq 0$. It may, however, sometimes be possible to extend the range of angular frequencies to arbitrarily negative values in (8), so that $|N|^2 = \sigma\sqrt{\pi}$, with an exponential small error, which makes some of the expressions more tractable and transparent. With our choice of beam parameters below, it turns out that such an approximation plays only a minor role with regard to our numerical evaluations where we make use of $\omega \geq 0$. It is now clear that $\langle f|\vec{E}(\vec{x},t)|f\rangle \equiv 0$, that is, a single-photon state has no average electromagnetic field.

There are, however, quantum states with a nonzero expectation value of the electromagnetic field operator $\vec{E}(\vec{x},t)$ are like coherent states. Conventional coherent states $|f\rangle_c$, as expressed in terms of the single-photon distribution f, can for example, be obtained using a multimode displacement operator (see e.g., [8–10]), that is,

$$|f\rangle_c = e^{(f,a^\dagger)-(f^*,a)}|0\rangle = e^{-\langle f|f\rangle/2}e^{(f,a^\dagger)}|0\rangle \tag{11}$$

such that

$$a(\omega)|f\rangle_c = (\delta\omega)^{1/2}f(\omega)|f\rangle_c, \tag{12}$$

and hence

$${}_c\langle f|\hat{N}|f\rangle_c = \sum_\omega \delta\omega|f(\omega)|^2. \tag{13}$$

We also observe that the expectation value of the single-component electromagnetic field operator $E^{(+)}(t)$ as defined in (5) is given by

$${}_c\langle f|E^{(+)}(t)(\vec{x},t)|f\rangle_c = i\sum_\omega \delta\omega\omega^{1/2}f(\omega)e^{i\phi(t)}, \tag{14}$$

which therefore corresponds to a classical electromagnetic field. Classical interference effects as expressed in terms of such expectation values, which are nonlinearly dependent of the quantum states considered, are, of course, not directly linked to quantum-mechanical interference effects which are linearly dependent of quantum states. The use of quantum optics clarifies the connection between these views on interference effects.

The input quantum states to be considered in Figure 1, will, of course, correspond to the replacements $a(\omega)\rightarrow a_0(\omega)$ and $a(\omega)\rightarrow a_1(\omega)$, respectively in the definition of the states considered above.

Transformation in the Beam Splitter

Next, we consider a dispersionless beam splitter with frequency-independent transmittance T and reflectance R. If we assume a prefect beam splitter, where all light is either reflected or transmitted, we have $R+T=1$. The input annihilation operators $a_0(\omega)$ and $a_1(\omega)$ of the beam splitter will then transform according to (see e.g., [15–17])

$$a_2(\omega) = \sqrt{T}a_0(\omega) + i\sqrt{R}a_1(\omega),$$

$$a_3(\omega) = \sqrt{T}a_1(\omega) + i\sqrt{R}a_0(\omega), \tag{15}$$

where $a_0(\omega)$ and $a_1(\omega)$ are the LO and signal port mode annihilation operators, and $a_2(\omega)$,$a_3(\omega)$ the output annihilation mode operators corresponding to the transmitted and reflected modes, respectively. The transformation rules for the corresponding creation operators follows from (15) by Hermitian conjugation. The two independent sets of output annihilation and creation operators will then obey the required canonical commutation relations $[a_2(\omega), a_2^\dagger(\omega)] = [a_3(\omega), a_3^\dagger(\omega)] = 1$. The phase shift between the transmitted and reflected part, described by the complex numbers in (15) will play an important role below as is also the case in the famous Hong-Ou-Mandel two-photon correlation experiment [31] and related investigations (see e.g., [32–35]).

A light beam arriving at the beam splitter after being reflected in the mirrors 1 and 2 will, in general, be phase shifted, that is, expressed in terms of mode operators this process corresponds to the propagation time replacements in (5):

$$a_2(\omega)e^{i\phi(t)} \longrightarrow a_2(\omega)e^{i\phi_2(t)},$$

$$a_3(\omega)e^{i\phi(t)} \longrightarrow a_3(\omega)e^{i\phi_3(t)}, \tag{16}$$

due to difference in optical path lengths with $\phi_2 = -\omega(t-\tau_2)$, $\phi_3 = -\omega(t-\tau_3)$ in terms of time-delays τ_2 and τ_3. The reflections at the identical mirrors in the MMI setup (Figure 1) will also introduce phase-shifts, but they are equal for the two light beams and can therefore be neglected all together. A light beam passing through the beam splitter after reflection at the mirrors will, again, transform according to (15) and we therefore, finally, obtain an expression for the mode operator describing incident light on the photon detector in (5), that is,

$$\begin{aligned} a_5(\omega)e^{i\phi(t)} &\longrightarrow \sqrt{T}a_2(\omega)e^{i\phi_2(t)} + i\sqrt{R}a_3(\omega)e^{i\phi_3(t)} \\ &= a_0(\omega)\left(Te^{i\phi_2(t)} - Re^{i\phi_3(t)}\right) \\ &\quad + a_1(\omega)\left(i\sqrt{TR}e^{i\phi_2(t)} + i\sqrt{TR}e^{i\phi_3(t)}\right). \end{aligned} \tag{17}$$

For reasons of clarity, we will consider a 50/50 beam splitter, which also makes interference effects dependent only on the phases present, that is, we make the choice $T=R=1/2$. The corresponding electromagnetic field operator to be used in Glauber's theory of photon detection, is therefore given by

$$\begin{aligned} E^{(+)}(t) &= i\sum_\omega \frac{1}{2}\sqrt{\delta\omega}\omega^{1/2} \\ &\quad \times \left(a_0(\omega)\left(e^{i\phi_2(t)} - e^{i\phi_3(t)}\right)\right. \\ &\quad \left. + ia_1(\omega)\left(e^{i\phi_2(t)} + e^{i\phi_3(t)}\right)\right). \end{aligned} \tag{18}$$

INTERFERENCE OF FOCK STATES AND COHERENT STATES

Let us now specifically consider single photons in the signal and local ports, that is, the following initial Fock state as follows:

$$|\text{in}\rangle = |f_s\rangle \otimes |f_{lo}\rangle, \tag{19}$$

with single-photon angular-frequency distributions according to (9). In the numerical results to be presented, we will use the same spectral widths σ but with $\omega\rightarrow\omega s=3\sigma$ for the signal port and $\overline{\omega} \rightarrow \overline{\omega}_s = 3\sigma$ or $\overline{\omega}_{lo} = 2.85\sigma$ for the local oscillator port. We now observe that the action of the electromagnetic field operator $E^{(+)}(t)$, as given by (18), on this initial state is given by

$$E^{(+)}(t)|\text{in}\rangle = i\sum_{\omega}\frac{1}{2}\delta\omega\omega^{1/2} \times \Big(if_s(\omega)e^{i\phi_2(t)}\big(1+e^{i\omega\tau}\big)|0\rangle\otimes|f_{lo}\rangle + f_{lo}(\omega)e^{i\phi_2(t)}\big(1-e^{i\omega\tau}\big)|f_s\rangle\otimes|0\rangle\Big), \quad (20)$$

where we have made use of the fact that $\phi_3(t) - \phi_2(t) = \omega(\tau_3 - \tau_2) \equiv \omega\tau$, which now defines the optical time-delay τ. The nonnormalized state $E^{(+)}(t)|\text{in}\rangle$ describes an entangled combination of the two orthogonal state vectors $|0\rangle\otimes|f_{lo}\rangle$ and $|f_s\rangle\otimes|0\rangle$ corresponding to the absorption of one photon from the signal source or from the local oscillator, respectively. Apart from an overall normalization constant, each expansion coefficient, that is, the corresponding probability amplitude, contains information about the single-photon interference for the signal or the local oscillator photon, respectively. As we have discussed above, the probability for the detection of one photon will, according to Glauber's theory of photon detection, be proportional to the matrix element $\langle\text{in}|E^{(-)}(t)E^{(+)}(t)|\text{in}\rangle$, that is, the absolute modulus square $|E^{(+)}(t)|\text{in}\rangle|^2$ of the nonnormalized state $E^{(+)}(t)|\text{in}\rangle$. A time average over the time t according to (6) with $T_{\text{int}} \gg 1/\sigma$ leads to a Dirac $(\omega-\omega')$-function and therefore makes any double-sum over frequencies into a single-sum. In the large-volume limit and for d=1, we then obtain

$$\langle I\rangle(\tau) = \frac{1}{2T_{\text{int}}}\int_0^{\infty} d\omega 2\pi \times \Big(\omega|f_s(\omega)|^2(1+\cos\omega\tau) + \omega|f_{lo}(\omega)|^2(1-\cos\omega\tau)\Big). \quad (21)$$

The single-photon detection intensity $\langle I\rangle(\tau)$ is then a sum of two single-photon interference patterns. We notice the appearance of a relative minus sign in the local oscillator photon contribution, which has its origin in the relative phase-shift between the transmission-reflection and transmission-transmission coefficients of a beamsplitter alluded to in connection with (15) and (17). If $\overline{\omega}_s = \overline{\omega}_{lo}$ then the detector cannot distinguish between a photon from the signal or the local source with the same spectral widths σ and the two interference effects will exactly cancel against each other with the result that $\langle I\rangle(\tau)$ will be independent of τ.

If $\overline{\omega}_s, \overline{\omega}_{lo} \gg \sigma$, then, within a good numerical approximation, we can first replace the linear ω dependence in (21) with $\overline{\omega}_s$ and $\overline{\omega}_{lo}$ in front of the corresponding angular-frequency distributions, and then extend the integration to include arbitrarily negative angular-frequencies. One then finds that $\langle I\rangle(0) \simeq 2\pi\overline{\omega}_s/T_{\text{int}}$ and

$$\frac{\langle I\rangle(\tau)}{\langle I\rangle(0)} \simeq \frac{1}{2}\Big(1+\frac{\overline{\omega}_{lo}}{\overline{\omega}_s} + e^{-\sigma^2\tau^2/4}\Big(\cos\tau\overline{\omega}_s - \frac{\overline{\omega}_{lo}}{\overline{\omega}_s}\cos\tau\overline{\omega}_{lo}\Big)\Big), \quad (22)$$

where the interference effects are exponentially sensitive to the spectral width σ of the single-photon angular-distributions in a fashion similar to the spectral width dependence in the famous Hong-Ou-Mandel two-photon experiment [31]. The asymptotic value of $\langle I\rangle(\tau)/\langle I\rangle(0)$ is given by $(1+\overline{\omega}_{lo}/\overline{\omega}_s)/2$. In Figure 2, we exhibit $\langle I\rangle(\tau)/\langle I\rangle(0)$ according to (21) with the choice as in (19). With a given reference distribution $|f_{lo}(\omega)|^2$ of the local oscillator one can, for example, now in principle infer the common spectral width σ of the single-photon sources as well as the corresponding angular-frequency $\overline{\omega}_s$. A related experimental situation is actually discussed in [36] for a general single-photon state, that is, not necessarily a pure quantum state.

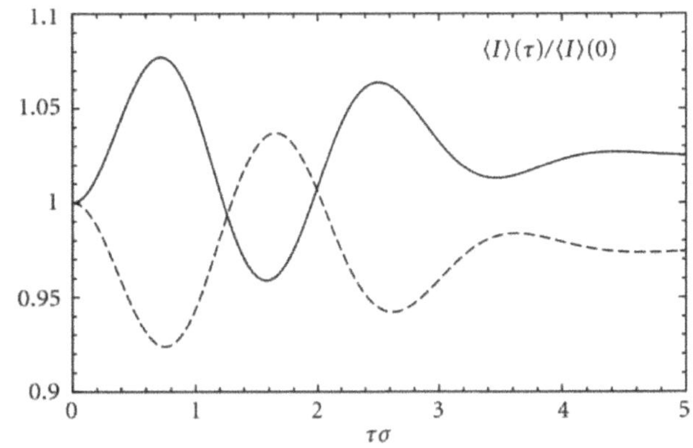

Figure 2: The normalized single-photon intensity $\langle I\rangle(\tau)/\langle I\rangle(0)$ as a function of the dimensionless time-delay $\tau\sigma$ for the case of single-photon states in the signal and the LO ports with the same spectral width σ but with different mean frequencies. Two different examples are plotted with $\overline{\omega}_s = 3\sigma$: $\overline{\omega}_{lo} = 3.15\sigma$ with (solid line) or $\overline{\omega}_{lo} = 2.85\sigma$ (dashed line). The asymptotic values of $\langle I\rangle(\tau)/\langle I\rangle(0)$ can be obtained from the expression (22) in the main text, that is, $(1+\overline{\omega}_{lo}/\overline{\omega}_s)/2$.

In the case of coherent states in the signal and local ports, with single-particle state angular-frequency distributions $f(\omega)$ and $f_{lo}(\omega)$ as above, (20) is now modified according to

$$E^{(+)}(t)|\text{in}\rangle = i\sum_{\omega}\frac{1}{2}\delta\omega\omega^{1/2} \times \Big(if_s(\omega)e^{i\phi_2(t)}\big(1+e^{i\omega\tau}\big) + f_{lo}(\omega)e^{i\phi_2(t)}\big(1-e^{i\omega\tau}\big)\Big)|f_s\rangle_c\otimes|f_{lo}\rangle_c, \quad (23)$$

which is not an entangled state. This is so since the removal of a photon from a coherent state does not change the state since it actually contains infinitely many photons, that is, $a(\omega)|f\rangle_c = (\delta\omega)^{1/2}f(\omega)|f\rangle_c$ according to (12). The intensity $\langle I\rangle(\tau)$ according to (21) is, for real-valued single-photon distributions $f_s(\omega)$ and $f_{lo}(\omega)$, therefore replaced by

$$\langle I\rangle(\tau) = \frac{1}{2T_{\text{int}}}\int_0^{\infty} d\omega\omega 2\pi\Big(|f_s(\omega)|^2(1+\cos\omega\tau) + |f_{lo}(\omega)|^2(1-\cos\omega\tau) - 2f_s(\omega)f_{lo}(\omega)\sin\omega\tau\Big), \quad (24)$$

which now contains an additional interference contribution as compared with (21). In Figure 3, we exhibit $\langle I\rangle(\tau)/\langle I\rangle(0)$ with coherent states generated by the same choice of single-photon states as in Figure 2. The interference pattern is, as in Figure 2, sensitive to the actual angular-frequency distributions. With a given reference distribution $|f_{lo}(\omega)|^2$ of the local oscillator, one can now infer the common spectral width σ of the coherent state sources as well as the corresponding angular-frequency $\overline{\omega}_s$. By a comparison with Figure 2, we conclude that the MMI setup is sensitive to the actual form of the initial quantum states despite the fact that we are only considering single-photon detection processes. With the same approximations leading to (22), we find an additional contribution to $\langle I\rangle(\tau)/\langle I\rangle(0)$, that is,

$$\frac{\langle I\rangle(\tau)}{\langle I\rangle(0)} \simeq \frac{1}{2}\left(\left(1+\frac{\overline{\omega}_{lo}}{\overline{\omega}_s}\right)\left(1-e^{-\delta\overline{\omega}^2/4\sigma^2-\sigma^2\tau^2/4}\sin\overline{\omega}\tau\right) + e^{-\sigma^2\tau^2/4}\left(\cos\tau\overline{\omega}_s-\frac{\overline{\omega}_{lo}}{\overline{\omega}_s}\cos\tau\overline{\omega}_{lo}\right)\right), \quad (25)$$

with $\delta\overline{\omega} \equiv (\overline{\omega}_s-\overline{\omega}_{lo})$ and $\overline{\omega} \equiv (\overline{\omega}_{lo}+\overline{\omega}_s)/2$. In particular, and more important, we notice that when $\overline{\omega} \equiv \overline{\omega}_{lo} = \overline{\omega}_s$ we can write

$$\frac{\langle I\rangle(\tau)}{\langle I\rangle(0)} \simeq 1-e^{-\sigma^2\tau^2/4}\sin\tau\overline{\omega}, \quad (26)$$

which now explicitly depends on τ in contrast to the initial Fock state considerations above in which case $\langle I\rangle(\tau)\simeq\langle I\rangle(0)$.

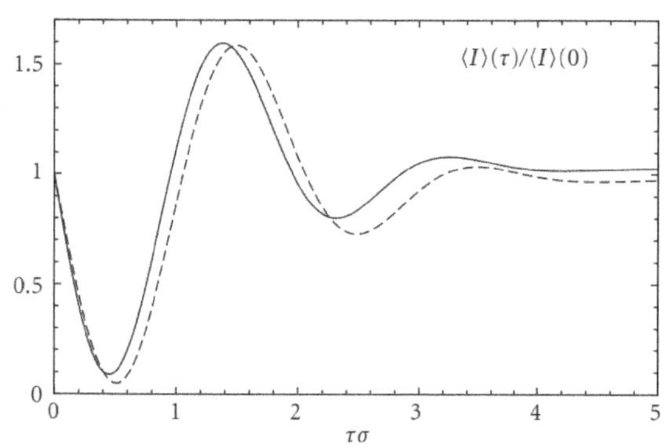

Figure 3: The normalized intensity $\langle I\rangle(\tau)/\langle I\rangle(0)$ as a function of the time-delay $\tau\sigma$ for the case of coherent states as generated by the same single-photon states as in Figure 2. Two different examples are plotted with $\overline{\omega}_s = 3\sigma$: $\overline{\omega}_{lo} = 3.15\sigma$ with (solid line) or $\overline{\omega}_{lo} = 2.85\sigma$ (dashed line). The asymptotic values of $\langle I\rangle(\tau)/\langle I\rangle(0)$ are as in Figure 2.

INTERFERENCE OF THERMAL LIGHT IN THE MICHELSON-MORLEY INTERFEROMETER

Thermal Light

For the readers convenience, let us first outline a simple and quantum mechanical description of thermal black body radiation at an absolute temperature T. Black body radiation is, basically, light with random phases. For a single mode with angular-frequency ω and if $|n\rangle = a^\dagger(\omega)^n|0\rangle/\sqrt{n!}$ denotes a n-photon state, the density matrix describing the thermal light is given by

$$\rho(\omega) = \sum_{n=0}^{\infty} p_n(\omega)|n_\omega\rangle\langle n_\omega|, \quad (27)$$

in terms of the Bose-Einstein distribution:

$$p_n(\omega) = \left(\frac{\overline{n}(\omega,T)}{1+\overline{n}(\omega,T)}\right)^n \frac{1}{1+\overline{n}(\omega,T)}, \quad (28)$$

with

$$\overline{n}(\omega,T) = \frac{1}{\exp(\hbar\omega/k_BT)-1}. \quad (29)$$

The state (ω) corresponds to an extreme value of the von Neumann entropy S, that is,

$$S = -k_BT\ \mathrm{Tr}[\rho(\omega)\ln\rho(\omega)], \quad (30)$$

subjected to the constraints (see e.g., [13, 22]):

$$\mathrm{Tr}[\rho(\omega)] = 1,$$
$$\mathrm{Tr}\left[a^\dagger(\omega)a(\omega)\rho(\omega)\right] = \overline{n}(\omega,T). \quad (31)$$

The random, or chaotic, nature of the quantum state (ω) corresponds to a phase-independent Glauber-Sudarshan $\mathcal{P}(\alpha$-representation (see e.g., [37–42]) in terms of a coherent state, that is,

$$\rho(\omega) = \int d^2\alpha\mathcal{P}(\alpha)|\alpha\rangle\langle\alpha|, \quad (32)$$

using a single-mode coherent state $|\alpha\rangle = \exp(\alpha a^\dagger(\omega) - \alpha^* a(\omega))|0\rangle$. For thermal light, one finds that

$$\mathcal{P}(\alpha) = \frac{1}{\pi\overline{n}(\omega,T)}\exp\left(-\frac{|\alpha|^2}{\overline{n}(\omega,T)}\right), \quad (33)$$

which obeys the normalization condition:

$$\mathrm{Tr}[\rho(\omega)] = \int d^2\alpha\mathcal{P}(\alpha) = 1, \quad (34)$$

as well as

$$\mathrm{Tr}\left[a^\dagger(\omega)a(\omega)\rho(\omega)\right] = \int d^2\alpha P(\alpha)|\alpha|^2 = \overline{n}(\omega,T). \quad (35)$$

A multimode system at thermal equilibrium is then described in terms of a tensor product $\rho(T) = \otimes_\omega\rho(\omega)$, where we have performed the replacement $\alpha\rightarrow\alpha(\omega)$ in (32) and (33). The Glauber-Sudarshan (α)-representation for the state $\rho(T)$ is now, in particular, useful in our considerations since the response in a single-photon detector can be obtained immediately from the previous results for coherent light using an average procedure.

Thermal Light in the Signal Port and Vacuum in the Local Oscillator Port

We are now in the position to consider a density matrix describing a system where we have thermal light in the signal port and vacuum in the LO port, that is, the initial density matrix ρ_{in} of the total system is given by

$$\rho_{\text{in}} = \rho(T)\otimes(|0\rangle\langle 0|)_{lo}. \quad (36)$$

By making use of [25], with $f_s(\omega) \rightarrow \alpha(\omega)$ and $f_{lo}(\omega) \rightarrow 0$, , and then performing an average over $\alpha(\omega)$according to (35), we immediately obtain the following result:

$$\langle I\rangle(\tau) = \frac{1}{2T_{\text{int}}}\int_0^\infty d\omega 2\pi\omega^d \overline{n}(\omega,T)(1+\cos\omega\tau), \tag{37}$$

in d space dimensions.

In passing, we notice the important fact that (37) is, in fact, valid for any physical quantum state of the form (36), where $\rho(T)$ is replaced with any density matrix $\rho(\omega)$ not necessarily thermal, due to the generality of the Glauber-Sudarshan $P(\alpha)$-representative equation (32). The interference effects as exhibited by the MMI setup, therefore, actually only depends on the, in general, angular-frequency dependent mean number $\text{Tr}[a^\dagger(\omega)a(\omega)\rho(\omega)] = n(\omega)$ and not on other features of the actual quantum state $\rho(\omega)$.

By a straightforward change of the variable of integration in (37), and with $a \equiv \tau k_B T/\hbar$, we then find for a thermal quantum state $\rho(T)$ that

$$\frac{\langle I\rangle(\tau)}{\langle I\rangle(0)} = \frac{1}{2}\left[1+\frac{1}{J(d)}\int_0^\infty dx\left(\frac{x^d\cos(ax)}{\exp(x)-1}\right)\right], \tag{38}$$

where $J(d)$ can be expressed in terms of gamma and Riemann's ζ functions, that is,

$$J(d) = \Gamma(1+d)\zeta(d+1) = \Gamma(1+d)\sum_{n=1}^{\infty}\frac{1}{n^{d+1}}. \tag{39}$$

Particular values are (1)=$\pi^2/6$ and $J(3)=\pi^4/15$. In the case of d=3, we, therefore, recover the well-known expression for $\langle I\rangle(\tau)/\langle I\rangle(0)$ [22] as also discussed in [21]. In, for example, d=3 it is actually possible to carry out the relevant integral in the expression equation (38) for $\langle I\rangle(\tau)/\langle I\rangle(0)$ analytically with the result

$$\frac{\langle I\rangle(\tau)}{\langle I\rangle(0)} = \frac{1}{2}\left[1+15\left(\frac{2+\cosh(2a\pi)}{\sinh(a\pi)^4}-\frac{3}{(a\pi)^4}\right)\right]. \tag{40}$$

Equation (40) shows that interference effects have a power-law sensitivity for larger a. In Figure 4, we show $\langle I\rangle(\tau)/\langle I\rangle(0)$ for varying values of a in the case with d=3 and one infers a characteristic coherence length lc of thermal light in the MMI of the form:

$$l_c \simeq 1.5\frac{\hbar c}{k_B T}, \tag{41}$$

as also discussed in [21].

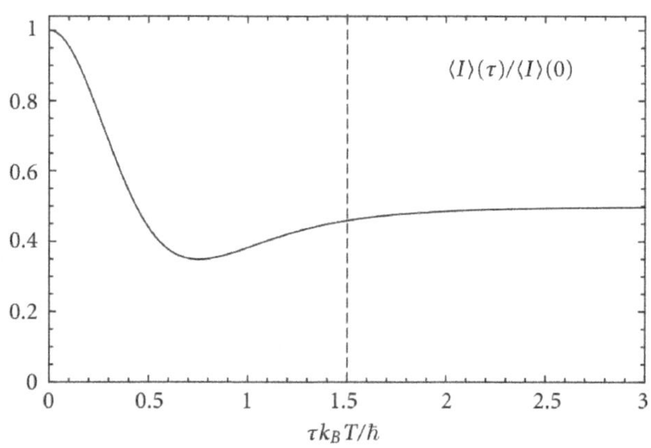

Figure 4: The normalized intensity $\langle I\rangle(\tau)/\langle I\rangle(0)$ as a function of $a \equiv \tau k_B T_0/\hbar$ for the case of thermal light in the signal port and vacuum in the LO port and for d=3. We infer a characteristic thermal coherence time τc in terms of $a\simeq1.5$ (dashed vertical line), that is, $\tau_c \simeq 1.5\hbar/k_B T$.

THERMAL LIGHT IN THE SIGNAL AND LOCAL PORTS

As we have seen above, with a vacuum in the LO port and with a signal thermal source the single-photon detection process exhibits an interference pattern. We now investigate what happens if we have thermal light with a temperature T_0 in the LO port and thermal light with temperature $T1$ in the signal port. For this setup, the corresponding initial density matrix becomes

$$\rho_{\text{in}} = \rho(T_1)\otimes\rho(T_0). \tag{42}$$

By making use of this density matrix as well as the same methods as described in Section 4.2 by performing independent averages over $f_s(\omega)$ and $f_{lo}(\omega)$ in (24) according to (35), we immediately obtain, in the large-volume limit, the following result:

$$\langle I\rangle(\tau) = \frac{1}{2T_{\text{int}}}\int_0^\infty d\omega\omega^d 2\pi(\overline{n}(\omega,T_1)(1+\cos\omega\tau) + \overline{n}(\omega,T_0)(1-\cos\omega\tau)). \tag{43}$$

In obtaining this result, we notice that the last term in (24), suitably extended to complex-valued $f_s(\omega)$ and $f_{lo}(\omega)$, averages to zero due to the chaotic nature, that is, phase-independence of thermal light according to (32) and (33). The relative intensity then takes the form (d=3) as follows:

$$\frac{\langle I\rangle(\tau)}{\langle I\rangle(0)} = \frac{1}{2}\left(1+\left(\frac{T_0}{T_1}\right)^4\right) + \frac{15}{2\pi^4}\int_0^\infty dx\frac{x^3}{\exp(x)-1} \times\left(\cos(a_1x)-\left(\frac{T_0}{T_1}\right)^4\cos(a_0x)\right), \tag{44}$$

where $a_0 = \tau k_B T_0/\hbar$ and $a_1 = \tau k_B T_1/\hbar$. The integrals in (44) can, again, be solved analytically in a fashion similar to the integral in (38), that is,

$$\frac{\langle I\rangle(\tau)}{\langle I\rangle(0)} = \frac{1}{2}\left(1+\left(\frac{T_0}{T_1}\right)^4\left(1-15\frac{2+\cosh(2a_0\pi)}{\sinh(a_0\pi)^4}\right)\right) + \frac{15}{2}\left(\frac{2+\cosh(2a_1\pi)}{\sinh(a_1\pi)^4}\right). \tag{45}$$

Due to the presence of hyperbolic functions in (45), we observe that $\langle I\rangle(\tau)/\langle I\rangle(0)$ approaches its asymptotic value $(1+(T_0/T_1)^4)/2$ exponentially fast as a function of the time-delay τ in contrast to the power-law dependence in (40). It is now of interest to study the behavior of (45) when the temperatures T_0 and T_1 of the local oscillator and signal, respectively are varied. In Figure 5, we exhibit the interference when T_0 is slightly smaller or larger than T_1 as a function of the parameter $\tau k_B T_1/\hbar$ for d=3. The corresponding interference, of course, disappears when the two temperatures are equal. The sensitivity of the interference pattern with regard to the difference in temperatures of the source and the reference temperature, that is, of the local source, constitutes the basic ingredient of the FIRAS setup. We also observe that the coherence length c for this MMI setup is roughly the same as in Section 4.2 since the two temperatures of the source considered are close to each other.

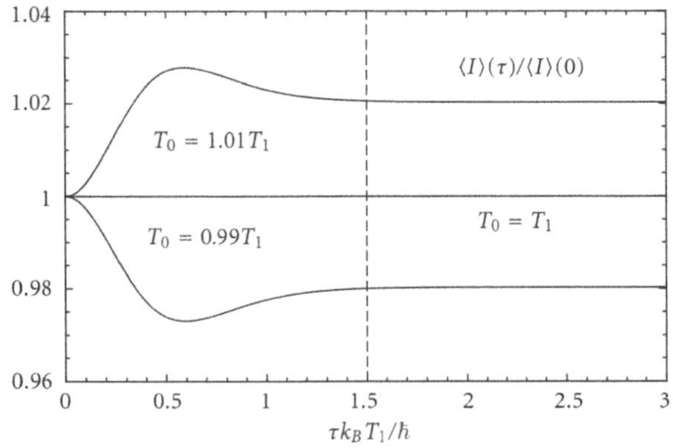

Figure 5: The same as in Figure 4 but with thermal light in both ports with T_1=1.01T_0(upper curve), T_1=0.99T_0 (lower curve), and T_1=T_0 (solid line). The vertical dashed line corresponds to the thermal coherence length $l_c = c\tau_c$ with $\tau_c k_B T_1/\hbar \simeq 1.5$.

CONCLUSIONS AND FINAL REMARKS

We have seen that the interference of thermal light in the Michelson-Morley interferometer can be described, in a straightforward manner, by making use of Glauber's theory of photon detection and elementary quantum theory of the electromagnetic field. Furthermore, we have seen the emergence of a natural coherence length $l_c \simeq 1.5\hbar c/k_B T$ of thermal light in the MMI. The result for thermal light in both the signal and the local oscillator ports shows that the interference pattern is sensitive to the difference in temperature of the two sources. This is the basic principle used by the FIRAS on board the COBE satellite in order to perform high-precision measurements of the temperature and the spectrum of the cosmic micro-wave background radiation.

It may first appear surprising that nonmonochromatic and chaotic/thermal light, with random phases, exhibits interference effects since it can be argued that a quantum state with a fixed number of photons has an undetermined "quantum phase" (see e.g., [43–48]). A mixed quantum state of such photon-number eigenstates therefore have no intrinsic and well-defined phase. As we, however, have seen the interference effects are naturally obtained in terms of the phase-dependent normal-mode expansion of the quantized electromagnetic field and its corresponding timeevolution rather than in terms of a Schrödinger picture and possible phases of quantum states (see e.g., the comments in [49]).

Since we have been considering initial quantum states in terms of a fixed number of photons as well as "classical" states, corresponding to coherent states with an infinite number of photons present, a quantum-mechanical language is mandatory. The signal and local ports of the MMI setup corresponds to independent input sources. It is, of course, a well-known experimental fact that independent photon sources can give rise to interference effects (see e.g., [50–55]). Despite the fact that such interference effects are well established, the interpretation of them can, nevertheless, gives rise to interesting issues regarding the very fundamental aspects of the quantum-mechanical world (see e.g., [56]) when considering, in particular, interference effects using single-photon sources.

We have seen that for multimode systems the quantum nature of these independent sources actually affects the nature of the single-photon intensity measurements. We have already mentioned that the angular-frequency distribution of a single photon can be measured using a similar experimental setup as the MMI considered in the present paper [36]. With a vacuum state in the local oscillator port and a single-photon signal angular-frequency distribution $f(\omega)$ of the form considered in (9), one finds, using (21), that$\langle I\rangle(\tau)\simeq 1\langle I\rangle$(0)2□1+$e-\sigma 2\tau 2/4\cos\tau\omega s$□.(46)We conclude by noticing that a characteristic exponential behavior as in (46) for single-photon interference has recently been observed by measuring the photoluminescence signal of a single quantum dot [57] using a Michelson-Morley interferometer (for a related study of interference effects of dissimilar photon sources, see [58]).

ACKNOWLEDGMENTS

This research has made use of NASA's Astrophysics Data System. The research was supported by the Research Council of Norway through Grants 170935/V30 and FRINAT-191564, and by NTNU. One of the authors (B.-S. Skagerstain) wishes to thank Professor F. G. Scholtz for a generous and stimulating hospitality during a joint NITheP and Stias, Stellenbosch (SA), workshop in 2009, Professors M. Reid and D. V. Ahluwalia, University of Christchurch, NZ, J. R. Klauder, University of Florida, Gainesville, USA, P. S. Riseborough, Temple University, USA., the TH Division at CERN, and the Centre for Advanced Study (CAS), Oslo, for hospitality when the present work was in progress.

REFERENCES

1. Nobel Prize Website, http://www.nobelprize.org/.
2. D. J. Fixsen and J. C. Mather, "The spectral results of the far-infrared absolute spectrophotometer instrument on COBE," Astrophysical Journal Letters, vol. 581, no. 2, pp. 817–822, 2002.
3. J. C. Mather, D. J. Fixsen, and R. A. Shafer, "Design for the COBE far-infrared absolute spectrophotometer," in Society of Photo-Optical Instrumentation Engineers (SPIE) Conference Series, M. S. Scholl, Ed., vol. 2019 of Proceedings of SPIE, pp. 168–179, 1993.
4. J. C. Mather, "The cosmic background explorer (COBE)," Optical Engineering, vol. 21, no. 4, pp. 769–774, 1982.
5. J. C. Mather, Section 2.8 in Questions of Modern Cosmology—Galileo's Legacy, Edited by M. D'Onofrio and C.Burigana, Springer, Berlin, Germany, 2009.
6. R. J. Glauber, "The quantum theory of optical coherence," Physical Review, vol. 130, no. 6, pp. 2529–2539, 1963.
7. R. J. Glauber, "Optical coherence and photon statistics," in Quantum Optics and Electronics, C. DeWitt, A. Blandin, and C. Cohen-Tannoudji, Eds., Les Houches, p. 621, Gordon and Breach, New York, NY, USA, 1965.
8. J. R. Klauder and B.-S. Skagerstam, Coherent States—Applications in Physics and Mathematical Physics, World Scientific, Singapore, 1985, (2nd edition in

Chinese, Beijing, 1986).

9. B.-S. Skagerstam, "Coherent states—some applications in quantum field theory and particle physics," inCoherent States: Past, Present, and the Future, D. H. Feng, J. R. Klauder, and M. R. Strayer, Eds., World Scientific, Singapore, 1994.
10. J. R. Klauder, "The current state of coherent states," in Proceedings of the the 7th ICSSUR Conference, Boston, Mass, USA, 2001.
11. J. R. Klauder and E. C. G. Sudarshan, Fundamentals of Quantum Optics, W.A. Benjamin, New York, NY, USA, 1968, and Dover, New York, NY, USA, 2006.
12. L. Mandel and E. Wolf, Optical Coherence and Quantum Optics, Cambridge University Press, Cambridge, UK, 1995.
13. M. O. Scully and M. S. Zubairy, Quantum Optics, Cambridge University Press, Cambridge, UK, 1997.
14. B. E. S. Saleh and M. C. Teich, Fundamentals of Photonics, Wiley, New York, NY, USA, 2nd edition, 2007.
15. B. Yurke, S. L. McCall, and J. R. Klauder, "SU(2) and SU(1,1) interferometers," Physical Review A, vol. 33, no. 6, pp. 4033–4054, 1986.
16. Z. Y. Ou and L. Mandel, "Derivation of reciprocity relations for a beam splitter from energy balance,"American Journal of Physics, vol. 57, pp. 66–67, 1989.
17. R. A. Campos, B. E. A. Saleh, and M. C. Teich, "Quantum-mechanical lossless beam splitter: SU(2) symmetry and photon statistics," Physical Review A, vol. 40, no. 3, pp. 1371–1384, 1989.
18. L. Mandel, "Quantum effects in one-photon and two-photon interference," Reviews of Modern Physics, vol. 71, no. 2, pp. S274–S282, 1999.
19. A. Zeilinger, "Experiment and the foundations of quantum physics," Reviews of Modern Physics, vol. 71, no. 2, pp. S288–S297, 1999.
20. A. Zeilinger, G. Weihs, T. Jennewein, and M. Aspelmeyer, "Happy centenary, photon," Nature, vol. 446, no. 7133, p. 342, 2007.
21. A. Donges, "The coherence length of black-body radiation," European Journal of Physics, vol. 19, no. 3, pp. 245–249, 1998.
22. L. Mandel and E. Wolf, "Coherence properties of optical fields," Reviews of Modern Physics, vol. 37, no. 2, pp. 231–287, 1965.
23. B.-S. Skagerstam, "Localization of massless spinning particles and the berry phase," in On Klauders Path: A Field Trip—Festschrift for John R. Klauder on Occasion of His 60th Birthday, G. G. Emch, G. C. Hegerfeldt, and L. Streit, Eds., pp. 209–222, World Scientific, 1994.
24. D. F. Walls, "A simple field theoretic description of photon interference," American Journal of Physics, vol. 45, pp. 952–956, 1977.
25. P. Grangier, G. Roger, and A. Aspect, "Experimental evidence for a photon anticorrelation effect on a beam splitter: a new light on single-photon interference," Europhysics Letters, vol. 1, no. 4, pp. 173–179, 1986.
26. A. Aspect and P. Grangier, "Wave-particle duality for single photons," Hyperfine Interactions, vol. 37, no. 1–4, pp. 1–17, 1987.
27. T. J. Herzog, J. G. Rarity, H. Weinfurter, and A. Zeilinger, "Frustrated two-photon creation via interference," Physical Review Letters, vol. 72, no. 5, pp. 629–632, 1994.
28. G. Scarcelli, A. Valencia, and Y. Shih, "Experimental study of the momentum correlation of a pseudothermal field in the photon-counting regime," Physical Review A, vol. 70, no. 5, Article ID 051802, 4 pages, 2004.
29. L. J. Wang, X. Y. Zou, and L. Mandel, "Induced coherence without induced emission," Physical Review A, vol. 44, no. 7, pp. 4614–4622, 1991.
30. P. W. Milonni, H. Fearn, and A. Zeilinger, "Theory of two-photon down-conversion in the presence of mirrors," Physical Review A, vol. 53, no. 6, pp. 4556–4566, 1996.
31. C. K. Hong, Z. Y. Ou, and L. Mandel, "Measurement of subpicosecond time intervals between two photons by interference," Physical Review Letters, vol. 59, no. 18, pp. 2044–2046, 1987.
32. Y. H. Shih and C. O. Alley, "New type of einstein-podolsky-rosen-bohm experiment using pairs of light quanta produced by optical parametric down conversion," Physical Review Letters, vol. 61, no. 26, pp. 2921–2924, 1988.
33. M. A. Steinberg, P. G. Kwiat, and R. Y. Chiao, "Measurement of the single-photon tunneling time,"Physical Review Letters, vol. 71, no. 5, pp. 708–711, 1993.
34. P. G. Kwiat, A. M. Steinberg, and R. Y. Chiao, "Observation of a quantum eraser: a revival of coherence in a two-photon interference experiment," Physical Review A, vol. 45, no. 11, pp. 7729–7739, 1992.
35. D. V. Strekalov, T. B. Pittman, and Y. H. Shih, "What we can learn about single photons in a two-photon interference experiment," Physical Review A, vol. 57, no. 1, pp. 567–570, 1998.
36. W. Wasilewski, P. Kolenderski, and R. Frankowski, "Spectral density matrix of a single photon measured," Physical Review Letters, vol. 99, no. 12, Article ID 123601, 4 pages, 2007.
37. E. C. G. Sudarshan, "Equivalence of semiclassical and quantum mechanical descriptions of statistical light beams," Physical Review Letters, vol. 10, no. 7, pp. 277–279, 1963.
38. R. J. Glauber, "Coherent and incoherent states of the radiation field," Physical Review, vol. 131, no. 6, pp. 2766–2788, 1963.
39. J. R. Klauder, J. McKenna, and D. G. Currie, "On "Diagonal" coherent-state representations for quantum-mechanical density matrices," Journal of Mathematical Physics, vol. 6, pp. 734–739, 1965.
40. C. L. Metha and E. C. G. Sudarshan, "Relation between quantum and semi-classical description of optical coherence," Physical Review, vol. 138, no. 1, pp. B274–B280, 1965.

41. J. R. Klauder, "Improved version of optical equivalence theorem," Physical Review Letters, vol. 16, no. 12, pp. 534–536, 1966.
42. J. R. Klauder and B. S. K. Skagerstam, "Generalized phase-space representation of operators," Journal of Physics A, vol. 40, no. 9, pp. 2093–2105, 2007
43. L. Susskind and J. Glogower, "Quantum mechanical phase and time operator," Physics, vol. 1, pp. 49–61, 1964.
44. S. M. Barnett and D. T. Pegg, "Phase in quantum optics," Journal of Physics A, vol. 19, pp. 3849–3862, 1986.
45. S. M. Barnett and D. T. Pegg, "On the hermitian optical phase operator," Journal of Modern Optics, vol. 36, no. 1, pp. 7–19, 1989.
46. S. M. Barnett and D. T. Pegg, "Phase properties of the quantized single-mode electro-magnetic field,"Physical Review A, vol. 39, no. 4, pp. 1665–1675, 1989.
47. S. M. Barnett and D. T. Pegg, "Quantum optical phase," Journal of Modern Optics, vol. 44, pp. 225–264, 1997.
48. Å. Bjørn and B.-S. Skagerstam, "On the quantum phase operator for coherent states," Physica Scripta, vol. 70, no. 1, pp. 26–32, 2004.
49. J. Xiong, D. Z. Cao, F. Huang, H.-G. Li, X.-J. Sun, and K. Wang, "Experimental observation of classical subwavelength interference with a pseudothermal light source," Physical Review Letters, vol. 94, no. 17, Article ID 173601, 2005.
50. G. Magyar and L. Mandel, "Interference fringes produced by superposition of two independent maser light beams," Nature, vol. 198, no. 4877, pp. 255–256, 1963.
51. R. L. Pfleegor and L. Mandel, "Further experiments on interference of independent photons beams at low light levels," JOSA, vol. 58, pp. 946–950, 1968.
52. L. Mandel, "Photon interference and correlation effects produced by independent quantum sources,"Physical Review A, vol. 28, no. 2, pp. 929–943, 1983.
53. H. Paul, "Interference between independent photons," Reviews of Modern Physics, vol. 58, no. 1, pp. 209–231, 1986.
54. F. Louradour, F. Reynaud, B. Colombeau, and C. Froehly, "Interference fringes between two separate lasers," American Journal of Physics, vol. 61, pp. 242–245, 1993.
55. R. K. Kaltenbaek, B. Blauensteiner, M. Zukowski, M. As-Pelmeyer, and A. Zeilinger, "Experimental interference of independent photons," Physical Review Letters, vol. 96, no. 24, Article ID 240502, 4 pages, 2006. ·
56. R. J. Glauber, "Diracs's fameous dictum on the interference: one photon or two photons?" American Journal of Physics, vol. 63, no. 12, 1995.
57. C. Kammerer, G. Cassabois, C. Voisin et al., "Interferometric correlation spectroscopy in single quantum dots," Applied Physics Letters, vol. 81, no. 15, pp. 2737–2739, 2002.
58. A. J. Bennett, R. B. Patel, C. A. Nicoll, D. A. Ritchie, and A. J. Shields, "Interference of dissimilar photon sources," Nature Physics, vol. 5, no. 10, pp. 715–717, 2009.

CHAPTER 13 RESOLVED-SIDEBAND LASER COOLING IN A PENNING TRAP

J. F. Goodwin[1], G. Stutter[1], R. C. Thompson[1], and D. M. Segal[1]

[1]Quantum Optics and Laser Science, Blackett Laboratory, Imperial College London, Prince Consort Road, London SW7 2AZ, United Kingdoma

ABSTRACT

We report the laser cooling of a single $^{40}Ca^+$ ion in a Penning trap to the motional ground state in one dimension. Cooling is performed in the strong binding limit on the 729-nm electric quadrupole $S_{1/2}\leftrightarrow D_{5/2}$ transition, broadened by a quench laser coupling the $D_{5/2}$ and $P_{3/2}$ levels. We find the final ground-state occupation to be 98(1)%. We measure the heating rate of the trap to be very low with $\dot{\bar{n}} \approx 0.3(2)\ \mathrm{s}^{-1}$ for trap frequencies from 150–400 kHz, consistent with the large ion-electrode distance.

INTRODUCTION

Cold, trapped ions are one of the leading systems with which to study processes that require excellent environmental isolation, including quantum computation [1], quantum simulation [2], and metrology [3,4]. Penning traps [5,6] are most widely used for precision measurements on fundamental particles and atomic ions [7–9] and have also found applications in quantum information[2,10,11]. Unlike radio frequency (rf) traps, Penning traps require no oscillating fields, making them suitable for trapping 2D and 3D Coulomb crystals [12] and ions in states sensitive to rf perturbation (e.g., Rydberg ions [13,14]).

Citation: J. F. Goodwin, G. Stutter, R.C. Thompson, and D.M. Segal, Resolved-Sideband Laser Cooling in a Penning Trap, DOI: 10.1103/PhysRevLett.116.143002

Many ion trap experiments require motional ground-state confinement, typically achieved via resolved-sideband cooling. Sideband cooling was first demonstrated in rf traps many years ago[15,16] but, due to the technical complexities associated with Penning traps, had not yet been realized in this system.

Ground-state cooling is of particular importance to quantum gates, and recent years have seen the application of these gates to precision measurement [17–19]. The ability to apply quantum logic spectroscopy to ions in Penning traps will greatly increase the precision of experiments that necessitate their use [20,21]. Coherent control of the motional state also underpins many experiments in quantum thermodynamics [22,23], a field where the very low heating rates achievable in Penning traps offer a distinct advantage.

In this Letter, we demonstrate the application of resolved optical sideband cooling to the Penning trap, cooling the axial motion of a calcium ion to its quantum ground state with

98% probability. We demonstrate our ability to coherently manipulate the ion's electronic state by observing its Rabi dynamics. Finally, we measure the ion heating rate, which we find to be the lowest reported in the literature to date for any type of trap [24]. The low heating rate is consistent with the large trap size, which has a characteristic dimension of d_0=1.32 cm, and distance to the nearest electroded=1.08 cm.

We have previously reported work on resolved-sideband spectroscopy of an ion in a Penning trap[26]. The experiments described here use a modified version of the same apparatus. We trap a $^{40}Ca^+$ ion in a Penning trap consisting of a stack of cylindrical electrodes, held in a 1.85 T axial field provided by a superconducting solenoid magnet. Doppler cooling in the axial and radial directions is performed using two lasers at 397 nm, tuned to two components of the $S_{1/2}\leftrightarrow P_{1/2}$ transition, with four laser frequencies around 866 nm applied along the trap axis to repump population in the $D_{3/2}$ states. At high magnetic fields, j-state mixing [27] provides a small branching ratio of $4.2\times10^{-7}B^2/T^2$ to the $D_{5/2}$ manifold, necessitating four additional repump laser frequencies around 854 nm. A laser at 729 nm, addressing the electric quadrupole $S_{1/2}\leftrightarrow D_{5/2}$ transition, is used for resolved-sideband spectroscopy via the electron shelving technique. Details of the trap, lasers, and spectroscopy scheme can be found in [26,28,29].

We have made two major changes to the experiment to enable us to perform sideband cooling. The first is an increase in the 729-nm laser power using a tapered amplifier. This increases the power at the ion from 4 to 40 mW, providing Rabi frequencies of up to $\Omega_0/2\pi\approx50$ kHz.

The second change is to introduce an oscillating quadrupolar "axialization" field [30] coupling the magnetron and modified cyclotron radial trap modes. Our previous work [26] was performed without any oscillating fields, using a radial cooling beam with an intensity gradient across the trap centre to cool the otherwise-unstable magnetron motion [31]. For higher axial frequencies, the intensity gradient necessary for effective cooling increases, and with our current optical system we are unable to reliably cool ions above axial frequencies of 200 kHz. The axialization technique [32], which had been used by the group in earlier laser cooling experiments [33–35], works at all trap frequencies but has the disadvantage of introducing an rf electric field and associated micromotion. Fortunately, the rf potential required for axialization is several orders of magnitude lower than that used for ponderomotive trapping in Paul traps, and in this experiment never exceeds 50 mV. For a typical trapping potential of 200 V, the micromotion amplitude is expected to be <10 nm for an ion situated10 μm from the rf null.

The ion is Doppler cooled, before one of the two 397-nm lasers is switched off to optically pump population into the $S_{1/2}(m_j=-\frac{1}{2})$ sublevel. For the axial trap frequencies used in this experiment ($\omega/2\pi$=150–400 kHz), the ion remains outside the Lamb-Dicke regime [after Doppler cooling. For the result presented below, at $\omega/2\pi$=389 kHz, this does not prevent efficient sideband cooling on the first red sideband. However, approximately 0.17% of the population is initially in oscillator states higher than n=150. As the first red sideband Rabi frequency is very close to zero for this state, any population cooled from higher phonon states will accumulate in a narrow distribution of states immediately above this level. To prevent such Fock-state population-trapping it is necessary to alternate cooling on the first-order and a higher-order sideband, as demonstrated by Poulsen *et al.* [36].

Sideband cooling is therefore performed by applying the 729-nm laser alternately to the first and second red sidebands of the $S_{1/2}(m_j=-\frac{1}{2})\leftrightarrow D_{5/2}(m_j=-\frac{3}{2})$ transition. The scattering rate is increased by using a weak, 854-nm quench laser to empty the $D_{5/2}$ level via $P_{3/2}$, which rapidly decays at 393-nm to $S_{1/2}$. For small quench laser saturation parameters, the $P_{3/2}$ level can be adiabatically eliminated and the system behaves like a two level system with a linewidth set by the properties of the quench laser and upper level [37]. The effective $D_{5/2}$ linewidth is measured spectroscopically and the quench laser intensity is adjusted to give $\tilde{\Gamma}/2\pi\approx50$ kHz. The second 397-nm laser and 866-nm repump lasers are applied continuously to ensure that population decaying on the $P_{3/2}\leftrightarrow S_{1/2}$ transition is optically pumped into the correct $(m_j=-\frac{1}{2})$ ground-state sublevel. After sideband cooling, electron shelving spectroscopy is performed as described in [26].

In the limit $\Omega0\ll\omega$, for the cooling cycle used in this experiment, the sideband cooling limit is given by [38]

$$\bar{n}_{\text{lim}}=\left(\frac{\tilde{\Gamma}}{2\omega}\right)^2\left[\frac{\tilde{\eta}^2+\eta_q^2+\eta_r^2+\tilde{\eta}_r^2}{\eta^2}+\frac{1}{4}\right], \tag{1}$$

where η and $\tilde{\eta}$ are Lamb-Dicke parameters associated with absorption on the 729-nm transition and emission on the 393-nm transition, η_q^2 is the Lamb-Dicke parameter associated with absorption at 854 nm from the quench laser, and η_r and $\tilde{\eta}_r$ are Lamb-Dicke parameters associated with emission and absorption at 397 nm during optical pumping. Geometric factors due to absorption and emission patterns are included in these Lamb-Dicke parameters.

To achieve the highest probabilities of ground-state occupation, especially at lower trap frequencies, we reduce the intensity of the 729-nm laser to 25% of its maximum during the final stage of cooling, so that $\Omega_0/2\pi\approx25$ kHz.

The ion is trapped with an axial frequency of 389 kHz and Doppler cooled for 5 ms. We measure an axial temperature of 0.45(2) mK, consistent with the Doppler cooling limit of 0.45 mK and corresponding to an average phonon number of $\bar{n}=24(1)$. Figure 1 shows a typical axial spectrum after this step, where each data point is the average of 400 repeats. We do not measure a radial temperature during these experiments, but independent measurements suggest this is several times higher than the axial temperature [26], typically 3 mK.

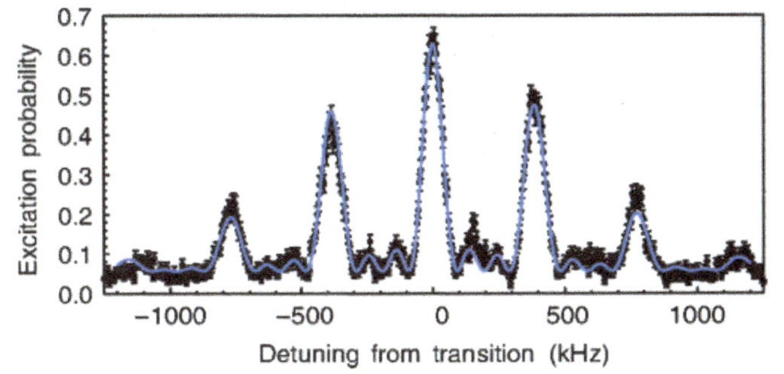

Figure. 1. Typical axial motional sideband spectrum after Doppler cooling, with a 10-μs probe pulse length. The solid line is a fit to the Rabi dynamics of a thermally distributed population, giving $\bar{n} = 24(1)$, equivalent to a temperature of 0.45(2) mK.

Figure 2 shows the first red and blue sidebands after 5, 5, and 10 ms of sideband cooling on the first red, second red, and first red sidebands, sequentially [39]. The red and blue sidebands are fitted simultaneously to Rabi sinc profiles on a constant background [40], with the background amplitude, Ω_0 and $\bar{n}$ as free parameters. The fit shows the average phonon number after sideband cooling to be $\bar{n}$ =0.02(1), consistent with the theoretical sideband cooling limit of $\bar{n}_{\text{lim}}$ =0.015 [Eq. (1)].

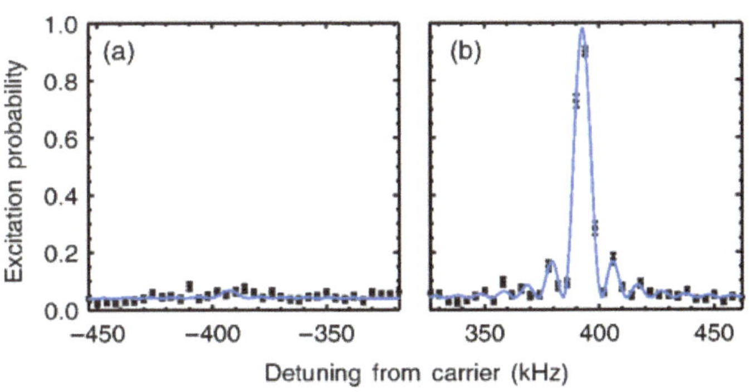

Figure. 2. (a) First red and (b) first blue sidebands after sideband cooling, with a trap frequency of ω/2π=389 kHz and a probe time of 100 μs. The solid line is a fit to the Rabi dynamics with a constant background, which gives $\bar{n} = 0.02(1)$.

Once cooled to the ground state, dephasing due to thermal effects becomes insignificant, allowing us to perform coherent qubit manipulations. After ground-state cooling, we probe the carrier transition and observe damped Rabi oscillations (Fig. 3). The source of the decoherence cannot be identified from these data so the function describing the damping is unknown. To extract the Rabi frequency and provide an approximate measure of the coherence time, we fit an exponentially decaying sinusoid, which gives $\Omega_0/2\pi$=28 kHz and τ≈0.7 ms, consistent with our spectroscopically measured linewidth of Δν=0.6(4) kHz [26]. As the number of oscillations increases, the data increasingly deviate from the theoretical fit. This is predominantly due to intensity noise on the 729-nm probe laser, on a characteristic time scale longer than that taken to record each data point. To reduce this effect we are currently developing a power-noise-eating feedback system for this laser.

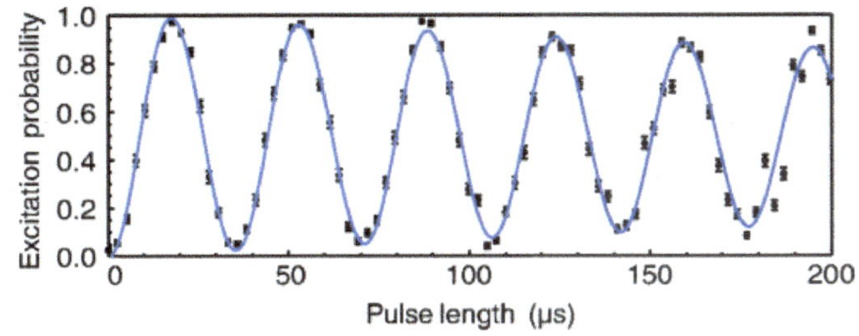

Figure. 3. Rabi oscillation on the carrier after ground-state cooling. The Rabi frequency is $\Omega_0/2\pi$≈28 kHz and the overall visibility decays with a coherence time of τ≈0.7 ms.

The heating rate $\dot{\bar{n}}$ of the trap is determined by inserting a delay period between ground-state cooling and spectroscopy, during which no cooling is applied, and measuring the increase in phonon number. We perform heating rate measurements with delays of 0, 50, and 100 ms, interleaved in a single experimental pulse sequence.

To investigate the source of the heating, we repeat this experiment at a range of axial trap frequencies, summarized in Fig. 4. In general, the heating rate for an ion of mass m due to electric field noise with spectral density $S_E(\omega)\propto\omega^{-\alpha}$ is given by $\dot{\bar{n}} = (e^2/4m\hbar\omega)S_E(\omega)$ [41]. Different sources of electric field noise lead to different characteristic frequency scalings: Johnson noise due to purely resistive elements is independent of frequency (α=0); inductive pickup (e.g., by trap supply cabling) of environmental electromagnetic interference (EMI) leads to α=−1 to 1 [41] as well as resonances due to local sources; while the patch-potential models developed to describe the d−4ion-electrode distance scaling observed in early experiments predict α=1 to 2 [42].

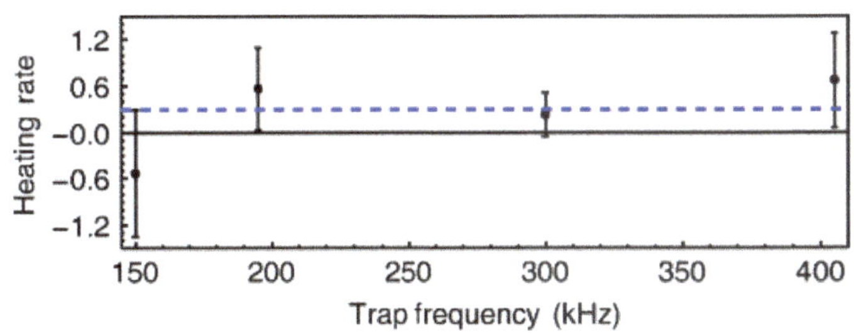

Figure. 4. Heating rate vs axial trapping frequency ω/2π. The data do not indicate a clear frequency scaling; a constant fit gives an average of $\dot{\bar{n}} = 0.3(2)\ \text{s}^{-1}$.

Considering the data in Fig. 4, it is apparent that the heating rate is very low across all trap frequencies, averaging $\dot{\bar{n}} = 0.3(2)\ \text{s}^{-1}$. The large uncertainties in this data set prevent us from determining a precise frequency dependence, although it appears very unlikely that an ω^{-2} or ω^{-3}scaling is present, as would be expected if fluctuating patch potentials were the dominant heating mechanism. The long delay periods required to measure heating rates of less than one phonon per second reduce the number of repeats it is possible to take for each spectroscopy point, limiting the precision of these measurements. No significant variation of the heating rate has been observed over several months with the current apparatus.

Voltage noise on the electrodes (such as is produced by Johnson noise and inductive pickup) produces field noise at the ion that scales as d^{-2}. When working with larger traps it is likely that such noise sources will eventually become more significant than patch-potential heating. These sources also show weaker scalings with frequency, as suggested by our data. For these reasons we believe that these are the most likely candidates for the source of the inferred electric field noise in our trap, $S_E \approx 5 \times 10^{-16}\ V^2 m^{-2} Hz^{-1}$.

Because of the symmetries of our trap, only differential-mode noise between the trap end caps or between the compensation electrodes can lead to significant heating of the axial mode. These end caps are individually connected via 3 m of cabling to a common voltage source. The resistance of this loop of cabling is approximately 1.5 Ω, leading to a Johnson-noise-induced $S_E = 4k_BTR/d^2 = 2 \times 10^{-16}\ \mathrm{V^2\,m^{-2}\,Hz^{-1}}$ at the ion, independent of frequency. Inductive pickup by this loop is reduced by routing the two cables side by side, minimizing the loop area. However, even very small linked fluxes can lead to significant field noise. Assuming a typical laboratory EMI noise figure of F_a=140 dB at 0.3 MHz with an ω^{-5} frequency scaling [43], the field noise at the ion due to inductive pickup by a loop of area A would be $S_E \approx A^2 \times 10^{-9}\ \mathrm{V^2\,m^{-6}\,Hz^{-1}}$, scaling with ω^{-1} [41]. This would exceed our inferred spectral noise for loop areas as small as 5 cm². While calculating the exact effects of EMI will require a measurement of the environmental noise in the vicinity of the experiment, it is certainly plausible that a combination of inductive pickup and Johnson noise could account for the observed heating rate. A simple way to reduce these effects by over 2 orders of magnitude would be to connect the two end caps at the location of the trap. Similar forms of voltage noise on the compensation electrodes could be minimized by moving the low-pass filters for these electrodes from the remote trap supply to the trap.

In Fig. 5 we compare our heating rate to those in a range of other traps, plotted against the distance to the nearest electrode. Here, we have taken the usual approach of plotting the phonon heating rate in terms of an inferred scaled noise spectral density $\omega S_E(\omega)$ that assumes α=1. Because of the large uncertainties of our individual measurements of $\dot{\bar{n}}$ and the absence of a clear frequency scaling, we have plotted the average of the noise densities calculated for each frequency. The list of measurements is not exhaustive; from those collated in Ref. [54], we have included only room temperature traps with three-dimensional electrode structures, and have additionally included the results from Poulsen *et al.* [36] as these were the lowest heating rates previously reported in any trap. Our values of $\omega S_E(\omega)$ are several times lower than those in Ref. [36]. This is not a particularly surprising result given the large ion-electrode distance in our trap.

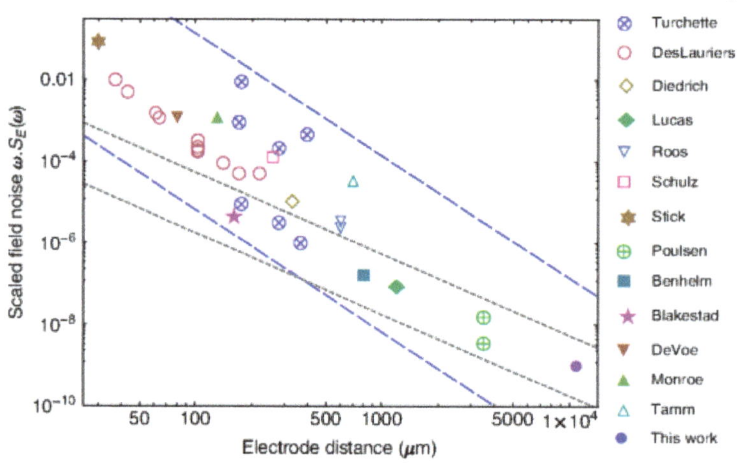

Figure. 5. The reported heating rates of a variety of room temperature, three-dimensional traps, shown as a frequency-scaled, noise spectral density plotted against the distance to the nearest electrode. The result reported in this Letter is shown in the lower right corner. The blue (dashed) and gray (dotted) lines are guides representing $1/d^2$ and $1/d^4$ distance scalings, respectively. (Data sources: Turchette [44], DesLauriers [45], Diedrich [15], Lucas [46], Roos [16], Schulz [47], Stick[48], Poulsen [36], Benhelm [49], Blakestad [50], DeVoe [51], Monroe [52], Tamm [53]).

We have included guidelines in this figure showing the slope of the expected d^{-4} scaling due to patch potential heating. Our data are approximately consistent with this scaling, although somewhat higher than might be expected. However, if voltage noise sources (e.g., Johnson noise) dominate for large traps, one would expect to see a transition to a d^{-2} scaling at some point, as indicated here by the second pair of guidelines. While there does appear to be some limited evidence of such a transition for d>1 mm, this could well be coincidental. To confirm such a crossover would require a carefully designed experiment, as variations in the supply electronics and cabling between apparatus would likely outweigh any distance scaling, and indeed would determine the distance at which the transition would occur. It is interesting to note that the results in Ref. [36] also suggest a heating rate that is constant in frequency, with a greater confidence than that given by our data. This would be consistent with inductive noise due to some forms of EMI.

Achieving three-dimensional ground-state confinement requires the magnetron mode to be cooled from a typical $\bar{n}$ after Doppler cooling of several thousand. This is further complicated by the nonseparability of the radial modes and the negative total energy associated with the magnetron motion. Work on this subject is ongoing, but the process is considerably more challenging than axial cooling.

We have demonstrated the application of resolved-sideband laser cooling to an ion in a Penning trap, achieving occupancy of the axial motional ground state with 98% probability. We have measured the heating rate of our ion trap and found it to be the lowest reported to date, although the uncertainties in the data prevent unambiguous identification of the source of the heating. These results pave the way for an exciting new range of Penning trap experiments in precision measurement [20], quantum information [10], and quantum thermodynamics [55].

ACKNOWLEDGEMENTS

This work was supported by the U.K. Engineering and Physical Sciences Research Council (Grant No. EP/D068509/1) and by the European Commission STREP PICC (FP7 2007-2013 Grant No. 249958). We gratefully acknowledge financial support towards networking activities from COST Action MP 1001—Ion Traps for Tomorrow's Applications.

REFERENCES

1. D. Nigg, M. Müller, E.A. Martinez, P. Schindler, M. Hennrich, T. Monz, M.A. Martin-Delgado, and R. Blatt, Science 345, 302 (2014).
2. J.W. Britton, B.C. Sawyer, A.C. Keith, C.C.J. Wang, J.K. Freericks, H. Uys, M.J. Biercuk, and J.J. Bollinger, Nature (London) 484, 489 (2012).
3. P. Gill, G.P. Barwood, H.A. Klein, G. Huang, S.A. Webster, P.J. Blythe, K. Hosaka, S.N. Lea, and H.S. Margolis, Meas. Sci. Technol. 14, 1174 (2003).
4. W. Quint, J. Dilling, S. Djekic, H. Häffner, N. Hermanspahn, H.-J. Kluge, G. Marx, R. Moore, D. Rodriguez, J. Schnfelder *et al.*, Hyperfine Interact. 132, 453 (2001).
5. R.C. Thompson, S. Donnellan, D.R. Crick, and D.M. Segal, J. Phys. B 42, 154003 (2009).
6. L.S. Brown and G. Gabrielse, Rev. Mod. Phys. 58, 233 (1986).
7. A. Mooser, S. Ulmer, K. Blaum, K. Franke, H. Kracke, C. Leiteritz, W. Quint, C.C. Rodegheri, C. Smorra, and J. Walz, Nature (London) 509, 596 (2014).
8. J. DiSciacca *et al.*, Phys. Rev. Lett. 110, 130801 (2013).
9. M. Vogel *et al.*, Phys. Scr. T166, 014066 (2015).
10. J.F. Goodwin, B.J. Brown, G. Stutter, H. Dale, R.C. Thompson, and T. Rudolph, Phys. Rev. A92, 032314 (2015).
11. D. Porras and J.I. Cirac, Phys. Rev. Lett. 96, 250501 (2006).
12. S. Mavadia, J.F. Goodwin, G. Stutter, S. Bharadia, D.R. Crick, D.M. Segal, and R.C. Thompson, Nat. Commun. 4, 2571 (2013).
13. M. Müller, L. Liang, I. Lesanovsky, and P. Zoller, New J. Phys. 10, 093009 (2008).
14. T. Feldker, P. Bachor, M. Stappel, D. Kolbe, R. Gerritsma, J. Walz, and F. Schmidt-Kaler, Phys. Rev. Lett. 115, 173001 (2015).
15. F. Diedrich, J.C. Bergquist, W.M. Itano, and D.J. Wineland, Phys. Rev. Lett. 62, 403 (1989).
16. C. Roos, T. Zeiger, H. Rohde, H.C. Nägerl, J. Eschner, D. Leibfried, F. Schmidt-Kaler, and R. Blatt, Phys. Rev. Lett. 83, 4713 (1999).
17. C.W. Chou, D.B. Hume, J.C.J. Koelemeij, D.J. Wineland, and T. Rosenband, Phys. Rev. Lett. 104, 070802 (2010).
18. M. Shi, P.F. Herskind, M. Drewsen, and I.L. Chuang, New J. Phys. 15, 113019 (2013).
19. F. Wolf, Y. Wan, J.C. Heip, F. Gebert, C. Shi, and P.O. Schmidt, Nature (London) 530, 457 (2016).
20. C. Smorra *et al.*, Eur. Phys. J. Spec. Top. 224, 3055 (2015).
21. J.M. Cornejo, P. Escobedo, and D. Rodríguez, Hyperfine Interact. 227, 223 (2014).
22. J.F. Poyatos, J.I. Cirac, and P. Zoller, Phys. Rev. Lett. 77, 4728 (1996).
23. J.P. Home, M.J. McDonnell, D.M. Lucas, G. Imreh, B.C. Keitch, D.J. Szwer, N.R. Thomas, S.C. Webster, D.N. Stacey, and A.M. Steane, New J. Phys. 8, 188 (2006).
24. This is the first direct measurement of the single-ion heating rate in a Penning trap; a recent measurement by Sawyer *et al.* [25] provided an indirect constraint via the heating rate scaling of large crystals.
25. B.C. Sawyer, J.W. Britton, and J.J. Bollinger, Phys. Rev. A 89, 033408 (2014).
26. S. Mavadia, G. Stutter, J.F. Goodwin, D.R. Crick, R.C. Thompson, and D.M. Segal, Phys. Rev. A 89, 032502 (2014).
27. D.R. Crick, S. Donnellan, D.M. Segal, and R.C. Thompson, Phys. Rev. A 81, 052503 (2010).
28. S. Mavadia, Ph.D. thesis, Imperial College London, 2013,http://www3.imperial.ac.uk/iontrap/pubs.
29. J.F. Goodwin, Ph.D. thesis, Imperial College London, 2015,http://www3.imperial.ac.uk/iontrap/pubs.
30. G. Savard, S. Becker, G. Bollen, H.-J. Kluge, R. Moore, T. Otto, L. Schweikhard, H. Stolzenberg, and U. Wiess, Phys. Lett. A 158, 247 (1991).
31. W.M. Itano and D.J. Wineland, Phys. Rev. A 25, 35 (1982).
32. Note that this technique is sometimes also referred to as "sideband cooling," but is unrelated to the laser cooling method described in this Letter.
33. H.F. Powell, D.M. Segal, and R.C. Thompson, Phys. Rev. Lett. 89, 093003 (2002).
34. E.S. Phillips, R.J. Hendricks, A.M. Abdulla, H. Ohadi, D.R. Crick, K. Koo, D.M. Segal, and R.C. Thompson, Phys. Rev. A 78, 032307 (2008).
35. R.J. Hendricks, E.S. Phillips, D.M. Segal, and R.C. Thompson, J. Phys. B 41, 035301 (2008).
36. G. Poulsen, Y. Miroshnychenko, and M. Drewsen, Phys. Rev. A 86, 051402 (2012).
37. I. Marzoli, J.I. Cirac, R. Blatt, and P. Zoller, Phys. Rev. A 49, 2771 (1994).
38. See Supplemental Material athttp://link.aps.org/supplemental/10.1103/PhysRevLett.116.143002 for derivation of the sideband cooling limit given in Eq. (1).
39. 4 ms of cooling is sufficient to reach $\bar{n} \approx 0$, but we use a much longer cooling time to ensure minimum $\bar{n}$.
40. A.M. Kaufman, B.J. Lester, and C.A. Regal, Phys. Rev. X 2, 041014 (2012).

41. M. Brownnutt, M. Kumph, P. Rabl, and R. Blatt, Rev. Mod. Phys. 87, 1419 (2015).

42. D. Hite, Y. Colombe, A. Wilson, D. Allcock, D. Leibfried, D. Wineland, and D. Pappas, MRS Bull. 38, 826 (2013).

43. I. Landa, A. Arrinda, I. Fernandez, and P. Angueira, IEEE Antennas Wireless Propag. Lett. 10, 17 (2011).

44. Q.A. Turchette *et al.*, Phys. Rev. A 61, 063418 (2000).

45. L. Deslauriers, S. Olmschenk, D. Stick, W.K. Hensinger, J. Sterk, and C. Monroe, Phys. Rev. Lett. 97, 103007 (2006).

46. D.M. Lucas, B.C. Keitch, J.P. Home, G. Imreh, M.J. McDonnell, D.N. Stacey, D.J. Szwer, and A.M. Steane, arXiv:0710.4421.

47. S.A. Schulz, U. Poschinger, F. Ziesel, and F. Schmidt-Kaler, New J. Phys. 10, 045007 (2008).

48. D. Stick, W.K. Hensinger, S. Olmschenk, M.J. Madsen, K. Schwab, and C. Monroe, Nat. Phys. 2, 36 (2006).

49. J. Benhelm, G. Kirchmair, C.F. Roos, and R. Blatt, Phys. Rev. A 77, 062306 (2008).

50. R.B. Blakestad, C. Ospelkaus, A.P. VanDevender, J.H. Wesenberg, M.J. Biercuk, D. Leibfried, and D.J. Wineland, Phys. Rev. A 84, 032314 (2011).

51. R.G. DeVoe and C. Kurtsiefer, Phys. Rev. A 65, 063407 (2002).

52. C. Monroe, D.M. Meekhof, B.E. King, S.R. Jefferts, W.M. Itano, D.J. Wineland, and P. Gould,Phys. Rev. Lett. 75, 4011 (1995).

53. C. Tamm, D. Engelke, and V. Bühner, Phys. Rev. A 61, 053405 (2000).

54. http://www.quantum.gatech.edu/heating_rate_plot.shtml.

55. A. Ruiz, D. Alonso, M.B. Plenio, and A. del Campo, Phys. Rev. B 89, 214305 (2014).

CHAPTER 14

A NOVEL FEMTOSECOND LASER SYSTEM FOR ATTOSECOND PULSE GENERATION

Jianqiang Zhu, Xinglong Xie, Meizhi Sun, Qunyu Bi, and Jun Kang

National Laboratory on High Power Laser and Physics, Shanghai Institute of Optics and Fine Mechanics, Chinese Academy of Sciences, Number 390, Qinghe Road, Jiading, Shanghai 201800, China

ABSTRACT

We report a novel ultrabroadband high-energy femtosecond laser to be built in our laboratory. A 7-femtosecond pulse is firstly stretched by an eight-pass offner stretcher with a chirp rate 15 ps/nm, and then energy-amplified by a two-stage optical parametric chirped pulse amplification (OPCPA). The first stage as preamplification with three pieces of BBO crystals provides the majority of the energy gain. At the second stage, a YCOB crystal with the aperture of ~50 mm is used instead of the KDP crystal as the gain medium to ensure the shortest pulse. After the completion, the laser will deliver about 8 J with pulse duration of about 10 femtoseconds, which should be beneficial to the attosecond pulse generation and other ultrafast experiments.

The attosecond pulse has been of interest in recent years and proved to be a powerful tool in studying the ultrafast phenomena, such as the chemical/biological transformations occurring on the femtosecond timescale and the evolution of the dynamics of the electrons in atoms or molecules irradiated by intense lasers. There are two ways to generate an attosecond pulse/pulse train. One is through the non-linear processes of the superposition of high order harmonics generated in the laser-gas atom interactions in which the laser intensity should be low so as to avoid the ionization of atoms. Up to now, the laser intensity is no more than 10^{16} W cm^{-2} and the efficiency of harmonic emission from the atoms is low. The other way is to generate the high order harmonics from the dense surface of a plasma created by the high-intensity femtosecond laser, which is promising for the intense attosecond pulse generation. Recent experiments and simulations have shown that attosecond pulses with high conversion efficiency, high photon energy, and excellent divergence can be generated [1–4].

The efficiency of harmonic generation from the laser-plasma surface relies on the laser intensity. In the experiments for the brilliant harmonic generation from the surface, the laser has energy of 0.1 J to less than 10 J with a pulse duration of 30–50 fs [5]. In this paper, we report a novel ultrabroadband ultrashort high-energy laser system based on optical parametric chirped pulse amplification (OPCPA). After completion, the laser can deliver 8.0 J with pulse width of about 10 fs, which is much shorter than those used in the previous intense

Citation: Jianqiang Zhu, Xinglong Xie, Meizhi Sun, Qunyu Bi, and Jun Kang, "A Novel Femtosecond Laser System for Attosecond Pulse Generation," Advances in Optical Technologies, vol. 2012, Article ID 908976, 6 pages, 2012. doi:10.1155/2012/908976

attosecond pulse generation experiments. The laser should be beneficial to the attosecond pulse generation and other ultrafast experiments.

OPCPA has been applied and becomes promising for constructing high-intensity lasers [5–9], particularly with the invention of new nonlinear crystals such as YCOB ($YCa_4O(BO_3)_3$, YCOB) [10–12]. Compared with other nonlinear crystals commonly utilized in OPCPA systems, YCOB crystal has several prominent properties: high damage threshold, moderate thermal conductivity, and nonlinear coupling [13]. It has been shown that the size of YCOB crystal can be as large as 7.5 cm in diameter and 25 cm in length [14], much larger than those currently used, such as LBO and BBO crystals. Chekhlov et al. reported their OPCPA laser system which used large aperture KDP crystals for power amplification. Their laser delivers output energy up to 35 J and pulse width of 84 fs, [15]. To our knowledge, YCOB crystal has not been used in OPCPA laser systems with wavelength centered at 808 nm and output energy up to several joules and pulse width of less than 10 fs.

To design the OPCPA-based ultrashort laser system which can deliver several joules within about 10 fs, we use YCOB instead of KDP or DKDP crystal as the gain medium for power amplification stage. Because YCOB has nonlinear coefficient four times of that of KDP and DKDP, it can provide broad spectral gain bandwidth. As indicated in Figure 1, our laser system consists of five parts: a signal laser pulse, an Offner stretcher, a multistage optical parametric amplifier (OPA), a compressor, and a pump laser system. The oscillator provides signal pulses of 7 fs in time duration and 2 nJ in energy. The signal pulses are chirped to 15 ps/nm by an eight-pass Offner stretcher with 10 percent transmission efficiency, and the energy of chirped signal pulse injected into OPA chain is measured to be about 0.2 nJ. The OPCPA amplification chain is composed of two stages. The first stage, as a preamplification, includes three pieces of BBO crystals that provide the majority of the energy gain. The pump pulses at 532 nm with 300 mJ for the first two pieces of BBO crystals are from a commercial single longitudinal mode Nd:YAG pump laser which operated at a repetition of 10 Hz. For the third piece of BBO crystal, 2.5 J second harmonic pulse, from an Nd:glass Subpicosecond Laser System (SPS), is utilized as the pump source [16]. The SPS laser is a two beam system, one beam is short with pulse width ~1.0 ps and pulse energy 100 J. The other beam is a 1053 nm beam with the pulse width of about 2.0 ns and maximum pulse energy about 150 J. The nanosecond beam is used as the pump source for our OPCPA system. The whole system works at one shot every thirty minutes. In the second stage, a piece of YCOB crystal with large aperture ~50 mm is utilized as the gain medium and the SPS provides pulses at 526.5 nm with the desired special-temporal shape and energy exceeding 50 J.

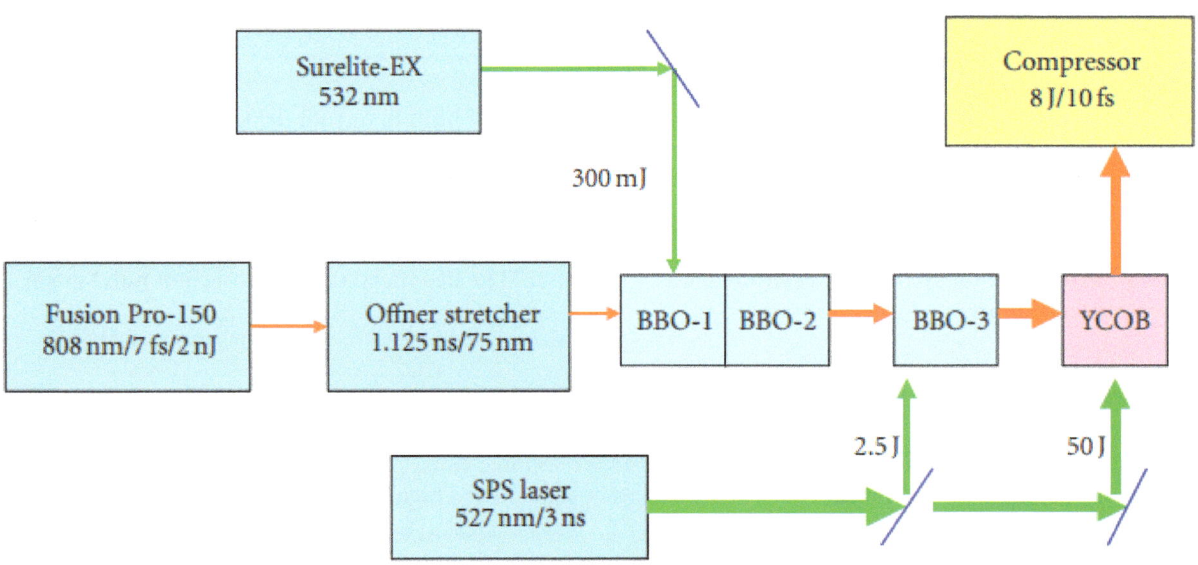

Figure 1: Optical schematic layout of OPCPA.

The signal is amplified in the XOZ principle of YCOB because the effective nonlinear coefficient (d_{eff}) in XOZ principle plane is much larger than that in XOY and YOZ principle planes [17, 18].

Noncollinear geometric configuration with type 1 phase matching $o + o \rightarrow e$ is utilized in all OPA stages. Compared with the collinear geometric configuration, the noncollinear geometric configuration can achieve larger parametric bandwidth and get higher conversion efficiency by compensating walk-off effects of the pump pulse with noncollinear angle. Wave vectors and parameters of noncollinear phase matching in biaxial crystals are shown in Figure 2. X, Y, and Z are principle axes. The refractive indexes along principle axes are ordered $nZ > nY > nX$. C_1 and C_2 are optical axes. Ω is optic axial angle, Φ the azimuth of k_p in XOZ principle plane, and k_s, k_i, and k_p wave vectors of signal, idle, and pump, respectively. Symbol of α refers to the noncollinear angle, θ the phase matching angle, ρ the walk-off angle, s_p the Poynting vector of pump pulse, and β the angle between k_i and k_p. Momentum conservation and energy conservation during the OPA process can be expressed as (1) and (2). The wave vector mismatch is expressed as (3):

$$\vec{k}_p = \vec{k}_i + \vec{k}_s, \tag{1}$$

$$\omega_p = \omega_i + \omega_s, \tag{2}$$

$$\Delta k = k_p - k_s \cos\alpha - k_i \cos\beta. \tag{3}$$

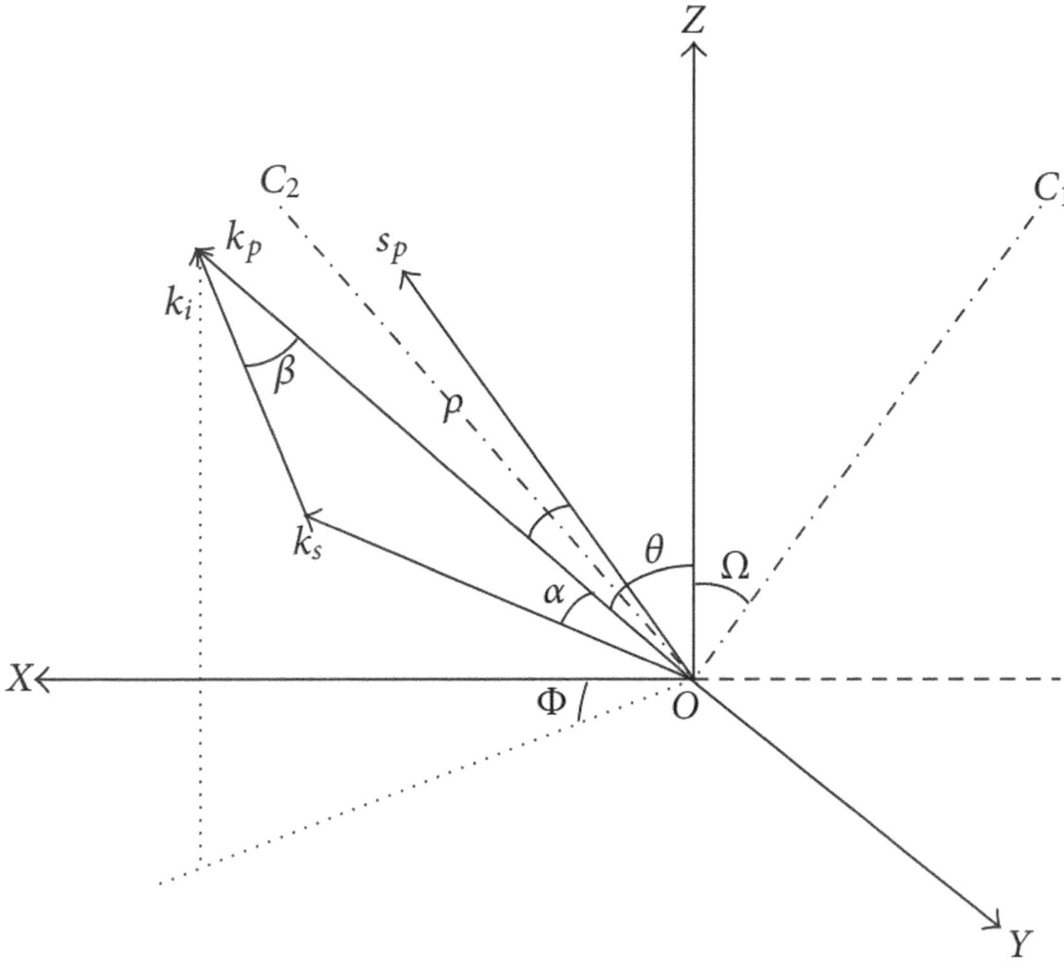

Figure 2: Noncollinear phase matching vectors in biaxial crystals.

The optimized noncollinear configuration can realize the highest parametric bandwidth (($\Delta\lambda_p$) for the ultrabroadband optical parametric amplifier. Generally speaking, noncollinear angle in such configuration is set to realize group velocity matching (GVM) between signal pulse and pump pulse [19]. However, according to numerical analysis, there exists a small difference between the GVM noncollinear angle and the one for realizing the highest parametric bandwidth [20]. Parametric bandwidths as well as gain bandwidths ($\Delta\lambda_g$) and acceptance angles ($\Delta\theta$) are calculated exactly according to the definitions without using approximation of Taylor series. Identical analytic method was utilized in reference [18].

Parametric bandwidth is defined as the interval of the signal wavelength that restricts the phase mismatching smaller than $\pm\pi$ under the condition that perfect phase matching is achieved at the center wavelength of OPA. It can be expressed as $|\Delta kL| \leq \pi$, here L is the crystal length. Figure 3(a) shows the curves of the parametric bandwidths versus noncollinear angles in 15 mm BBO crystal, 15 mm YCOB crystal, and 35 mm DKDP. As shown in Figure 3(a), parametric bandwidths of 149.6 nm and 108 nm are achieved in YCOB and BBO when noncollinear angle are 2.763° and 2.358°, respectively. The value is 46 nm in DKDP when noncollinear angle is 3.821°. Parametric bandwidth is consistent with the gain bandwidth to a large extent. Gain bandwidth is defined as the interval of the signal wavelength that restricts the gain bigger than half of that achieved at the center wavelength of signal pulse. As shown in Figure 3, the largest gain bandwidth of 133.6 nm is achieved at noncollinear angle of 2.775° in YCOB, 112.4 nm at 2.345° in BBO, and 46.6 nm at 3.819° in DKDP. As a conclusion, gain bandwidths of YCOB and BBO are broad enough for the amplification of chirped pulses, while DKDP is confided to the amplification of chirped pulses with original pulse duration exceeding 21 fs at 808 nm.

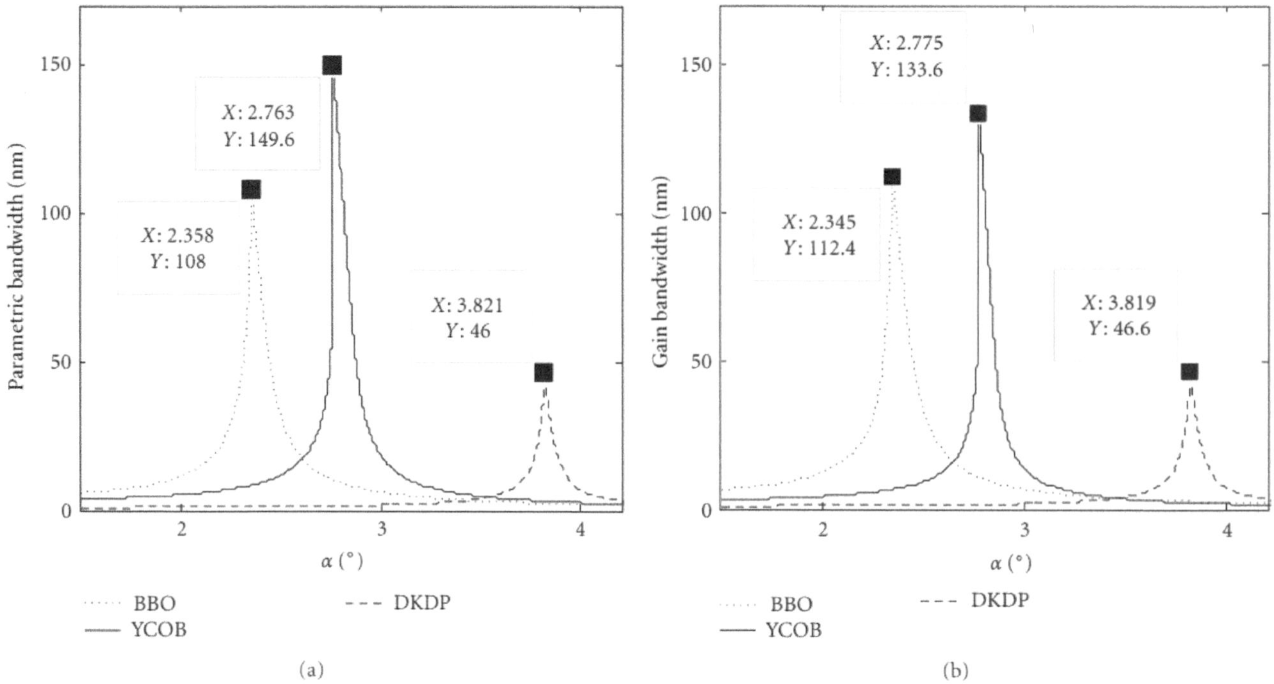

Figure 3: Parametric bandwidth (a) and gain bandwidth (b) versus noncollinear angles in BBO (dotted), in the XOZ principle plane of YCOB (solid), and in DKDP (dashed).

The walk-off effect results from the separation of Poynting vector and wave vector of light transmitting in nonlinear crystals (inhomogeneous medium) and leads to spatial displacement between the signal pulse and pump pulse and the degradation of conversion efficiency in OPCPA. In addition, walk-off effect plays an important role in the impact of wavefront phase distortion of pump on the beam quality of signal in OPA [21,22]. The acceptance angle is defined as the interval of θ that restricts the wave vector mismatch smaller than $\pm\pi/L$ around the perfect phase matching angle (θ_{pm}) at 808 nm. A large acceptance angle means a stable OPA. The walk-off angles, acceptance angles, and phase-matching angles versus noncollinear angles are shown in Figure 4. Corresponding to the optimal noncollinear angle, the walk-off angle in YCOB crystal is 1.134°, as well as acceptance angle ~0.06° and phase matching-angle 26.4°. The values of YCOB, BBO, and DKDP are listed in Table 1 and that the YCOB has bigger acceptance angle and smaller walk-off angle than BBO and DKDP.

Table 1: Noncollinear phase matching parameters of BBO, YCOB, and DKDP

	α (°)	θ_{pm} (°)	$\Delta\lambda_p$ (nm)	d_{eff} (pm/V)	ρ (°)	$\Delta\theta$ (°)	$\Delta\lambda_g$ (nm)	I_p (GW/cm^2)	L (m)
BBO-1/2	2.345	23.78	108	2.096	−3.288	0.0214	112.4	0.4	0.015
BBO-3	2.4	23.86	132	2.095	−3.295	0.0214	127	0.4	0.015
YCOB	2.775	(26.4, 180)	149.6	0.935	1.134	0.0599	133.6	0.5	0.015
DKDP	3.819	52.76	46	0.295	−2.913	0.0113	46.6	2	0.035

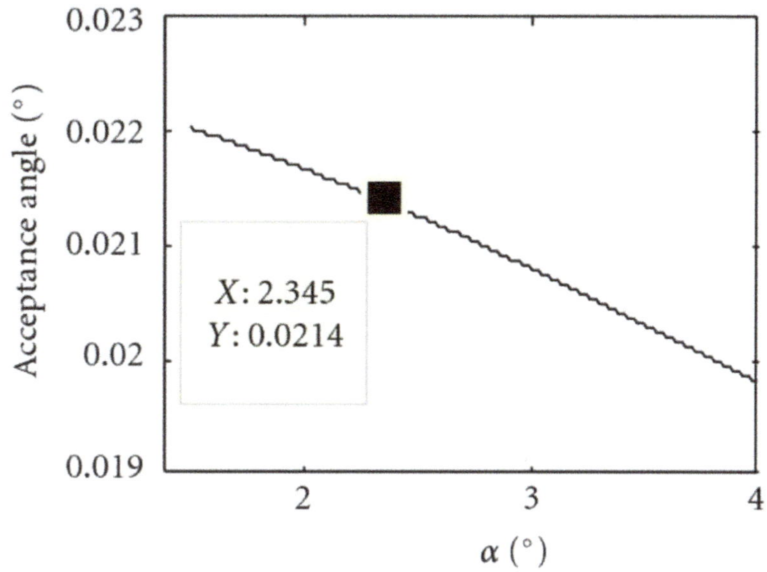
0.023
0.022
0.021
0.02
0.019
Acceptance angle (°)
X: 2.345
Y: 0.0214
2
3
4
α (°)

(a)

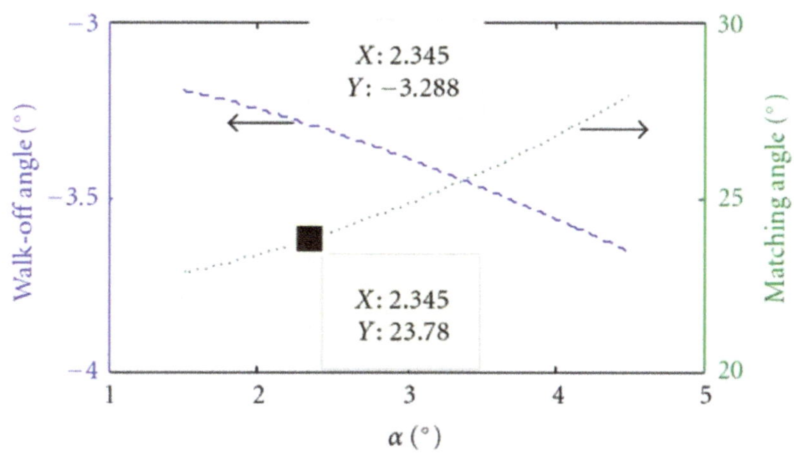
−3
−3.5
−4
Walk-off angle (°)
30
25
20
Matching angle (°)
X: 2.345
Y: −3.288
X: 2.345
Y: 23.78
1
2
3
4
5
α (°)

(b)

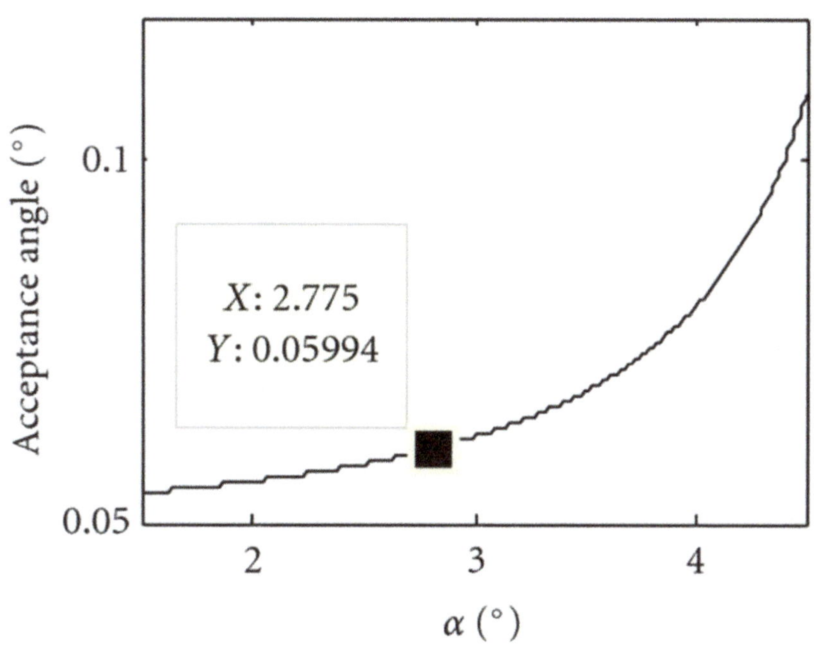
0.1
0.05
Acceptance angle (°)
X: 2.775
Y: 0.05994
2
3
4
α (°)

(c)

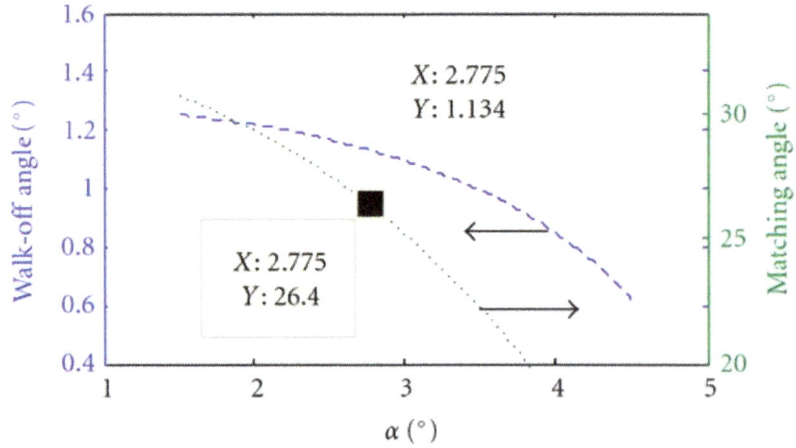

(d)

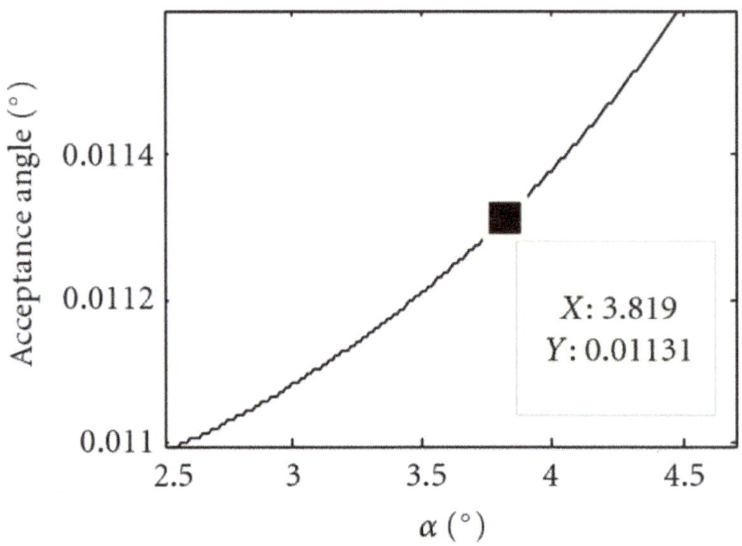

(e)

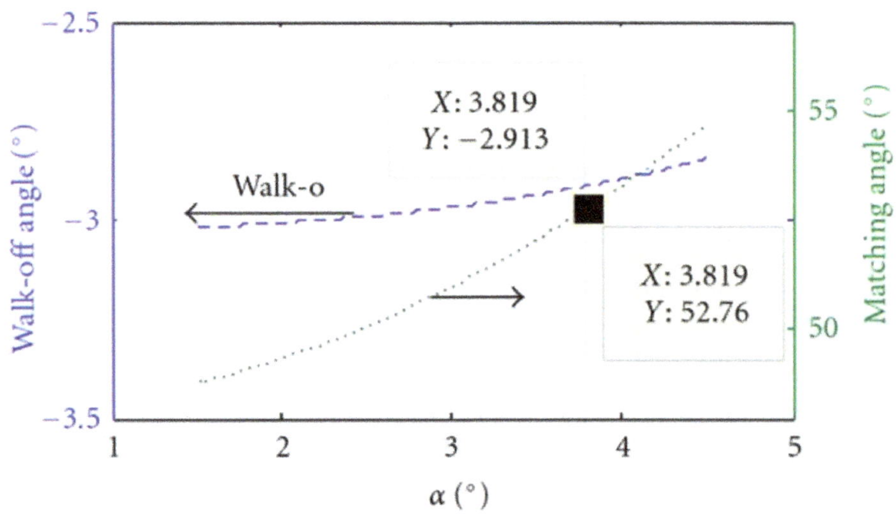

(f)

Figure 4: Acceptance angle (solid), walk-off angle (dashed), and phase-matching angle (dotted) versus noncollinear angle in BBO (a and b), the XOZ principle plane of YCOB (c and d), and DKDP (e and f).

As shown in Figure 5(a), for chirped signal pulse pumped by 532 nm pulse at intensity () of 0.4 GW/cm^2 in 15 mm BBO crystal, gain of 2732 can be realized. In the first two pieces of BBO crystals, unsaturated gain of 1000 × 1000 can be achieved by pumping with 532 nm pulse of 300 mJ at an intensity of 0.4 GW/cm^2. In the third piece of BBO crystal, the injected signal pulse is amplified by 526.5 pump pulse with energy of 2.5 J at an intensity of 0.4 GW/cm^2. The phase matching parameters listed in Table 1 are calculated with the same method. There are not big differences with those pumped by pulse at 532 nm. The BBO crystal utilized in this stage is 13 mm (aperture) × 15 mm (length) to realize saturated gain of ~2000 so that energy of signal pulse is amplified up to ~0.4 J. The majority of energy is provided by the second stage. Figure 5(b) shows the gain curve in the XOZ principle plane of YCOB with 526.5 nm pump pulse. For YCOB crystal of 15 mm in length, saturated gain of ~20 can be realized with pump pulse at intensity of 0.5 GW/cm^2 and energy of 50 J. In the power amplification stage, signal pulse of ~10 J can be achieved with broad spectral bandwidth. Moreover, saturated amplification in both stages would greatly improve the stability of the system [23]. Normalized gain curve in 35 mm DKDP is also shown in Figure 5(a) when pumped by 526.5 nm pulse at intensity of 2 GW/cm^2. DKDP is a good choice for amplifying pulses centered at 808 nm with duration exceeding 20 fs.

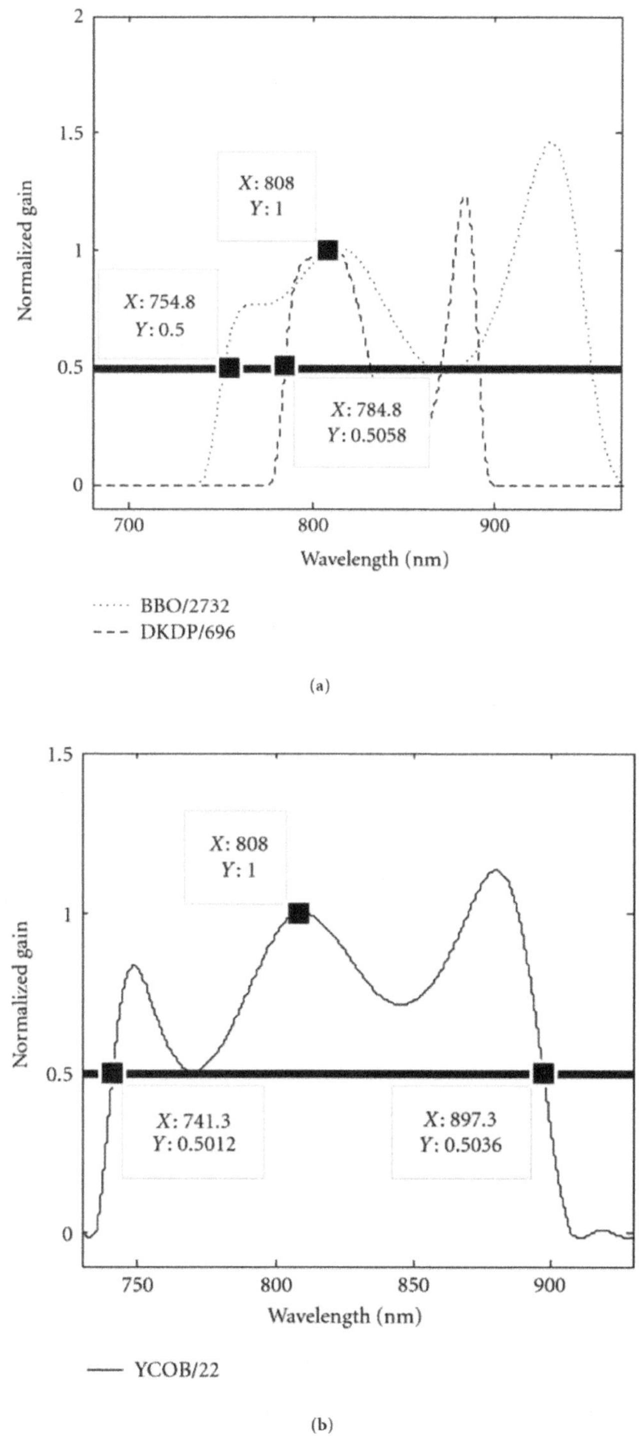

Figure 5: Normalized gain curves in BBO (a) (dotted), DKDP (a) (dashed), and XOZ principle plane of YCOB crystal (b) (solid).

As the temporal profile is an important factor in the ultrashort laser mater interaction experiment, a spectrum shaper or a Dazzler with the resolution of 0.6 nm is used within the first OPCPA stage to control the pulse envelop. A 39-element deformable mirror with diameter of 50 mm is used before the pulse compression to ensure a good beam quality and the shortest pulse width.

The compressor is consisted of a pair of gratings and a piece of reflector. They are arranged in tandem in a four-pass configuration with Littrow incident angle of 36.3° to match with the stretcher so that the chirped pulse can be compressed to Fourier transform limitation. The transmission efficiency of the compressor is ~75% and the energy of output ultrashort pulse would be ~8 J.

In conclusion, a new type of OPCPA system to realize the amplification of pulse centered at 808 nm is analyzed, and noncollinear phase matching parameters of YCOB is compared with those of BBO and DKDP in Table 1. Numerical analysis shows that YCOB rather than DKDP can be utilized to amplify ultrabroadband chirped pulse so that the signal pulse can be compressed to about 10 fs after the compressor. With YCOB as the gain medium of OPCPA, a total gain of $\sim 4 \times 10^{10}$ is achieved and the energy of the amplified signal pulse is up to several joules.

REFERENCES

1. Y. Nomura, R. Hörlein, P. Tzallas et al., "Attosecond phase locking of harmonics emitted from laser-produced plasmas," Nature Physics, vol. 5, no. 2, pp. 124–128, 2009.
2. F. Quere, "Attosecond plasma optics," Nature Physics, vol. 5, pp. 93–94, 2009.
3. D. Charalambidis, P. Tzallasl, E. P. Benis et al., "Exploring intense attosecond pulses," New Journal of Physics, vol. 10, Article ID 025018, 2008.
4. G. D. Tsakiris, K. Eidmann, J. Meyer-ter-Vehn, and F. Krausz, "Route to intense single attosecond pulses," New Journal of Physics, vol. 8, Article ID 025018, 2006.
5. E. Hugonnot, G. Deschaseaux, O. Hartmann, N. Beck, and H. Coïc, "Optical parametric chirped pulse amplification for the PETAL front-end: design and first results," in Proceedings of the Advanced Solid-State Photonics (ASSP '07), OSA Technical Digest Series (CD), p. MB4, Optical Society of America, Vancouver, Canada, 2007.
6. M. Sun, J. Kang, A. Guo et al., "Effect of single-mode fiber on the contrast of chirped-pulse," Acta Optica Sinica, vol. 30, no. 5, pp. 1234–1237, 2010.
7. E. Gaul, M. Martinez, J. Blakeney et al., "Activation of a 1.1 Petawatt hybrid, OPCPA-Nd: glass laser," inProceedings of the Conference on Lasers and Electro-Optics and the Conference on Quantum Electronics and Laser Science Conference (CLEO/QELS '09), June 2009.
8. Y. Kitagawa, H. Fujita, R. Kodama et al., "Prepulse-free Petawatt laser for a fast ignitor," IEEE Journal of Quantum Electronics, vol. 40, no. 3, pp. 281–293, 2004.
9. C. N. Danson, P. A. Brummitt, R. J. Clarke et al., "Vulcan Petawatt—an ultra-high-intensity interaction facility," Nuclear Fusion, vol. 44, no. 12, pp. S239–S246, 2004.
10. R. Norrestam, M. Nygren, and J. O. Bovin, "Structural investigations of new calcium-rare earth (R) oxyborates with the composition $Ca_4RO(BO_3)_3$," Chemistry of Materials, vol. 4, no. 3, pp. 737–743, 1992.
11. Z. Wang, K. Fu, X. Xu et al., "The optimum configuration for the third-harmonic generation of 1.064μm in a YCOB crystal," Applied Physics B, vol. 72, no. 7, pp. 839–842, 2001.
12. M. Iwai, T. Kobayashi, H. Furuya, Y. Mori, and T. Sasaki, "Crystal growth and optical characterization of rare-earth (Re) calcium oxyborate $ReCa_4O(BO_3)_3$ (Re = Y or Gd) as new nonlinear optical material," Japanese Journal of Applied Physics: Part 2, vol. 36, no. 3, pp. L276–L279, 1997.
13. C. A. Ebbers, A. J. Bayramian, R. W. Campbell, et al., "High average power frequency conversion with large aperture YCOB," in Proceedings of the Advanced Solid-State Photonics (ASSP '08), OSA Technical Digest Series (CD), p. WD3, Optical Society of America, Nara, Japan, January 2008.
14. Z. M. Liao, I. Jovanovic, C. A. Ebbers, Y. Fei, and B. Chai, "Energy and average power scalable optical parametric chirped-pulse amplification in yttrium calcium oxyborate," Optics Letters, vol. 31, no. 9, pp. 1277–1279, 2006.
15. O. V. Chekhlov, J. L. Collier, I. N. Ross et al., "35 J broadband femtosecond optical parametric chirped pulse amplification system," Optics Letters, vol. 31, no. 24, pp. 3665–3667, 2006.
16. X. L. Xie, J. Q. Zhu, F. Q. Liu et al., "20 TW sub-picosecond laser system (SPS) applied for the neutron experiment," Chinese Journal of Lasers, vol. 30, no. 10, pp. 865–872, 2003.
17. M. V. Pack, D. J. Armstrong, A. V. Smith, G. Aka, B. Ferrand, and D. Pelenc, "Measurement of the X(2)tensor of $GdCa_4O(BO_3)_3$ and $YCa_4O(BO_3)_3$ crystals," Journal of the Optical Society of America B, vol. 22, no. 2, pp. 417–425, 2005.
18. M. Sun, Q. Bi, F. Zhang, J. Kang, X. Xie, and Z. Lin, "Study on noncollinear phase matching in YCOB crystal," Acta Optica Sinica, vol. 31, no. 1, Article ID 0119001, 2011.
19. T. Wilhelm, J. Piel, and E. Riedle, "Sub-20-fs pulses tunable across the visible from a blue-pumped single-pass noncollinear parametric converter," Optics Letters, vol. 22, no. 19, pp. 1494–1496, 1997.
20. J. Wang, J. Q. Yao, X. F. Li, Y. Z. Yu, and X. C. Shi, "Amendatory calculation of the acceptance parameters in three-wave interactions," Acta Optica Sinica, vol. 21, no. 2, pp. 139–141, 2001.
21. F. Zhang, Y. Wang, M. Sun, Q. Bi, X. Xie, and Z. Lin, "Numerical simulations of the impact of wavefront phase distortions of pump on the beam quality of OPA," Chinese Optics Letters, vol. 8, no. 2, pp. 217–220, 2010.
22. X. Wei, L. Qian, P. Yuan, H. Zhu, and D. Fan, "Optical parametric amplification pumped by a phase-aberrated

beam," Optics Express, vol. 16, no. 12, pp. 8904–8915, 2008.

23. J. Kang, S. H. Chen, J. Q. Zhu, and H. Wei, "Design of a highly stable, high-conversion-efficiency pumping source for optical parametric amplifier by extending efficient crystal length," Optics and Laser Technology, vol. 39, no. 5, pp. 1084–1088, 2007.

CHAPTER 15 NONLINEAR PULSE SHAPING IN FIBRES FOR PULSE GENERATION AND OPTICAL PROCESSING

Sonia Boscolo[1] and Christophe Finot[2]

[1]Photonics Research Group, School of Engineering and Applied Science, Aston University, Birmingham B4 7ET, UK

[2]Laboratoire Interdisciplinaire Carnot de Bourgogne, UMR 6303 CNRS, Université de Bourgogne, 21078 Dijon, France

ABSTRACT

The development of new all-optical technologies for data processing and signal manipulation is a field of growing importance with a strong potential for numerous applications in diverse areas of modern science. Nonlinear phenomena occurring in optical fibres have many attractive features and great, but not yet fully explored, potential in signal processing. Here, we review recent progress on the use of fibre nonlinearities for the generation and shaping of optical pulses and on the applications of advanced pulse shapes in all-optical signal processing. Amongst other topics, we will discuss ultrahigh repetition rate pulse sources, the generation of parabolic shaped pulses in active and passive fibres, the generation of pulses with triangular temporal profiles, and coherent supercontinuum sources. The signal processing applications will span optical regeneration, linear distortion compensation, optical decision at the receiver in optical communication systems, spectral and temporal signal doubling, and frequency conversion.

INTRODUCTION

The use of photonic technologies for data processing in the all-optical domain has a strong potential for a variety of interesting applications in such diverse areas as optical telecommunications, metrology, optical sensing, microwave engineering, advanced microscopy image processing, optical computing, and many others. Advantages of processing the information in the all-optical domain include the large available bandwidth and the (potential) parallelism intrinsic to the optical approach, which translate into high-processing speeds. However, today electronic techniques of signal manipulation are advanced compared to all-optical processing devices, which are still at the research stage rather than under commercial development. A key feature of electronics, enabling many different applications, is the capability of generating electrical waveforms with arbitrary temporal profiles by use of simple integrated circuits. This capability is so far unmatched in the optical frequency range, where bulky and complex devices are required to shape the light fields. Hence, considerable knowledge still has to be accumulated and new methodologies need to be explored before a true breakthrough can be achieved in this field, allowing the range of functions and operations

Citation: Sonia Boscolo and Christophe Finot, "Nonlinear Pulse Shaping in Fibres for Pulse Generation and Optical Processing," International Journal of Optics, vol. 2012, Article ID 159057, 14 pages, 2012. doi:10.1155/2012/159057

currently accomplished electronically, to be performed in the optical domain. Furthermore, by using optical rather than electronic processing additional functionality may be possible.

In order to realize such all-optical processing, nonlinear photonics is seen as a key technology. Optical fibre materials exhibit a nonlinear response to strong electric fields, such those of optical signals confined within the small fibre core. In high-speed optical communications such nonlinear effects generally degrade the integrity of the transmitted signal but the same effects can be used to realize a variety of optical functions that have practical applications in the field of lightwave technology. In this paper, we provide a snapshot of recent results and advances in the use of nonlinear effects in optical fibres for optical waveform generation and pulse shaping and in the applications of advanced pulse profiles in all-optical signal processing. We would like to note that the main attention in this paper will be focused on results obtained in our groups, and that it is not our intention here to comprehensively cover all the possible examples of fibre-based pulse shaping and signal processing.

FIBRE NONLINEARITIES

The third-order $\chi^{(3)}$ optical nonlinearity in silica-based single-mode (SM) fibres is one of the most important effects that can be used for all-optical signal processing. This happens not only because the third-order nonlinearity provides ultra-fast response times in the femtosecond range, but also because it is responsible for a wide range of phenomena such as third-harmonic generation, nonlinear refraction (Kerr nonlinearity), and stimulated Raman and Brillouin scattering [1]. The rapid recent developments of microstructured fibres with extremely small effective core areas and exhibiting enhanced nonlinear characteristics [2] and of fibres using materials with refractive indexes higher than that of the silica glass [3] have enabled dramatic reduction of the required fibre lengths for nonlinear processing compared to conventional fibres. Such progress has paved the way for the integration of fibre-based nonlinear processing functions on photonic chips.

The intensity dependence of the refractive index of a fibre gives rise to different effects depending on the shape of the input signal; the most widely studied effects are self-phase modulation (SPM) and cross-phase modulation (XPM). Considering the localized optical pulse evolution in a fibre medium and neglecting higher order effects in the fibre, pulse propagation can be described by the nonlinear Schrödinger (NLS) equation [4]:

$$\psi_z = -i\frac{\beta_2}{2}\psi_{tt} + i\gamma|\psi|^2\psi, \qquad (1)$$

where $\psi(z,t)$ is the slowly varying pulse envelope in the comoving system of coordinates, β_2 and $\gamma = 2\pi n_2/(\lambda A_{\mathrm{eff}})$ are the respective group-velocity (second-order) dispersion (GVD) and Kerr nonlinearity coefficients of the fibre, is the nonlinear-index coefficient, λ the central wavelength of the pulse, and A_{eff} the effective core area. As a result of the nonlinear term in (1), upon propagation in the fibre the pulse acquires an intensity-dependent nonlinear phase shift as $\phi(z,t) = \gamma|\psi(0,t)|^2 z$, namely, the frequency chirp $-\varphi_t$. When the length scale associated with the pulse where the GVD effects take place: $L_{\mathrm{D}} = T_0^2/|\beta_2|$ (with being some temporal characteristic value of the initial pulse) is much larger compared with both the fibre length and the nonlinear length $L_{\mathrm{NL}} = 1/(\gamma P_0)$ (with P_0 being the peak power of the initial pulse), the dispersive term in (1) can be neglected. In this case, the presence of a chirp causes a nonlinear broadening of the pulse spectrum. When two (or more) pulses copropagate inside a fibre, one still obtains an NLS equation for each pulse but these equations are coupled through XPM [4]. In this case, the nonlinear phase evolution for each pulse depends also on the power of the other pulses according to the expression $\phi_i(z,t) = \gamma_i(|\psi_i(0,t)|^2 + 2\sum_{j\neq i}|\psi_j(0,t)|^2)z$. XPM occurs only when the pulses overlap. Similar to the SPM case, XPM leads to additional nonlinear chirping and spectral shaping.

When the effects of chromatic dispersion are considered in combination with the Kerr nonlinearity, rich pulse dynamics arise from the interplay between dispersive and nonlinear effects depending on the sign of the dispersion and the relative magnitudes of the associated length scales. A well-known and fascinating example is the formation of optical solitons [5, 6] in the anomalous dispersion regime of a fibre ($\beta_2 < 0$) as a result of a cooperation between GVD and SPM. On the contrary, in the regime of normal dispersion pulse dynamics are highly affected by the phenomenon of optical wave breaking [7–9].

In the next Section we will discuss some newly emerged qualitative features of the optical pulse evolution in a fibre under the combined action of dispersion and nonlinearity.

OPTICAL PULSE GENERATION AND SHAPING USING FIBRE NONLINEARITIES

Techniques for generating, controlling, manipulating, and measuring ultrashort optical pulses and specialized waveforms have become increasingly important in many scientific areas including, amongst others, ultrahigh-speed optical communications [10], optical signal processing, and biophotonics. To date, conventional picosecond and femtosecond pulse shaping in the optical domain have been implemented using devices, such as liquid crystal spatial light modulators [11, 12], wavelength-to-time mapping [13, 14], acoustooptic modulators, and electrooptical phase arrays, that impart user-specified spectral amplitude and phase to the pulse. Though powerful and flexible, this approach has the drawback of requiring a rather bulky and expensive apparatus, entailing a level of cost and complexity that is not often commensurate to the used laser system and target application. In order to fulfill telecommunication requirements, more compact techniques for pulse shaping have been developed, which include the use of super-structured fibre Bragg gratings [15], long-period fibre grating filters [16], and arrayed waveguide gratings that are well suited for line by line processing of high-repetition periodic pulse shapes [17, 18]. However, in all these pulse shapers using linear techniques, the bandwidth of the input pulse determines the maximum bandwidth of the output. Indeed, a linear manipulation cannot increase the pulse bandwidth, and so to create shorter pulses nonlinear effects must be used. This limitation of the linear approaches becomes very stringent when synthesizing advanced pulse shapes such as rectangular pulses, for which a large optical bandwidth is required to reproduce accurately the spectral side lobes linked to the temporal compactness of the pulse. Moreover, in those applications that require spectral narrowing, linear filtering introduces a power penalty that is at least proportional to the ratio of the target spectrum to the input one.

The combination of third-order nonlinear processes and chromatic dispersion in optical fibres can provide efficient new solutions to overcome the aforementioned limitations. The pulse-shaping examples discussed below highlight some of the advantages offered by a passive, all-optical nonlinear fibre-based approach to pulse shaping.

Ultrahigh Repetition Rate Pulse Train Generation

As a first illustration of the possible nonlinear pulse shaping mechanisms in optical fibres, we review here the generation of ultrashort pulse trains with very high repetition rates resulting from the passive evolution of optical pulses in an anomalously dispersive fibre in the presence of Kerr nonlinearity. As mentioned in the previous Section, in the anomalous GVD regime of a fibre, the NLS equation (1) admits families of soliton solutions. While fundamental solitons are characterized by an exact balance between the GVD and SPM effects ($N = \sqrt{L_D/L_{NL}} = 1$), in the case of higher order solitons ($N > 1$) SPM dominates initially but GVD soon catches up and leads to pulse contraction. This stage of temporal compression can be used for generating pulses with ultrashort durations [19–21]. However, the compressed pulses exhibit a low-amplitude structure outside the central lobe. An alternative technique that prevents sidelobe formation relies on the progressive evolution of the peak power and temporal duration of a fundamental soliton in a distributed amplifier or a dispersion decreasing (DD) fibre [22, 23]. Step-like or comb-like [24] dispersion profiled structures based on carefully chosen sets of fibres have been successfully used to emulate continuous DD fibre profiles.

The generation of high repetition rate pulse sources has largely benefited from the nonlinear compression effect taking place in the anomalous dispersion region, and several works have demonstrated the nonlinear reshaping of a sinusoidal beat-signal into well-separated pulses, the pulse repetition rate being simply determined by the frequency separation between the two continuous-wave (CW) laser sources. This technique has been successfully demonstrated with various experimental setups including DD fibres, adiabatic Raman compression in standard fibres, and step-like and comb-like dispersion profiled fibres. More recently, this nonlinear compression effect has been observed through a multiple four-wave mixing (FWM) process in a single anomalously dispersive fibre, and repetition rates ranging from a few tens gigahertz to 1 terahertz have been achieved [25].

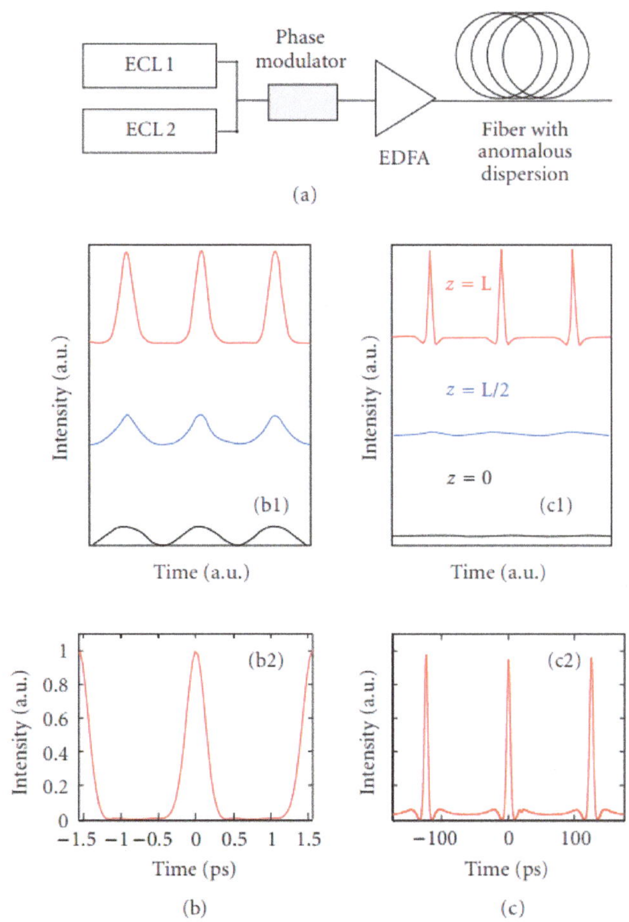

Figure 1: (a) Experimental setup for high repetition rate pulse train generation. (b, c) Results obtained for the optimum output shape and for the best compression, respectively. The temporal intensity profiles at different stages as predicted by numerical simulations (subplots 1) are compared to the experimental results (subplots 2) obtained by the FROG technique (b2—results adapted from [25]) or by an optical sampling oscilloscope (c2—results adapted from [31]).

In the frequency domain, such nonlinear compression can be viewed as the result of a modulation instability [26] experienced by the initial beat signal during propagation in the fibre, with the generation of new sidebands via FWM: two pump photons at and interact to generate two first-order sideband photons at and , being the pump frequency detuning. High-order sidebands are then generated due to multiple FWM interactions. We note that a powerful alternative technique for realizing high-repetition-rate short-pulse sources relies on harmonic passive mode locking of fibre lasers [27–30]. The advantage of laser cavities as such compared to nonlinear pulse shaping in one-stage fibre systems is the possibility of achieving lower pulse train duty cycles. However, laser cavities have the drawbacks that they require mode locking, they cannot be referenced to an external clock directly and are not very suitable for the generation of pulses with ultrahigh repetition rates above a hundred gigahertz.

A typical experimental setup for the generation of high repetition rate pulse trains at telecommunication wavelengths is sketched in Figure 1(a). The initial beat signal is easily obtained via temporal superposition of two CWs with slightly different wavelengths delivered by external cavity lasers or a single CW directly modulated by an intensity modulator. A phase modulator driven at a frequency of a few hundreds of megahertz is used to avoid the deleterious effects of Brillouin back scattering. The beat signal is then amplified up to the Watt level by an erbium-doped fibre amplifier (EDFA) and propagated through a suitable optical fibre with anomalous dispersion to realize the nonlinear reshaping. According to the pulse source parameters that are chosen (pulse repetition rate, initial peak power, nonlinearity and dispersion coefficients of the compression fibre, and fibre length), various output pulse shapes can be obtained. Intensity and phase measurements based on the frequency resolved optical gating (FROG) technique [32] show that high-quality compressed pulses with a Gaussian intensity profile and nearly constant phase can be achieved at a typical duty cycle of 1/5 [25, 33] (Figure 1(b)). Lower duty cycles can still be obtained with a single fibre, but at the expenses of increased sidelobes in the wings of the pulses [34]. Such sidelobes can be partly avoided by using specially designed arrangements of segments of fibres [35–37] or other comb-like structures.

We would like to note that the reshaping of a sinusoidal beat signal into ultrashort structures can be understood in terms of the nonlinear dynamics of "Akhmediev breathers" [38, 39]. In this context, the extreme spatiotemporal localization that can be observed (see, e.g., Figure 1(c)) is explained as a result of the formation of Peregrine solitons [31, 40], mathematical solutions of the standard NLS equation that have been first derived in 1983 in the hydrodynamics context [41] but have never been demonstrated experimentally before 2010, and using experiments in optical fibres. These observations offer a promising way to better understand the problem of the occurrence of rogue waves in optics and hydrodynamics [42, 43].

Generation of Parabolic Pulses in Active and Passive Fibres

Strong pulse reshaping can also occur in the normal regime of dispersion of a fibre. An important example that has generated a great deal of attention in the last decade is the occurrence of self-similar dynamical effects in the nonlinear propagation of ultrashort pulses in normally dispersive (ND) optical fibre amplifying media [44]. Results have demonstrated a fundamentally new operating regime where the interaction of Kerr nonlinearity, dispersion, and optical gain is exploited to generate a particular class of pulses with a parabolic intensity profile and a strictly linear frequency chirp. These pulses then propagate in a self-similar manner, holding certain relations (scaling) between pulse power, duration, and chirp parameter. The results for the self-similar parabolic pulses are obtained by assuming that a fibre amplifier can be described by the NLS equation with gain

$$\psi_z = -i\frac{\beta_2}{2}\psi_{tt} + i\gamma|\psi|^2\psi + \frac{g}{2}\psi, \tag{2}$$

where g is the distributed gain coefficient. Using a self-similar ansatz and standard amplitude/phase decomposition, one can find self-similar parabolic pulse solutions in the quasiclassical limit of (2) $(\beta_2|(|\psi|)_{tt}|/(2\gamma|\psi|^3) \ll 1)$, whose dynamic evolution is summarized as [45, 46]

$$\begin{aligned} |\psi(z,t)| &= a(z)\sqrt{1 - t^2/\tau(z)^2}\,\theta(\tau(z) - |t|), \\ \arg\psi(z,t) &= b(z)t^2 + \phi_0(z), \qquad a(z) = a_0 e^{gz/3}, \\ \tau(z) &= \tau_0 e^{gz/3}, \qquad b(z) = -\frac{g}{6\beta_2}, \end{aligned} \tag{3}$$

in the case of constant gain. Here, θ(x) is the Heaviside function. The amplitude term indeed reveals a parabolic intensity profile, and the phase term a parabolic phase modulation or linear chirp with chirp rate b/2. The peak amplitude and width of the pulse scale from their initial values, and the initial values depend only on the initial energy and amplifier parameters. These solutions possess the remarkable property of being a global attractor to the system for arbitrary initial conditions [47, 48]. By analogy with the well-known stable dynamics of solitary waves or solitons, these self-similar parabolic pulses have come to be known as "similaritons". In contrast to solitons, similaritons can tolerate strong nonlinearity without wave breaking. The normal GVD effectively linearizes the accumulated phase of the pulse allowing for the spectral bandwidth to increase without destabilizing the pulse [49]. The unique properties of similariton pulses have motivated many theoretical and experimental studies (see, e.g., [46, 48, 50–53]).

Experimental demonstrations of similariton generation relying on amplification from either rare-earth doping (with ytterbium [45, 56–59] or erbium [51] dopants) or Raman scattering [50, 54, 60] have been achieved, confirming the potential of this method especially when dealing with the generation of high-power ultrashort pulses. Thanks to the high level of linearity of the chirp developed during parabolic amplification and as long as the impact of third-order fibre dispersion is moderate [52, 61, 62], it is possible to efficiently compensate for the chirp slope, so that compressed pulses with very low substructures and temporal durations much shorter than the initial seed pulse can be obtained [45, 51, 56–59]. As an illustration, Figures2(a) and 2(b) show the setup and results of an experiment where similaritons are generated through Raman amplification at telecommunication wavelengths [54].

Moreover, recent fibre lasers that use self-similar pulse shaping in the normal dispersion regime have been demonstrated to achieve high-energy pulses [63–66].

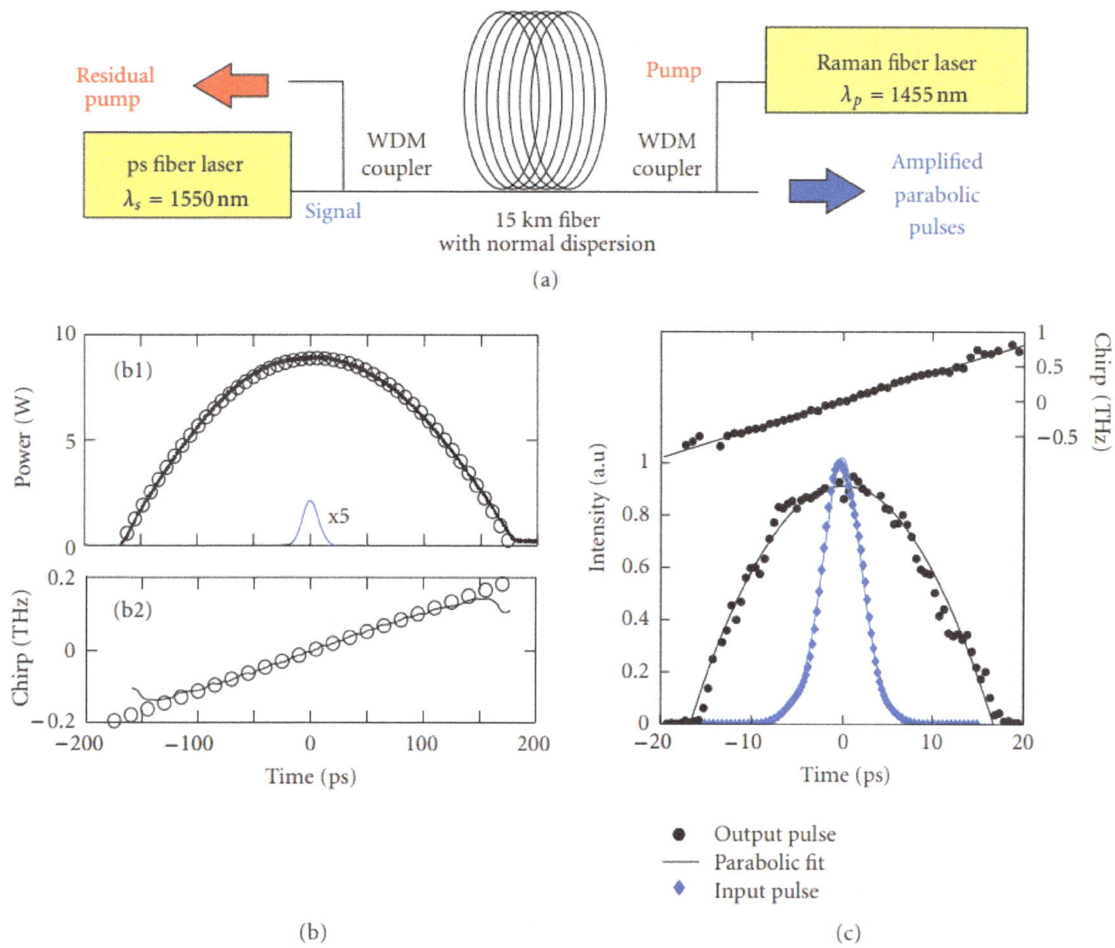

Figure 2: (a) Experimental setup for parabolic pulse generation through Raman amplification at telecommunication wavelengths. (b) Results obtained in the Raman amplifier (adapted from [54]): temporal intensity profile (subplot b1) and chirp profile (subplot b2) of the output pulse from the amplifier. The experimental results (solid black lines) are compared to a parabolic and a linear fit (open black circles), respectively. The intensity profile of the initial pulse is also shown (solid blue line). (c) Results obtained in a ND decreasing fibre (adapted from [55]): experimental temporal intensity and chirp profiles at the fibre output (black circles) compared to a parabolic and a linear fit (solid black lines), respectively. The initial intensity profile is also shown (blue diamonds).

In addition to fibre amplifiers and lasers, stable similariton pulses can be generated in passive fibres provided a suitable longitudinal variation of the dispersion is introduced [55, 67–69] (Figure 2(c)). This approach is based on the observation that a longitudinal decrease of the normal dispersion is formally equivalent to linear gain. Recently, a simple approach to the generation of parabolic pulses that uses progressive nonlinear pulse reshaping in a ND fibre with fixed dispersion has been demonstrated [70]. However, in contrast with the asymptotic similariton solutions obtained in fibre amplifiers, the generated parabolic waveforms represent transient states of the nonlinear pulse evolution in the passive fibre medium [71]. As such, they have a finite life distance that depends sensibly on the initial conditions (pulse shape, energy, and chirp profile). Nevertheless, stabilization of the parabolic features is possible by use of a second propagation stage in a fibre with specially adjusted nonlinear and dispersive characteristics relative to the first fibre [70, 71].

Generation of Triangular Pulses

Parabolic shapes are not the only pulse waveforms that can be generated in a passive ND fibre. Recently, the combination of pulse prechirping and nonlinear propagation in a section of ND fibre has been introduced as a method for passive nonlinear pulse shaping, which provides a simple way of generating various advanced field distributions, including flat-top- and triangular-profiled pulses with a linear chirp [71]. In this scheme, Kerr nonlinearity and GVD lead to various reshaping processes of an initial conventional pulse (e.g., a Gaussian pulse) according to the chirping value and power level at the entrance of the fibre. In particular, triangular pulses can be generated for a positive initial chirp parameter (using the definition for the phase profile) and sufficiently high energies.

These theoretical results have been confirmed experimentally by intensity and phase measurements of the generated pulses [72] (Figure 3(b)). In the experimental setup used (Figure 3(a)), the control of the pulse prechirping value was realized by propagation through different lengths of standard SM fibre with anomalous GVD, which imposed a positive chirp parameter on the pulse. The prechirped pulses were amplified to different power levels using an EDFA and then propagated through a ND fibre to realize the pulse reshaping.

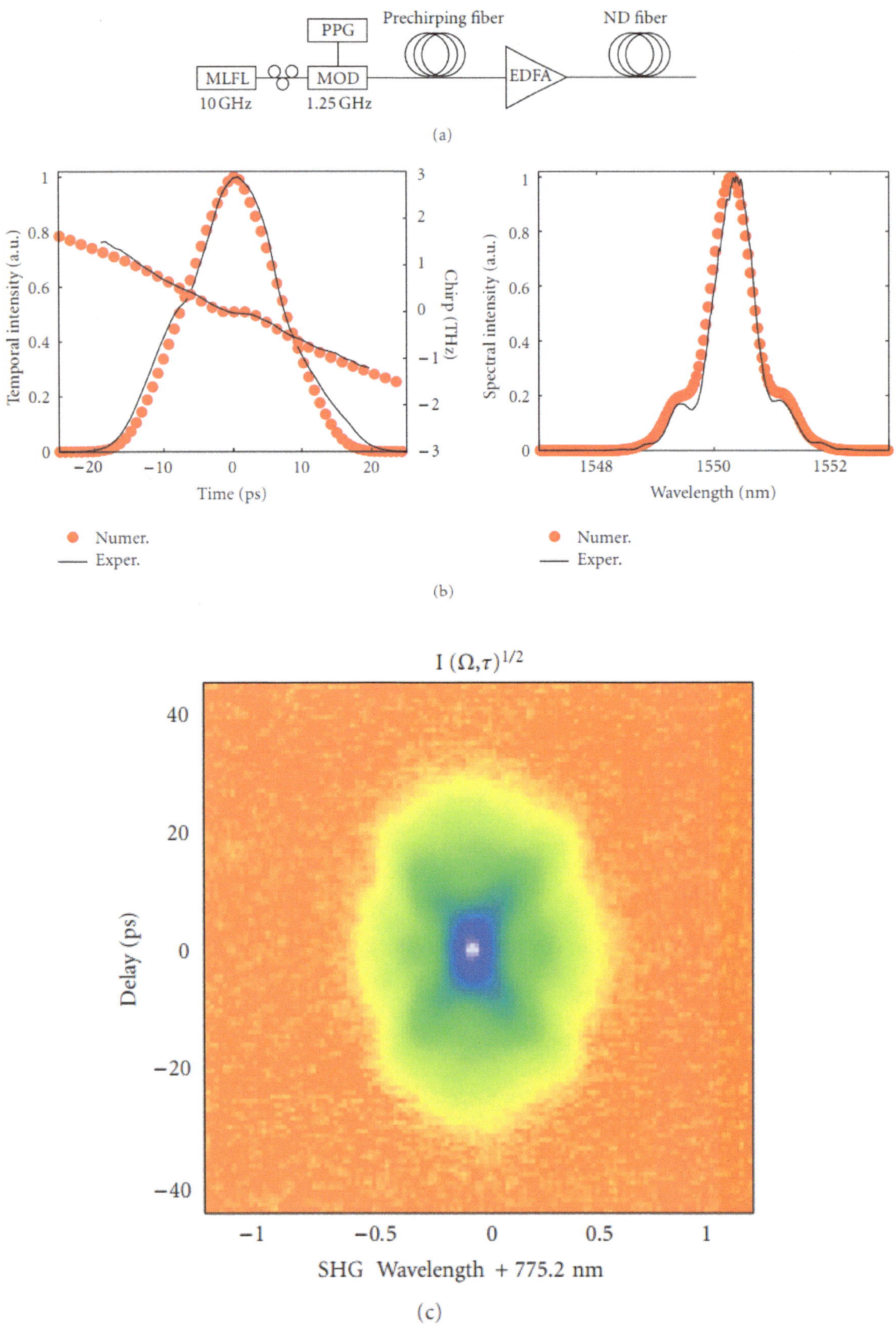

Figure 3: (a) Experimental setup for triangular pulse generation. (b) Experimental and numerical temporal intensity and chirp (φ_t) profiles retrieved from second-harmonic generation (SHG) FROG (left) and optical spectra (right) of the generated pulses for an example length of prechirping fibre. (c) SHG-FROG spectrogram of the pulse in (b). Results adapted from [72].

Different nonlinear dynamics may lead to the generation of triangular pulses in a passive ND fibre. Indeed, it has been recently shown qualitatively and numerically that temporal triangular intensity profiles can result from the progressive reshaping of initially parabolic pulses driven by the fourth-order dispersion (FOD) of the fibre [53]. The overall temporal effect of FOD on parabolic pulse propagation is to stretch and enhance the power reduction in the pulse wings, leading to a triangular profile. Furthermore, the possibility of triangular pulse shaping in mode-locked ring-cavity fibre lasers has been first reported in [73]. It has been numerically demonstrated that for normal net dispersion, formation of two distinct steady-state solutions of stable single pulses can be obtained in a laser cavity in different regions of the system parameter space: the previously known similariton [63] and a triangular-profiled pulse with a linear frequency chirp.

Coherent Continuums for Optical Telecommunications

One of the most well-known examples of pulse reshaping in the spectral domain is in supercontinuum (SC) generation. SC generation in optical fibres, and more specifically, in photonic crystal fibres with enhanced nonlinearity, is an established technique for producing broadband light sources in a wide range of research fields including metrology, biophotonics, and optical telecommunications [74, 75]. In telecommunications, the broadened spectrum can be spectrally sliced to generate wavelength multiplexed pulse trains [76–78]. The principle of operation of a multiwavelength source is illustrated in Figure 4(a). An initial high-power mode-locked picosecond or femtosecond pulse train with a repetition rate of several gigahertz is spectrally broadened through SPM in a kilometer-long highly nonlinear (HNL) fibre. At the fibre output, the spectrum is sliced by adequate bandpass filters (BPFs) to obtain pulse trains that have the same repetition rate as the initial train but new central frequencies.

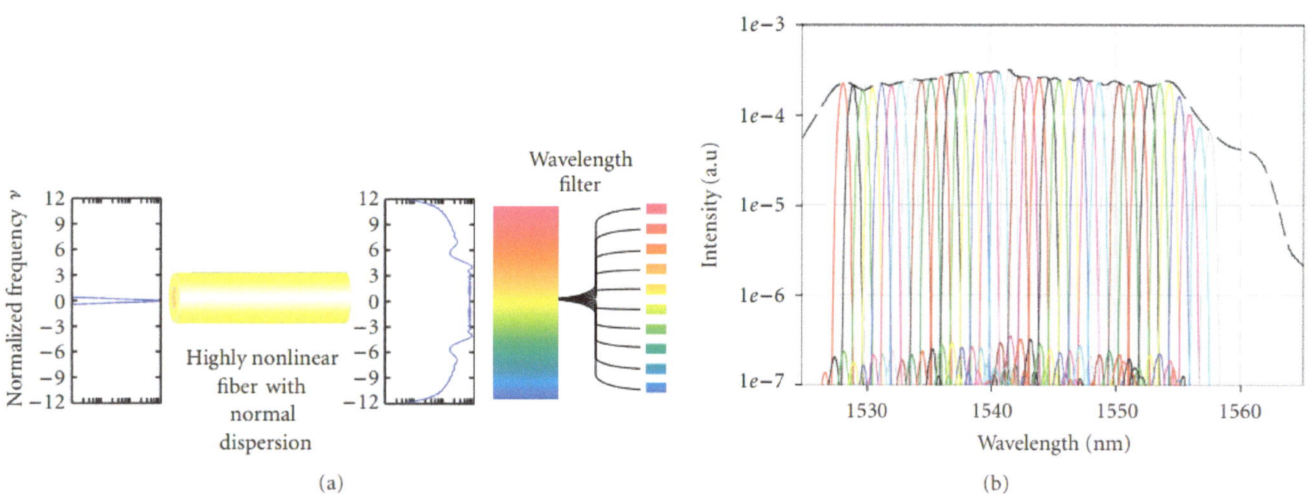

Figure 4: (a) Principle of operation of a multiwavelength picosecond source. (b) Experimental spectrum of a parabolic-shaped pulse after propagation in a ND-HNL fibre (black dashed line) and superposition of the measured spectra of 38 sliced channels. Results adapted from [79].

Typically the broadest spectra are generated in anomalously dispersive fibres, where the broadening mechanism is strongly influenced by soliton dynamics [80]. However, solitonic effects and modulation instability lead to decreased temporal coherence and spectral flatness of the continuum generated in the anomalous dispersion regime [81]. By contrast, the nonlinear pulse evolution in ND fibres leads to flatter spectra and improved temporal coherence, but at the expenses of relatively narrower spectra [82, 83]. Hence frequency broadening using a ND fibre is an attractive choice whenever a moderate amount of broadening alongside good spectral and temporal stability of the output pulses are targeted. However, the main limits to spectral pulse quality in the normal dispersion regime are the spectral ripple arising from SPM of conventional laser pulses, and the effects of optical wave breaking which may lead to significant change in the temporal pulse shape and severe energy transfer into the wings of the spectrum. These effects can in principle be avoided by using preshaped input pulses with a parabolic temporal intensity profile, which would preserve their shape whilst propagating within the fibre and, thus, result in spectrally flat, highly coherent pulses. Continuum with low-spectral ripple and high-energy density in the central part is indeed achievable through parabolic pulse propagation in ND-HNL fibre [79], and low-noise multiwavelength picosecond sources covering the whole C-band of optical telecommunications have been demonstrated based on this technique [12, 16, 79] (Figure 4(b)).

APPLICATIONS OF PARABOLIC AND TRIANGULAR PULSES IN SIGNAL PROCESSING

Applications of parabolic pulses have not been limited to ultrashort high-power pulse generation and highly coherent continuum sources, but several optical processing techniques have taken advantage of their specific features. These include optical regeneration [84, 85] including pulse retiming [86], prereceiver nonlinear processing in the optical domain [87, 88], and mitigation of linear waveform distortions [89, 90]. The simple intensity profile of triangular pulses is also highly desired for a range of processing applications, including time domain add-drop multiplexing [91, 92], wavelength conversion [93–95], optical signal doubling [96], time-to-frequency mapping of multiplexed signals [97], and enhanced spectral compression [98, 99]. In this Section, we will review some of these applications.

2R Regeneration

As a first example, here we discuss the usage of parabolic pulses in 2R (reamplification, retiming) signal regeneration in an extension of the technique proposed by Mamyshev [100]. The method relies on SPM in a nonlinear medium (fibre) and subsequent offset spectral filtering to suppress the noise in the zero-bit slots and the amplitude fluctuations in the one-bit pulses of return-to-zero (RZ) optical data streams. In the schematic of the Mamyshev regenerator depicted in Figure 5(a), the signal to be regenerated is first amplified to the optimum power level, and an optical BPF is also used to reduce the amplified spontaneous emission noise power. Then spectral broadening is achieved in a HNL fibre, and the broadened spectrum is partially sliced by a second BPF, which is shifted with respect to the input signal frequency in order to transmit SPM-generated frequencies and reject the original spectrum. The central frequency and output pulse shape of the regenerated signal are determined by the central frequency and spectral profile of the offset filter, respectively. The SPM phase shift experienced in the HNL fibre depends on the instantaneous power (cf. Section 2). Hence, a high-intensity pulse induces more SPM phase shift and spectrally broadens more than the low-intensity noise [100]. Filtering the new SPM-generated frequencies with the offset filter leads to a power transfer function for the pulses and the noise which is close to a binary one [101].

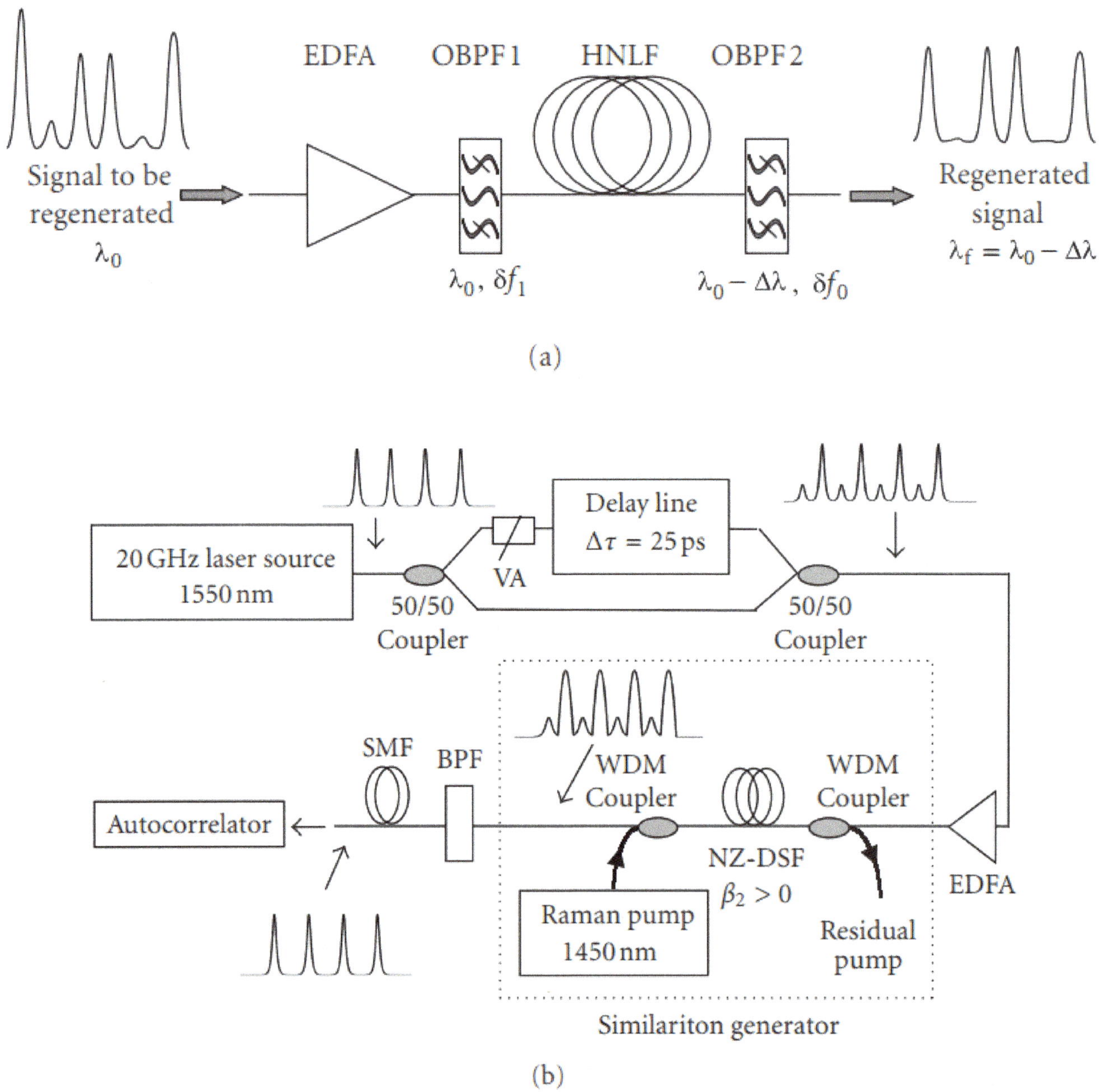

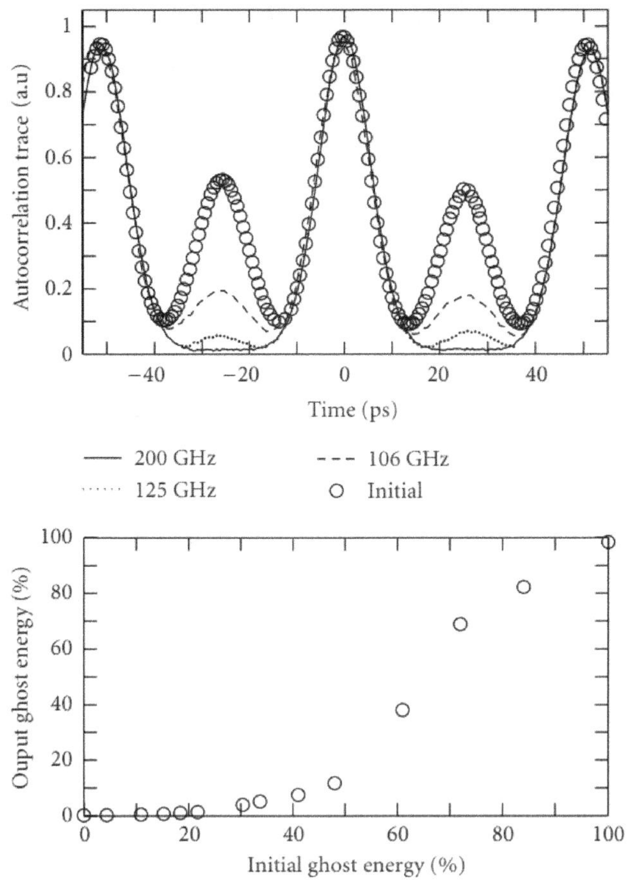

(c)

Figure 5: (a) Schematic of an optical regenerator based on SPM and filtering (after [100]). (b) Experimental setup for 40 Gbit/s similariton-based regeneration. (c) Top, autocorrelation traces of the pulse train at the Raman amplifier input (circles) and output for different values of the filter offset (dashed, dotted, and solid curves). Bottom, output versus input proportion of the spurious noise pulse energy with respect to the ones energy for a fixed filter offset. Results adapted from [84].

Spectral filtering of similariton pulses can improve the ability of the Mamyshev regenerator to eliminate spurious noise pulses as well as to simultaneously reduce any fluctuations in the signal one-bit level [84]. This ability relies on the specific features of self-similar amplification. Indeed, a spectrum broadened through self-similar amplification has a smooth profile (cf. Section 3.4), whereas a SPM-broadened spectrum is generally accompanied by oscillatory structures covering the entire frequency range. Moreover, the spectrum and chirp obtained by SPM both depend on the details of the initial pulse (shape, chirp, and energy), leading to some amplitude and time fluctuations of the regenerated pulses. In contrast, the similariton-based spectral broadening and similariton-induced chirp depend only on the initial pulse energy and/or amplifier parameters, respectively (cf. Section 3.2), so that both amplitude and timing jitter are reduced. In particular, the energy-dependent spectral broadening is responsible for efficient discrimination of high-energy pulses (ones) from low-energy pulses (zeros) with the offset BPF. An implementation of the similariton-based regeneration technique at 40 Gbit/s data rate where similariton pulses are generated through distributed Raman amplification in a ND fibre is illustrated in Figures 5(b) and 5(c). It is worth noting here that a distributed amplification is helpful in increasing the energy yield of the regenerator which is a highly dissipative element [102].

Linear-Distortion Compensation

As a second example of the application of parabolic pulses in signal processing, we review here the realization of a time-domain all-optical Fourier transformation (OFT) technique. Time-domain OFT in which the spectral profile is converted into a waveform in the time domain makes it possible to eliminate the waveform distortion caused by linear perturbations in optical fibres, such as higher order dispersion, jitter and polarization-mode dispersion [103–106]. Time-domain OFT is realized ideally by applying a quadratic temporal phase modulation (time lens) to a distorted pulse: $\psi_{\mathrm{chirp}}(t) = \psi(t)\exp(ibt^2)$ and passing the linearly chirped pulse through a dispersive medium such as a short piece of fibre. When the accumulated dispersion of the medium $D = \beta_2 L$ equates the inverse of the chirp coefficient b (i.e., when the chirp is fully compensated), the output waveform is proportional to the spectral amplitude before OFT: $\chi(t) = \sqrt{i/(2\pi D)}\exp(-ibt^2)\tilde{\psi}(t/D)$, where $\tilde{\psi}(\omega)$ is the Fourier transform of $\psi(t)$, and the frequency ω is converted into time t through the relationship $\omega = t/D$. Since generally the spectral envelope profile has no distortion even if its time-domain waveform is distorted by linear perturbations, this frequency-to-time mapping indicates that a distortion-free pulse waveform can be obtained in the time domain after OFT. In [103–106], this compensation technique was investigated by using electrooptic phase modulators. But this way the quadratic modulation region was restricted to around the center of the sinusoidal modulation imparted by the phase modulator. Thus, any parts of the pulse extending beyond the quadratic region could not be properly transformed, and only the section of the target pulse sitting within the quadratic region could be compensated (Figure 6).

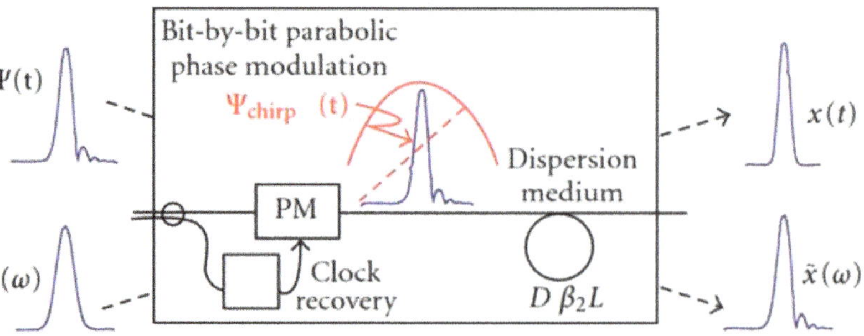

Figure 6: Schematic diagram of time-domain OFT (after [90]).

The use of XPM with parabolic pulses as a time lens has been recently proposed as a promising way to overcome the aforementioned limitation [89, 90]. Indeed, the chirp induced by XPM on the target pulse is proportional to the gradient of the intensity profile of the parabolic control pulse: $-\phi_t = -2bt, b = \mp 2\gamma a^2 L/\tau^2$, so it is exactly linear and thus ideal OFT can be achieved by complete chirp compensation. Here, a and τ are the respective peak amplitude and characteristic width of the parabolic pulse (cf. Section 3.2), and the sign ∓ is for either a standard "bright" or a dark parabolic pulse [17]. The parabolic XPM-based OFT technique has been successfully experimentally demonstrated to achieve cancellation of the third-order dispersion-induced distortion of short-optical pulses [89, 90].

Optical Decision at the Receiver

In this section, we review a possible application of parabolic/flat-top pulses to the improvement of the signal bit-error rate (BER) in an optical communication system by using optical decision before the receiver. An attractive approach to the design of the optical receiver for high-speed fibre communication systems using RZ data formats is based on employing an additional all-optical decision element (ODE) just before the conventional receiver [87, 107–109]. In this way, the first decision is carried out in the optical domain, thus improving the quality of the received signal. The attractive possibility of improving the BER by using optical devices has been studied in some recent works [88, 110, 111]. The conceptual basis of the BER improvement by optical devices comes from the information theory: an intrinsic BER improvement is possible whenever the decision performed by the device in the optical domain is based on information that is different from the information used by the decision circuit of the (suboptimal) receiver in the electrical domain. In [88] an application of this general concept was demonstrated, based on using an advanced optical receiver enhanced by a nonlinear ODE. The example of an ODE scheme used in [88] exploits the nonlinear mechanism that leads to the formation of parabolic pulses and the resultant flattening of the signal waveform (cf. Section 3.2). The method is based on the nonlinear reshaping of a pulse towards a broad and flat temporal waveform that can occur upon propagation in a ND fibre [9] and subsequent slicing of the pulse temporal profile by an optical temporal gate (e.g., an amplitude modulator with a sharp nonlinear transfer function or a nonlinear optical loop mirror provided with a clock) [85] or XPM-based techniques.

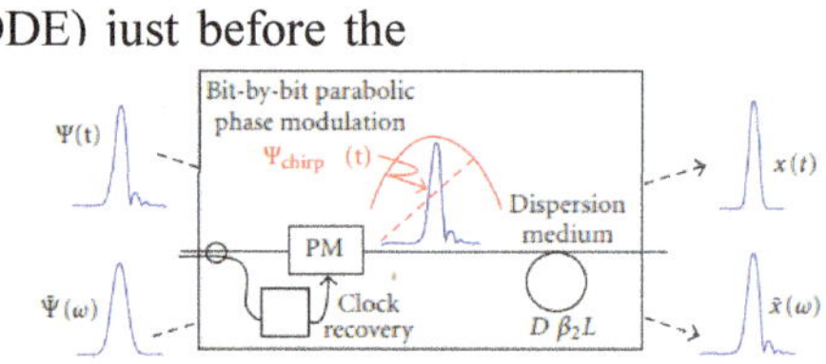

FIGURE 6: Schematic diagram of time-domain OFT (after [90]).

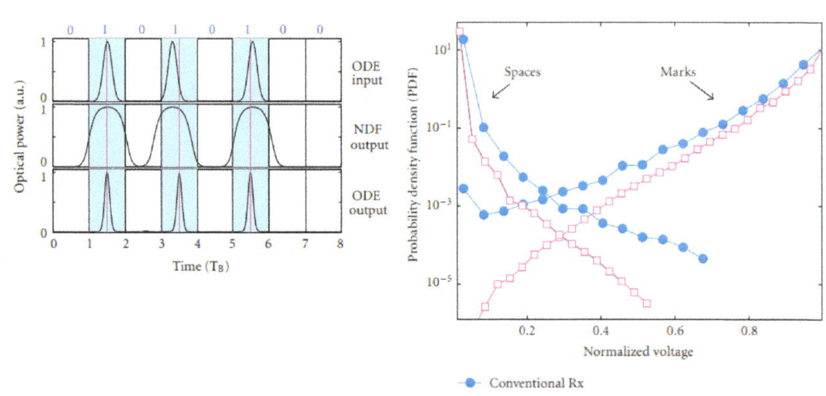

(a) (b)

Figure 7: (a) Schematic illustration of the optical decision at an ODE-enhanced receiver. (b) Numerically calculated probability density functions for the voltage of the marks and spaces in the conventional and ODE-enhanced receivers for a 40 Gbit/s RZ input signal with a random jitter of the pulse temporal position. Results adapted from [88].

The basic mechanism that is responsible for BER improvement in this scheme can be understood in the following way. The pulse temporal broadening and flattening in the ND fibre permits to recover the center of the bit slot for those pulses that are significantly shifted from the bit slot center in the input pulse train because of timing jitter (Figure 7(a)). Such strongly time-shifted pulses, once they are converted into electrical pulses and arrive at the decision point, might be missed by the threshold detector in those conventional receivers that use a relatively small fraction of the bit period. In contrast, the bit slot recovery in the ND fibre enables such pulses to be recognized as marks in the ODE-enhanced receiver. The slicing of the central portion of the broadened pulse temporal profile by the temporal gate enables efficient suppression of the timing jitter of the pulses which, in turn, improves the receiver sensitivity [87]. This approach is illustrated in Figure 7(b), which shows that BER improvement over the conventional scheme, indeed, can be achieved using nonlinear signal processing in the optical domain before the conventional receiver. It is worth noting that the same processing technique based on temporal slicing of nonlinearly flattened pulse waveforms can also be used in all-optical 3R (2R + re-timing) signal regeneration [85].

Optical Signal Doubling and Frequency Conversion

Here we discuss the use of triangular pulses in the context of optical signal doubling and frequency conversion. A technique of copying optical pulses in both the frequency and time domains based on a combination of either XPM with a triangular pump pulse or SPM of a triangular pulse in a nonlinear Kerr medium and subsequent propagation in a dispersive medium was introduced in [96] and experimentally demonstrated in [95]. In this scheme, spectral doubling of an optical signal is made possible by the XPM or SPM phase shift generated by a temporal intensity profile with a linear gradient: $\phi(t) = \phi_0(1 - |t/\tau|)$, $\phi_0 \propto a^2\gamma L$, (a and τ being the peak amplitude and characteristic width of a triangular pulse, resp.), which translates into a constant and distinct (opposite sign) frequency chirp induced onto the leading (downshifted) and trailing (upshifted) edges of the signal pulse. As a result of this constant and distinct chirp, under certain conditions the pulse spectrum develops a structure consisting of two equal peaks. The separation between the two spectral peaks increases with the maximum nonlinear phase shift ϕ_0, while their form is almost preserved. Under subsequent propagation of the modulated signal pulse in a linear dispersive medium, the produced spectral separation of the pulse components leads to their subsequent separation in time, thus allowing temporal replication of the pulse. In the case of normal dispersion, the pulse develops into a structure consisting of two identical pulses where the original pulse splits down the middle. The two pulse parts separate in time while experiencing dispersive broadening. The relative velocity of separation is proportional to the separation between the two peaks of the nonlinearly modulated pulse spectrum. Pulse splitting and separation occur also in the case of anomalous dispersion after an initial stage of pulse compression. Exact analytic expressions for the nonlinearly modulated pulse spectrum and the pulse temporal intensity distribution clearly show the features described above [96]. Applying spectral filtering to the XPM/SPM modulated signal is also possible to realize efficient and clean frequency conversion [93–95].

Figure 8 shows results of the experiment reported in [95]. Similarly to [72], triangular pulses were generated by passive nonlinear reshaping in a ND fibre. Spectral pulse doubling was realized through the SPM experienced by the triangular pulses over their life distance in a HNL fibre. The output SPM-shaped spectrum from the HNL fibre was sent into either a linear standard SM fibre segment to realize temporal pulse doubling or a wavelength demultiplexer filter for frequency conversion processing.

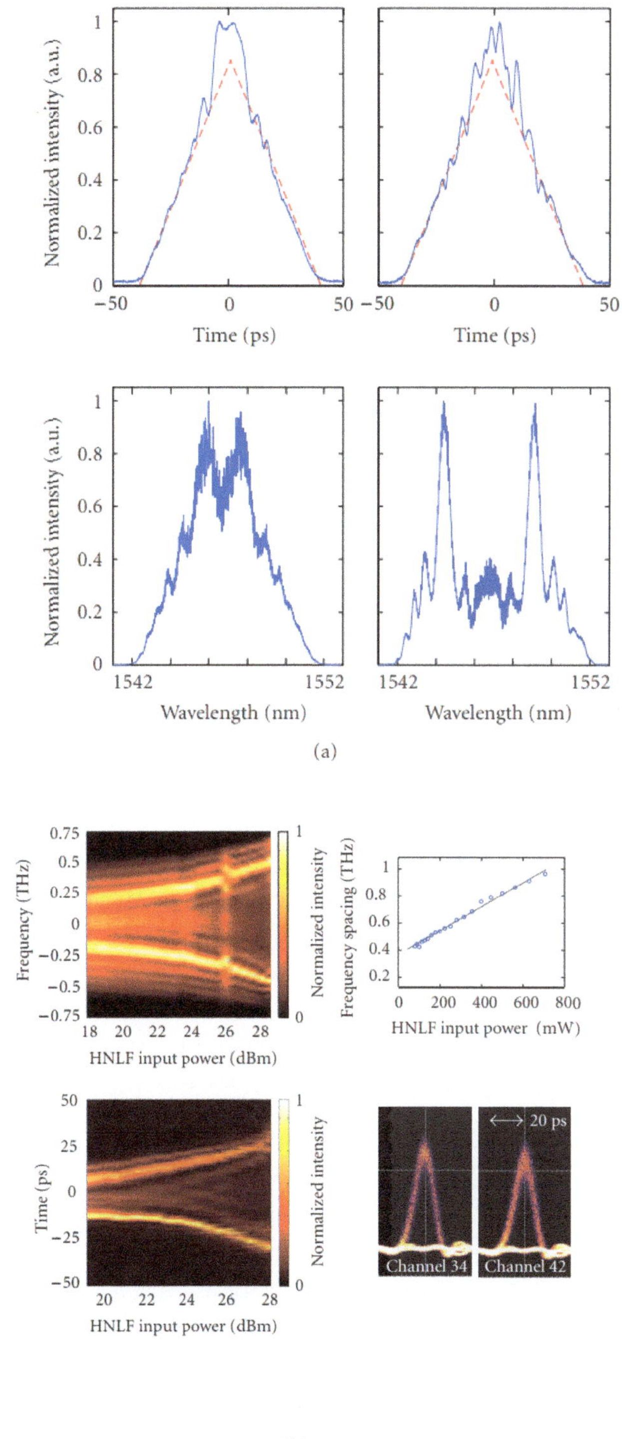

Figure 8: (a) Temporal (top) and spectral (bottom) intensity profiles after the ND reshaping fibre (left) and after the HNL fibre (right). (b)

Top, evolution of the output spectrum from the HNL fibre with the input power (left) and dependence of the frequency spacing between the two spectral peaks on the input power (right). Bottom, evolution of the temporal intensity profile at the output of the linear SM fibre with the input power to the HNL fibre (left) and eye diagrams of a 10 Gbit/s pulse train obtained after spectral filtering for the ITU channels 34 and 42. Results adapted from [95].

CONCLUSION

We have provided an overview of several recent examples of the use of nonlinear phenomena in optical fibres for the generation and shaping of optical pulses. We have discussed the generation of ultrashort pulse trains with very high-repetition rates in the anomalous dispersion regime of a fibre, the generation of parabolic-shaped pulses in active and passive ND fibres, the generation of pulses with triangular intensity profiles in passive ND fibres, and the generation of coherent continuums for optical telecommunications in ND fibres. Examples of new techniques or improvements in existing techniques of all-optical nonlinear fibre-based signal processing enabled by specialized pulse waveforms have been reviewed. Specifically, we have presented applications of parabolic pulses in optical signal regeneration, compensation of linear waveform distortions, and signal postprocessing in optical communication systems. Applications of triangular pulses in optical signal doubling and frequency conversion have also been discussed.

ACKNOWLEDGMENTS

The authors would like to acknowledge and thank our colleagues, namely, Prof. Sergei K. Turitsyn, Dr. Paul Harper, Dr. Anton I. Latkin, Prof. Keith J. Blow, Dr. Julien Fatome, Dr. Kamal Hammani, Dr. Bertrand Kibler, Prof. Guy Millot, and Prof. John M. Dudley and the Optoelectronics Research Center (Southampton, UK). The authors also acknowledge support from the British Council (Alliance Grant 10.002), the Ministère des Affaires Étrangères et Européennes (Alliance Grant 22836ZJ), the Agence Nationale de la Recherche, and the Conseil Regional de Bourgogne (PARI Photcom).

REFERENCES

1. Y. R. Shen, Principles of Nonlinear Optics, John Wiley & Sons, New York, NY, USA, 1984.
2. J. C. Knight, "Photonic crystal fibres," Nature, vol. 424, no. 6950, pp. 847–851, 2003.
3. B. J. Eggleton, B. Luther-Davies, and K. Richardson, "Chalcogenide photonics," Nature Photonics, vol. 5, no. 3, pp. 141–148, 2011.
4. G. P. Agrawal, Nonlinear Fiber Optics, Academic Press, San Diego, Calif, USA, 4th edition, 2006.
5. A. Hasegawa and F. Tappert, "Transmission of stationary nonlinear optical pulses in dispersive dielectric fibers I. Anomalous dispersion," Applied Physics Letters, vol. 23, no. 3, pp. 142–144, 1973.
6. L. F. Mollenauer and J. P. Gordon, Solitons in Optical Fibers: Fundamentals and Applications, Academic Press, San Diego, Calif, USA, 2006.
7. W. J. Tomlinson, R. H. Stolen, and A. M. Johnson, "Optical wave breaking of pulses in nonlinear optical fibers," Optics Letters, vol. 10, no. 9, pp. 457–459, 1985.
8. D. Anderson, M. Desaix, M. Lisak, and M. L. Quiroga–Teixeiro, "Wave breaking in nonlinear-optical fibers," Journal of the Optical Society of America B, vol. 9, no. 8, pp. 1358–1361, 1992.
9. C. Finot, B. Kibler, L. Provost, and S. Wabnitz, "Beneficial impact of wave-breaking for coherent continuum formation in normally dispersive nonlinear fibers," Journal of the Optical Society of America B, vol. 25, no. 11, pp. 1938–1948, 2008.
10. J. Azaña, L. K. Oxenløwe, E. Palushani, et al., "In-fiber subpicosecond pulse shaping for nonlinear optical telecommunication data processing at 640 Gbit/s," International Journal of Optics, vol. 2012, Article ID 895281, 16 pages, 2012.
11. A. M. Weiner, "Femtosecond pulse shaping using spatial light modulators," Review of Scientific Instruments, vol. 71, no. 5, pp. 1929–1960, 2000.
12. A. M. Clarke, D. G. Williams, M. A. F. Roelens, and B. J. Eggleton, "Reconfigurable optical pulse generator employing a Fourier-domain programmable optical processor," Journal of Lightwave Technology, vol. 28, no. 1, pp. 97–103, 2010.
13. D. Nguyen, M. U. Piracha, D. Mandridis, and P. J. Delfyett, "Dynamic parabolic pulse generation using temporal shaping of wavelength to time mapped pulses," Optics Express, vol. 19, no. 13, pp. 12305–12311, 2011.
14. J. Ye, L. S. Yan, W. Pan, et al., "Photonic generation of triangular-shaped pulses based on frequency-to-time conversion," Optics Letters, vol. 36, no. 8, pp. 1458–1460, 2011.
15. P. Petropoulos, M. Ibsen, A. D. Ellis, and D. J. Richardson, "Rectangular pulse generation based on pulse reshaping using a superstructured fiber Bragg grating," Journal of Lightwave Technology, vol. 19, no. 5, pp. 746–752, 2001.
16. D. Krčmařík, R. Slavík, Y. G. Park, and J. Azaña, "Nonlinear pulse compression of picosecond parabolic-like pulses synthesized with a long period fiber grating filter," Optics Express, vol. 17, no. 9, pp. 7074–7087, 2009.
17. T. Hirooka, M. Nakazawa, and K. Okamoto, "Bright and dark 40 GHz parabolic pulse generation using a picosecond optical pulse train and an arrayed waveguide grating," Optics Letters, vol. 33, no. 10, pp. 1102–1104, 2008.
18. K. Kashiwagi, H. Ishizu, and T. Kurokawa, "Fiber transmission characteristics of parabolic pulses generated by optical pulse synthesizer," Japanese Journal of Applied Physics, vol. 50, no. 9, Article ID 092501, 2011.
19. B. Kibler, R. Fischer, P. A. Lacourt, et al., "Optimised one-step compression of femtosecond fibre laser soliton pulses around 1550 nm to below 30 fs in highly nonlinear fibre," Electronics Letters, vol. 43, no. 17, pp. 915–916, 2007.
20. A. A. Amorim, M. V. Tognetti, P. Oliveira, et al., "Sub-two-cycle pulses by soliton self-compression in highly nonlinear photonic crystal fibers," Optics Letters, vol.

34, no. 24, pp. 3851–3853, 2009.

21. L. F. Mollenauer, R. H. Stolen, and J. P. Gordon, "Experimental observation of picosecond pulse narrowing and solitons in optical fibers," Physical Review Letters, vol. 45, no. 13, pp. 1095–1098, 1980.

22. S. V. Chernikov and P. V. Mamyshev, "Femtosecond soliton propagation in fibers with slowly decreasing dispersion," Journal of the Optical Society of America B, vol. 8, no. 8, pp. 1633–1641, 1991.

23. S. V. Chernikov, D. J. Richardson, E. M. Dianov, and D. N. Payne, "Picosecond soliton pulse compressor based on dispersion decreasing fibre," Electronics Letters, vol. 28, no. 19, pp. 1842–1844, 1992.

24. S. V. Chernikov, J. R. Taylor, and R. Kashyap, "Comblike dispersion-profiled fiber for soliton pulse train generation," Optics Letters, vol. 19, no. 8, pp. 539–541, 1994.

25. J. Fatome, S. Pitois, and G. Millot, "20-GHz-to-1-THz repetition rate pulse sources based on multiple four-wave mixing in optical fibers," IEEE Journal of Quantum Electronics, vol. 42, no. 10, Article ID 01703695, pp. 1038–1046, 2006.

26. K. Tai, A. Hasegawa, and A. Tomita, "Observation of modulational instability in optical fibers," Physical Review Letters, vol. 56, no. 2, pp. 135–138, 1986.

27. B. Ortaç, A. Hideur, G. Martel, and M. Brunel, "2-GHz passive harmonically mode-locked Yb-doped double-clad fiber laser," Applied Physics B, vol. 81, no. 4, pp. 507–509, 2005.

28. A. Komarov, H. Leblond, and F. Sanchez, "Passive harmonic mode-locking in a fiber laser with nonlinear polarization rotation," Optics Communications, vol. 267, no. 1, pp. 162–169, 2006.

29. A. Komarov, A. Haboucha, and F. Sanchez, "Ultrahigh-repetition-rate bound-soliton harmonic passive mode-locked fiber lasers," Optics Letters, vol. 33, no. 19, pp. 2254–2256, 2008.

30. J. Schröder, S. Coen, F. Vanholsbeeck, and T. Sylvestre, "Passively mode-locked Raman fiber laser with 100 GHz repetition rate," Optics Letters, vol. 31, no. 23, pp. 3489–3491, 2006.

31. K. Hammani, B. Kibler, C. Finot, et al., "Peregrine soliton generation and breakup in standard telecommunications fiber," Optics Letters, vol. 36, no. 2, pp. 112–114, 2011.

32. R. Trebino, Frequency-Resolved Optical Gating: The Measurement of Ultrashort Laser Pulses, Kluwer Academic, Norwell, Mass, USA, 2000.

33. J. Fatome, S. Pitois, C. Fortier, et al., "Multiple four-wave mixing in optical fibers: 1.5-3.4-THz femtosecond pulse sources and real-time monitoring of a 20-GHz picosecond source," Optics Communications, vol. 283, no. 11, pp. 2425–2429, 2010.

34. S. Pitois, C. Finot, J. Fatome, B. Sinardet, and G. Millot, "Generation of 20-GHz picosecond pulse trains in the normal and anomalous dispersion regimes of optical fibers," Optics Communications, vol. 260, no. 1, pp. 301–306, 2006.

35. C. Finot, J. Fatome, S. Pitois, and G. Millot, "All-fibered high-quality low duty-cycle 20-GHz and 40-GHz picosecond pulse sources," IEEE Photonics Technology Letters, vol. 19, no. 21, pp. 1711–1713, 2007.

36. C. Fortier, B. Kibler, J. Fatome, C. Finot, S. Pitois, and G. Millot, "All-fibered high-quality low duty-cycle 160-GHz femtosecond pulse source," Laser Physics Letters, vol. 5, no. 11, pp. 817–820, 2008.

37. I. E. Mansouri, J. Fatome, C. Finot, S. Pitois, and M. Lintz, "All-fibered high-quality stable 20- and 40-GHz picosecond pulse generators for 160-Gb/s OTDM applications," IEEE Photonics Technology Letters, vol. 23, no. 20, pp. 1487–1489, 2011.

38. N. Akhmediev and V. I. Korneev, "Modulation instability and periodic solutions of the nonlinear Schrödinger equation," Theoretical and Mathematical Physics, vol. 69, no. 2, pp. 1089–1093, 1986.

39. K. Hammani, B. Wetzel, B. Kibler, et al., "Spectral dynamics of modulation instability described using Akhmediev breather theory," Optics Letters, vol. 36, no. 11, pp. 2140–2142, 2011.

40. B. Kibler, J. Fatome, C. Finot et al., "The Peregrine soliton in nonlinear fibre optics," Nature Physics, vol. 6, no. 10, pp. 790–795, 2010.

41. D. H. Peregrine, "Water waves, nonlinear Schrödinger equations and their solutions," The Journal of the Australian Mathematical Society B, vol. 25, no. 1, pp. 16–43, 1983.

42. J. M. Dudley, C. Finot, G. Millot, et al., "Extreme events in optics: challenges of the MANUREVA project," European Physical Journal, vol. 185, no. 1, pp. 125–133, 2010.

43. D. R. Solli, C. Ropers, P. Koonath, and B. Jalali, "Optical rogue waves," Nature, vol. 450, no. 7172, pp. 1054–1057, 2007.

44. J. M. Dudley, C. Finot, G. Millot, and D. J. Richardson, "Self-similarity in ultrafast nonlinear optics,"Nature Physics, vol. 3, no. 9, pp. 597–603, 2007.

45. M. E. Fermann, V. I. Kruglov, B. C. Thomsen, J. M. Dudley, and J. D. Harvey, "Self-similar propagation and amplification of parabolic pulses in optical fibers," Physical Review Letters, vol. 84, no. 26, pp. 6010–6013, 2000.

46. S. Boscolo, S. K. Turitsyn, V. Y. Novokshenov, and J. H. B. Nijhof, "Self-similar parabolic optical solitary waves," Theoretical and Mathematical Physics, vol. 133, no. 3, pp. 1647–1656, 2002.

47. V. L. Kruglov and J. D. Harvey, "Asymptotically exact parabolic solutions of the generalized nonlinear Schrödinger equation with varying parameters," Journal of the Optical Society of America B, vol. 23, no. 12, pp. 2541–2550, 2006.

48. C. Finot and L. Wu, "Influence of the initial phase profile on the asymptotic self-similar parabolic dynamics," Journal of Nonlinear Optical Physics and Materials, vol. 18, no. 4, pp. 709–721, 2009.

49. D. Anderson, M. Desaix, M. Karlsson, M. Lisak, and M. L. Quiroka-Teixeiro, "Wave-breaking-free pulses in nonlinear-optical fibers," Journal of the Optical Society

of America B, vol. 10, no. 7, pp. 1185–1190, 1993.

50. C. Finot, G. Millot, and J. M. Dudley, "Asymptotic characteristics of parabolic similariton pulses in optical fiber amplifiers," Optics Letters, vol. 29, no. 21, pp. 2533–2535, 2004.

51. C. Billet, J. M. Dudley, N. Joly, and J. C. Knight, "Intermediate asymptotic evolution and photonic bandgap fiber compression of optical similaritons around 1550 nm," Optics Express, vol. 13, no. 9, pp. 3236–3241, 2005.

52. B. G. Bale and S. Boscolo, "Impact of third-order fibre dispersion on the evolution of parabolic optical pulses," Journal of Optics A, vol. 12, no. 1, Article ID 015202, 2010.

53. B. G. Bale, S. Boscolo, K. Hammani, and C. Finot, "Effects of fourth-order fiber dispersion on ultrashort parabolic optical pulses in the normal dispersion regime," Journal of the Optical Society of America B, vol. 28, no. 9, pp. 2059–2065, 2011.

54. K. Hammani, C. Finot, S. Pitois, J. Fatome, and G. Millot, "Real-time measurement of long parabolic optical similaritons," Electronics Letters, vol. 44, no. 21, pp. 1239–1240, 2008.

55. C. Finot, B. Barviau, G. Millot, A. Guryanov, A. Sysoliatin, and S. Wabnitz, "Parabolic pulse generation with active or passive dispersion decreasing optical fibers," Optics Express, vol. 15, no. 24, pp. 15824–15835, 2007.

56. J. P. Limpert, T. Schreiber, T. Clausnitzer, et al., "High-power femtosecond Yb-doped fiber amplifier,"Optics Express, vol. 10, no. 14, pp. 628–638, 2002.

57. A. Malinowski, A. Piper, J. H. V. Price, et al., "Ultrashort-pulse Yb^{3+}-fiber-based laser and amplifier system producing > 25-W average power," Optics Letters, vol. 29, no. 17, pp. 2073–2075, 2004.

58. P. Dupriez, C. Finot, A. Malinowski, et al., "High-power, high repetition rate picosecond and femtosecond sources based on Yb-doped fiber amplification of VECSELs," Optics Express, vol. 14, no. 21, pp. 9611–9616, 2006.

59. D. N. Papadopoulos, Y. Zaouter, M. Hanna, et al., "Generation of 63 fs 4.1 MW peak power pulses from a parabolic fiber amplifier operated beyond the gain bandwidth limit," Optics Letters, vol. 32, no. 17, pp. 2520–2522, 2007.

60. C. Finot, G. Millot, S. Pitois, C. Billet, and J. M. Dudley, "Numerical and experimental study of parabolic pulses generated via Raman amplification in standard optical fibers," IEEE Journal on Selected Topics in Quantum Electronics, vol. 10, no. 5, pp. 1211–1218, 2004.

61. S. Zhang, C. Jin, Y. Meng, X. Wang, and H. Li, "Propagation of high-power parabolic pulses in cubicon fiber amplifiers," Journal of the Optical Society of America B, vol. 27, no. 6, pp. 1272–1278, 2010.

62. Vladimir I. Kruglov, Claude Aguergaray, and John D. Harvey, "Propagation and breakup of pulses in fiber amplifiers and dispersion-decreasing fibers with third-order dispersion," Physical Review A, vol. 84, no. 2, Article ID 023823, 14 pages, 2011.

63. F. Ö. Ilday, J. R. Buckley, W. G. Clark, and F. W. Wise, "Self-similar evolution of parabolic pulses in a laser," Physical Review Letters, vol. 92, no. 21, Article ID 213902, 4 pages, 2004.

64. F. W. Wise, A. Chong, and W. H. Renninger, "High-energy femtosecond fiber lasers based on pulse propagation at normal dispersion," Laser and Photonics Reviews, vol. 2, no. 1-2, pp. 58–73, 2008.

65. C. Aguergaray, D. Méchin, V. I. Kruglov, and J. D. Harvey, "Experimental realization of a mode-locked parabolic Raman fiber oscillator," Optics Express, vol. 18, no. 8, pp. 8680–8687, 2010.

66. B. Oktem, C. Ülgüdür, and F. O. Ilday, "Soliton-similariton fibre laser," Nature Photonics, vol. 4, no. 5, pp. 307–311, 2010.

67. T. Hirooka and M. Nakazawa, "Parabolic pulse generation by use of a dispersion-decreasing fiber with normal group-velocity dispersion," Optics Letters, vol. 29, no. 5, pp. 498–500, 2004.

68. B. Kibler, C. Billet, P. A. Lacourt, R. Ferriere, L. Larger, and J. M. Dudley, "Parabolic pulse generation in comb-like profiled dispersion decreasing fibre," Electronics Letters, vol. 42, no. 17, pp. 965–966, 2006.

69. A. Plocky, A. A. Sysoliatin, A. I. Latkin, et al., "Experiments on the generation of parabolic pulses in waveguides with length-varying normal chromatic dispersion," JETP Letters, vol. 85, pp. 319–322, 2007.

70. C. Finot, L. Provost, P. Petropoulos, and D. J. Richardson, "Parabolic pulse generation through passive nonlinear pulse reshaping in a normally dispersive two segment fiber device," Optics Express, vol. 15, no. 3, pp. 852–864, 2007.

71. S. Boscolo, A. I. Latkin, and S. K. Turitsyn, "Passive nonlinear pulse shaping in normally dispersive fiber systems," IEEE Journal of Quantum Electronics, vol. 44, no. 12, pp. 1196–1203, 2008.

72. H. Wang, A. I. Latkin, S. Boscolo, P. Harper, and S. K. Turitsyn, "Generation of triangular-shaped optical pulses in normally dispersive fibre," Journal of Optics, vol. 12, no. 3, Article ID 035205, 2010.

73. S. Boscolo and S. K. Turitsyn, "Pulse shaping in mode-locked ring-cavity fibre lasers," in Proceedings of Lasers and Electro-Optics Europe and 12th European Quantum Electronics Conference (CLEO EUROPE/EQEC ‹11), Munich, German, May 2011, paper EH4.6.

74. R. R. Alfano, The Supercontinuum Laser Source, Springer, New York, NY, USA, 2006.

75. J. M. Dudley and J. R. Taylor, Supercontinuum Generation in Optical Fibers, Cambridge University Press, New York, NY, USA, 2010.

76. L. Boivin and B. C. Collings, "Spectrum slicing of coherent sources in optical communications," Optical Fiber Technology, vol. 7, no. 1, pp. 1–20, 2001.

77. Y. Takushima and K. Kikuchi, "10-GHz, over 20-channel multiwavelength pulse source by slicing supercontinuum spectrum generated in normal-dispersion fiber," IEEE Photonics Technology Letters, vol. 11, no. 3, pp. 322–324, 1999.

78. Z. Yusoff, P. Petropoulos, K. Furusawa, T. M. Monro, and D. J. Richardson, "A 36-Channel x 10-GHz spectrally sliced pulse source based on supercontinuum generation in normally dispersive highly nonlinear holey fiber," IEEE Photonics Technology Letters, vol. 15, no. 12, pp. 1689–1691, 2003.

79. F. Parmigiani, C. Finot, K. Mukasa, et al., "Ultra-flat SPM-broadened spectra in a highly nonlinear fiber using parabolic pulses formed in a fiber Bragg grating," Optics Express, vol. 14, no. 17, pp. 7617–7622, 2006.

80. J. M. Dudley, G. Genty, and S. Coen, "Supercontinuum generation in photonic crystal fiber," Reviews of Modern Physics, vol. 78, no. 4, pp. 1135–1184, 2006.

81. M. Nakazawa, K. Tamura, H. Kubota, and E. Yoshida, "Coherence degradation in the process of supercontinuum generation in an optical fiber," Optical Fiber Technology, vol. 4, no. 2, pp. 215–223, 1998.

82. A. M. Heidt, "Pulse preserving flat-top supercontinuum generation in all-normal dispersion photonic crystal fibers," Journal of the Optical Society of America B, vol. 27, no. 3, pp. 550–559, 2010.

83. L. E. Hooper, P. J. Mosley, A. C. Muir, W. J. Wadsworth, and J. C. Knight, "Coherent supercontinuum generation in photonic crystal fiber with all-normal group velocity dispersion," Optics Express, vol. 19, no. 6, pp. 4902–4907, 2011.

84. C. Finot, S. Pitois, and G. Millot, "Regenerative 40 Gbit/s wavelength converter based on similariton generation," Optics Letters, vol. 30, no. 14, pp. 1776–1778, 2005.

85. S. Boscolo and S. K. Turitsyn, "All-optical signal regeneration by temporal slicing of nonlinearly flattened optical waveform," IEEE Photonics Technology Letters, vol. 17, no. 6, pp. 1235–1237, 2005.

86. F. Parmigiani, P. Petropoulos, M. Ibsen, and D. J. Richardson, "Pulse retiming based on XPM using parabolic pulses formed in a fiber Bragg grating," IEEE Photonics Technology Letters, vol. 18, no. 7, pp. 829–831, 2006.

87. S. Boscolo, S. K. Turitsyn, and K. J. Blow, "Time domain all-optical signal processing at a RZ optical receiver," Optics Express, vol. 13, no. 16, pp. 6217–6227, 2005.

88. I. O. Nasieva, S. Boscolo, and S. K. Turitsyn, "Bit error rate improvement by nonlinear optical decision element," Optics Letters, vol. 31, no. 9, pp. 1205–1207, 2006.

89. T. T. Ng, F. Parmigiani, M. Ibsen, Z. Zhang, P. Petropoulos, and D. J. Richardson, "Compensation of linear distortions by using XPM with parabolic pulses as a time lens," IEEE Photonics Technology Letters, vol. 20, no. 13, pp. 1097–1099, 2008.

90. T. Hirooka and M. Nakazawa, "All-optical 40-GHz time-domain Fourier transformation using XPM with a dark parabolic pulse," IEEE Photonics Technology Letters, vol. 20, no. 22, pp. 1869–1871, 2008.

91. J. Li, B. E. Olsson, M. Karlsson, and P. A. Andrekson, "OTDM add-drop multiplexer based on XPM-induced wavelength shifting in highly nonlinear fiber," Journal of Lightwave Technology, vol. 23, no. 9, pp. 2654–2661, 2005.

92. F. Parmigiani, P. Petropoulos, M. Ibsen, P. J. Almeida, T. T. Ng, and D. J. Richardson, "Time domain add-drop multiplexing scheme enhanced using a saw-tooth pulse shaper," Optics Express, vol. 17, no. 10, pp. 8362–8369, 2009.

93. F. Parmigiani, M. Ibsen, T. T. Ng, L. Provost, P. Petropoulos, and D. J. Richardson, "An efficient wavelength converter exploiting a grating-based saw-tooth pulse shaper," IEEE Photonics Technology Letters, vol. 20, no. 17, pp. 1461–1463, 2008.

94. F. Parmigiani, M. Ibsen, P. Petropoulos, and D. J. Richardson, "Efficient all-optical wavelength-conversion scheme based on a saw-tooth pulse shaper," IEEE Photonics Technology Letters, vol. 21, no. 24, Article ID 5291729, pp. 1837–1839, 2009.

95. N. Verscheure and C. Finot, "Pulse doubling and wavelength conversion through triangular nonlinear pulse reshaping," Electronics Letters, vol. 47, no. 21, pp. 1194–1196, 2011.

96. A. I. Latkin, S. Boscolo, R. S. Bhamber, and S. K. Turitsyn, "Doubling of optical signals using triangular pulses," Journal of the Optical Society of America B, vol. 26, no. 8, pp. 1492–1496, 2009.

97. R. S. Bhamber, S. Boscolo, A. I. Latkin, and S. K. Turitsyn, "All-optical TDM to WDM signal conversion and partial regeneration using XPM with triangular pulses," in Proceedings of the 34th European Conference on Optical Communication (ECOC ‹08), pp. 1–2, Brussels, Belgium, 2008, paper Th.1.B.2.

98. E. R. Andresen, J. M. Dudley, D. Oron, C. Finot, and H. Rigneault, "Transform-limited spectral compression by self-phase modulation of amplitude-shaped pulses with negative chirp," Optics Letters, vol. 36, no. 5, pp. 707–709, 2011.

99. C. Finot, F. Parmigiani, P. Petropoulos, and D. J. Richardson, "Parabolic pulse evolution in normally dispersive fiber amplifiers preceding the similariton formation regime," Optics Express, vol. 14, no. 8, pp. 3161–3170, 2006.

100. P. V. Mamyshev, "All-optical data regeneration based on self-phase modulation effect," in Proceedings of the 24th European Conference on Optical Communication (ECOC ‹98), pp. 475–476, Madrid, Spain, September 1998.

101. L. Provost, C. Finot, P. Petropoulos, K. Mukasa, and D. J. Richardson, "Design scaling rules for 2R-optical self-phase modulation-based regenerators," Optics Express, vol. 15, no. 8, pp. 5100–5113, 2007.

102. C. Finot, J. Fatome, S. Pitois, G. Millot, and E. Pincemin, "Active Mamyshev regenerator," Optical Review, vol. 18, no. 3, pp. 257–263, 2011.

103. M. Romagnoli, P. Franco, R. Corsini, A. Schiffini, and M. Midrio, "Time-domain Fourier optics for polarization-mode dispersion compensation," Optics Letters, vol. 24, no. 17, pp. 1197–1199, 1999.

104. L. A. Jiang, M. E. Grein, H. A. Haus, E. P. Ippen, and H. Yokoyama, "Timing jitter eater for optical pulse trains," Optics Letters, vol. 28, no. 2, pp. 78–80, 2003.

105. M. Nakazawa and T. Hirooka, "Distortion-free optical transmission using time-domain optical Fourier transformation and transform-limited optical pulses," Journal of the Optical Society of America B, vol. 22, no. 9, pp. 1842–1855, 2005.

106. T. Hirooka and M. Nakazawa, "Optical adaptive equalization of high-speed signals using time-domain optical Fourier transformation," Journal of Lightwave Technology, vol. 24, no. 7, pp. 2530–2540, 2006.

107. N. Yoshikane, I. Morita, T. Tsuritani, A. Agata, N. Edagawa, and S. Akiba, "Benefit of SPMbased all-optical reshaper for long-haul DWDM transmission systems," Journal of Selected Topics in Quantum Electronics, vol. 10, pp. 412–420, 2004.

108. M. Matsuura, N. Kishi, and T. Miki, "Performance improvement of optical RZ-receiver by utilizing an all-optical waveform converter," Optics Express, vol. 13, no. 13, pp. 5074–5079, 2005.

109. P. Ghelfi, M. Secondini, M. Scaffardi, F. Fresi, A. Bogoni, and L. Poti, "Impact of an additional all-optical decision element in band-limited receivers for RZ systems," Journal of Lightwave Technology, vol. 25, no. 7, pp. 1728–1734, 2007.

110. M. Rochette, J. N. Kutz, J. L. Blows, D. Moss, J. T. Mok, and B. J. Eggleton, "Bit-error-ratio improvement with 2R optical regenerators," IEEE Photonics Technology Letters, vol. 17, no. 4, pp. 908–910, 2005.

111. M. Rochette, J. L. Blows, and B. J. Eggleton, "3R optical regeneration: an all-optical solution with BER improvement," Optics Express, vol. 14, no. 14, pp. 6414–6427, 2006.

CHAPTER 16 EFFICIENT NONLINEAR LIGHT EMISSION OF SINGLE GOLD OPTICAL ANTENNAS DRIVEN BY FEW-CYCLE NEAR-INFRARED PULSES

T. Hanke, G. Krauss, D. Tra¨utlein, B. Wild, R. Bratschitsch, and A. Leitenstorfer

Department of Physics and Center for Applied Photonics, University of Konstanz, D-78457 Konstanz, Germany

ABSTRACT

Individual nanometer-sized plasmonic antennas are excited resonantly with few-cycle laser pulses in the near infrared. Intense third-harmonic emission of visible light prevails for fundamental photon energies below 1.1 eV. Interband luminescence and second harmonic generation occur solely at higher driving frequencies. We attribute these findings to multiphoton resonances with the d-band transitions of gold. The strong third-order signal allows direct measurement of a subcycle plasmon dephasing time of 2 fs, highlighting the efficient radiation coupling and broadband response of the devices.

INTRODUCTION

Metal nanoantennas [1,2] have recently been demonstrated to work as optical analogues to their ubiquitous counterparts at radio and microwave frequencies. They have been used to couple visible radiation to single nanoscopic light emitters such as semiconductor quantum dots [3] or molecules[4]. Although the linear response of optical antennae has been investigated in detail [5–8], little is yet known about their nonlinear properties. Emission attributed to coherent supercontinuum and high harmonics generation [1,9], as well as incoherent two-photon photoluminescence [2,10] have been observed. Up to now, experiments were performed by exciting nanoantennae with ultrashort laser pulses originating from Ti:sapphire lasers operating at photon energies around 1.5 eV. Typically, large arrays of antennas have been employed to overcome small conversion efficiencies. In this work, we exploit ultrabroadband Er:fiber technology delivering tunable laser pulses in the few-cycle regime [11] to investigate single gold nanoantennae under excitation in the near infrared. We observe a variety of strong nonlinear signals in the visible spectral range. The efficiencies depend critically on the photon energy of the excitation.

Well-defined arrays of linear, bowtie, and elliptically shaped structures are fabricated via electron beam lithography on fused silica substrates. The thicknesses of the gold film, chromium adhesion layer, and dielectric substrate amount to 40 nm, 2 nm, and 170 μm, respectively. Antenna size and gap

Citation: T. Hanke, G. Krauss, D. Tra¨utlein, B. Wild, R. Bratschitsch, and A. Leitenstorfer, Efficient Nonlinear Light Emission of Single Gold Optical Antennas Driven by Few-Cycle Near-Infrared Pulses, DOI: 10.1103/PhysRevLett.103.257404

dimensions are systematically varied. The data featured here represent fundamental phenomena observed for all antenna geometries under investigation.

Our femtosecond Er:fiber system operates at a repetition rate of 40 MHz. It delivers pulses of a duration of 7.8 fs with a spectral bandwidth extending from 0.85 to 1.4 eV [11]. To single out individual nanoantennae, we focus to a spot diameter of 2μm via all-reflective optics, minimizing dispersion and achromatism. The linear polarization of the excitation is set parallel to the main antenna axis. The emission is collected in transmission geometry with an objective lens of a numerical aperture of NA=0.55. Antenna samples may be raster scanned relative to the confocal region with a piezoelectric translation stage. Figure 1 shows the spectrally integrated visible emission as a function of lateral position in an array of 48 linear nanoantennae of varying arm length l and feed gap g. We excite with the ultrabroadband spectrum centered at a wavelength of 1.2μm and an average intensity set to 0.6 MW/cm^2. Note that the blue light emanating from individual nanoantennae is easily visible with the bare human eye [12]. The average power generated by the brightest devices is in the order of 0.7 nW. For structures with feed gaps wider than g=40 nm, high emission intensities are observed when the excitation bandwidth includes wavelengths corresponding to a quarter of a single antenna arm, i.e., for l≈300 nm [6,7]. As soon as the two arms begin to couple electrodynamically for feed gaps g around 40 nm, the redshift of the dipole resonance [13] leads to resonant conditions at shorter overall antenna lengths (see dashed line in Fig. 1). This shift finally evolves into the $\lambda/4$ resonance of a single rod at the onset of close contact of the two antenna arms (nominal feed gaps g of 0 nm and 10 nm). The observation of this systematic behavior in the antenna array demonstrates the quality of the fabrication process and the reproducibility of our structures.

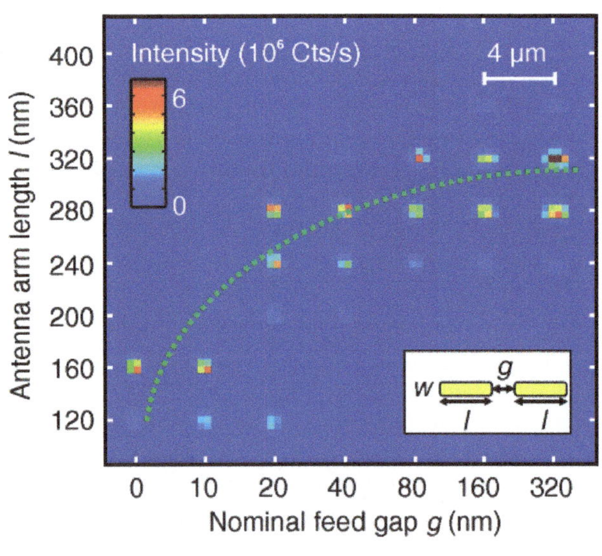

Figure. 1. Spectrally integrated nonlinear emission from 48 individual linear gold antennae of different arm length l and feed gap g. The arm width w is kept constant at 60 nm. Lateral and vertical distances between antennas are set to 4 μm and 3 μm, respectively. The nanostructures are raster scanned through the focus of a 7.8-fs laser pulse with a spectrum extending from 0.85 to 1.4 eV. The visible intensity is depicted in a color scale as a function of position. The dashed line is a guide to the eye illustrating the regions with antennae in resonance with the driving field. Inset: schematic drawing of the antenna geometry studied in this data set.

To spectrally resolve the radiation from each nanoantenna, we use a monochromator and a Peltier-cooled CCD camera. Figure 2 depicts a typical nonlinear emission spectrum from a single bowtie resonant device excited at a pump intensity of 0.5 MW/cm^2. We observe strong third-harmonic generation (THG) together with a weak second harmonic signal (SHG). The spectral positions of THG and SHG point to the same fundamental frequency determined by the plasmon resonance of the metal nanostructure in the near infrared. We find a cubic and quadratic dependence of the emission intensity on excitation power for the THG and SHG, respectively. The high intensity of the third harmonic may be attributed to the strong $\chi^{(3)}$ nonlinearity of Au [14]. The smaller efficiency for SHG is consistent with the centrosymmetric character of bulk gold and fused silica [15]. In the forward direction, frequency doubling might occur solely due to hyper-Rayleigh scattering at surface imperfections [16]. Most likely, the SHG we observe is due to excitation of nonlocal electric dipole and local electric quadrupole modes [17] or electromagnetic effects in the confined plasma system [18]. In our geometry, these components get collected due to the high numerical aperture of our detection optics.

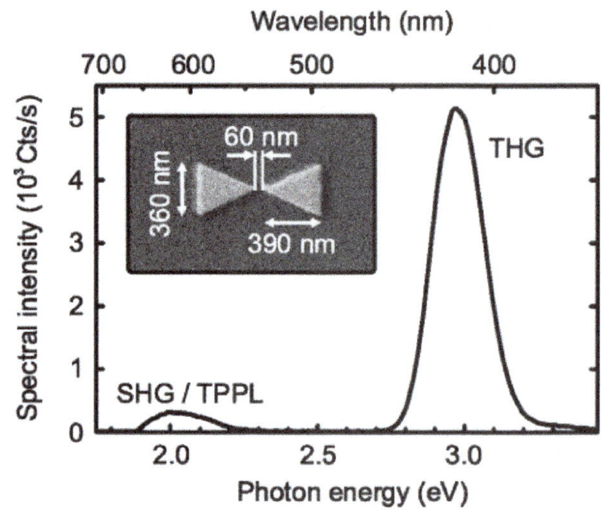

Figure. 2. Typical nonlinear emission spectrum from a single gold optical antenna excited with the total bandwidth of the 7.8-fs pulse. SHG: second harmonic generation, TPPL: two-photon photoluminescence, THG: third-harmonic generation. Inset: scanning electron microscope image of the antenna. A bowtie-shaped geometry is featured in this figure which is composed of two opposing nanotriangles (triangle height: 390 nm, triangle base: 360 nm, gap: 60 nm).

To elaborate the microscopic origin of the nonlinear emission components we tune the photon energy of the pump while keeping the antennas resonant by proper control of geometry: A strong third-harmonic emission is observed for excitation with a 20-fs laser pulse centered at 1 eV [Fig. 3(a)]. When the driving pulse contains photon energies beyond 1.1 eV, we obtain a signal at the second harmonic frequency [see Fig. 3(b)]. At the same time, the SHG peak develops a low-energy shoulder. We attribute this feature to two-photon photoluminescence (TPPL). This component is known from experiments with Ti:sapphire lasers [2,19]. It has been used previously to map the electric field around single metal nanoantennas [10]. A tentative explanation for the drastic change of the dominating nonlinear process when tuning the excitation above 1.1 eV is motivated by the synchronous appearance of SHG and TPPL: As long as the photon energy is too low to induce two-photon transitions of electrons out of the d-bands across the Fermi energy and into the empty states in the sp-type conduction band, the antenna response is dominated by the strong third-order nonlinearity and a clean THG signal is emitted. This situation is sketched in the band structure scheme of Fig. 3(c). As soon as the two-photon resonance of the d-band transition is reached, TPPL sets in and at the same time, second-order nonlinear effects may be resonantly enhanced [see Fig. 3(d)]. Interestingly, the THG emission from single antennae becomes hardly discernible for pump photon energies beyond 1.1 eV, although it falls well into our detection window [see Fig. 3(b)]. The explanation for this behavior may be found in a combination of two effects: (i) THG may be enhanced at lower driving frequencies due to a three-photon resonance with the d-band transitions of bulk Au [see Fig. 3(c)]. (ii) The two-photon resonance occurring at shorter pump wavelengths [Fig. 3(d)] may render the third-order nonlinear response predominantly imaginary, effectively quenching THG which is linked to the real part of $\chi^{(3)}$.

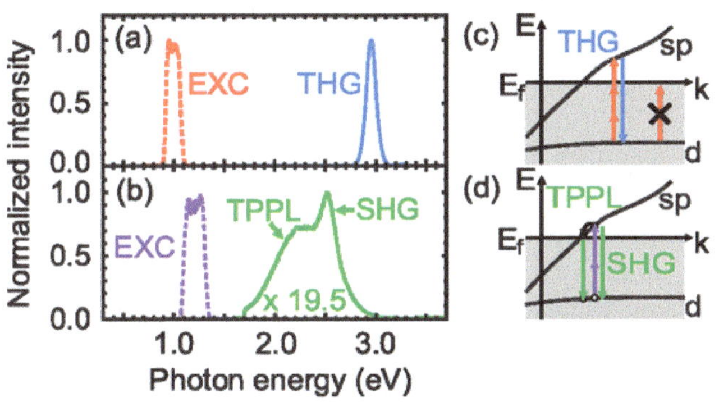

Figure. 3. Nonlinear emission from resonant gold nanoantennas depending on excitation photon energy. (a) Third-harmonic emission (THG, blue line) of a linear device with arm length l=320 nm, width w=60 nm and feed gap g=40 nm, excited with a pump spectrum (EXC) confined below 1.1 eV (dashed red line). (b) The two-photon photoluminescence (TPPL) and second harmonic generation (SHG) from a specimen with l=240 nm, w=60 nm and g=40 nm, excited with a pump photon energy above 1.1 eV (magenta dashed line) is depicted as a green line. The peak spectral intensity of the TPPL/SHG signal is weaker by a factor of 19.5 as compared to the THG emission in (a). (c) and (d) Sketches of the band structure of gold illustrating the relevant pump and emission processes for the conditions in (a) and (b), respectively. Arrow lengths indicate photon energies in excitation (up) and emission (down). Colors correspond to the processes indicated in (a) and (b).

The efficient THG emission we find for pump photon energies below 1.1 eV is now exploited to directly measure the ultrafast plasmon dephasing time of a single nanoantenna. To this end, we excite collinearly with two identical transients derived from a Mach-Zehnder interferometer and record the emitted THG signal as a function of time delay between the pulses. Frequency-resolved optical gating (FROG) measurements [20–22] are performed with the spectrometer and the CCD camera. To achieve the highest time resolution available combined with an unchirped driving field, we restrict the spectrum of the excitation pulse to the region where it features a flat phase relationship. This strategy results in a temporal duration close to the bandwidth limit and a clean pulse shape. We characterize the transient by recording a FROG trace from the nonresonant signal generated with the bare fused silica surface next to the antenna structure. Figure 4(a) shows the emitted THG intensity color coded as a function of delay time and frequency position. The interference fringes due to the collinear geometry were removed via Fourier filtering. Careful FROG analysis [23] yields a pulse duration of 9.8 fs, a spectral bandwidth of 51 THz and a carrier frequency of 248 THz, equivalent to a central wavelength of 1210 nm. The accuracy of our characterization is demonstrated by the close coincidence of the measured third-harmonic autocorrelation (dotted line in Fig. 4(c)] with the trace calculated from the amplitude and phase spectra obtained via FROG (red line in Fig. 4(c)]. With exactly the same pump pulse we now resonantly excite an elliptical Au nanoantenna. The Fourier-filtered FROG data are depicted in Fig. 4(b). Note that the THG intensity is 2 orders of magnitude higher than from the bare substrate. The spectral full width at half maximum (FWHM) of the signal decreases from 31.1 THz in the nonresonant measurement of Fig. 4(a) to 26.5 THz for the antenna emission in Fig. 4(b). Correspondingly, the temporal width of the spectrally integrated third-order autocorrelation [dotted line in Fig. 4(d)] increases significantly as compared to the instantaneous response of the substrate [Fig. 4(c)]. To quantify the plasmon dephasing of the nanoantenna system, we fit the interferometric autocorrelation by assuming a fundamental plasmon oscillator with an exponential damping term and a perturbative third-order nonlinearity [24]. The driving electric field is retrieved from the FROG trace in Fig. 4(a). If this method is applied integrating over an array of metal nanostructures, a lower limit for the individual dephasing times is obtained due to the potential influence of inhomogenous broadening [24]. In contrast, our study of a single specimen gives direct access to the precise value of the damping constant. The THG autocorrelation simulated assuming a plasmon dephasing time of 2 fs is depicted as a red line in Fig. 4(d). It is in excellent agreement with the experimental result [dotted line in Fig. 4(d)]. By comparison with simulated autocorrelation traces for 1.5 fs and 2.5 fs dephasing times, we estimate the experimental error of this measurement to be as small as ±500 attoseconds.

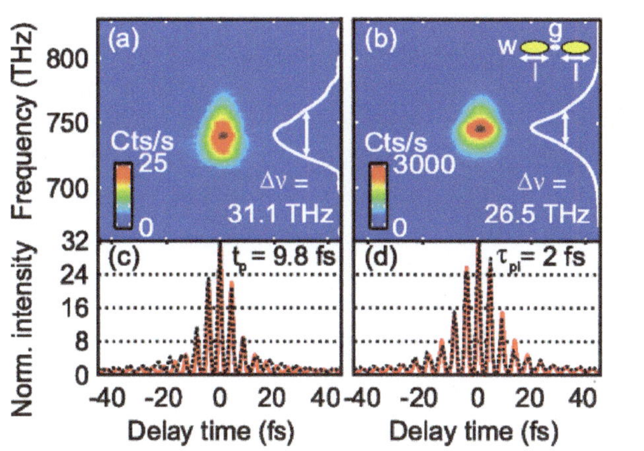

Figure. 4. Dephasing of a single resonantly driven gold nanoantenna: Fourier-filtered third-harmonic emission intensity as a function of delay time and frequency for (a) the bare fused silica substrate and (b) an elliptical gold nanoantenna (arm length l=250 nm, width w=100 nm, and feed gap g=40 nm). Integrated emission spectra are depicted as white lines. Interferometric third-order autocorrelation traces from (c) the substrate and (d) the resonant antenna. Careful analysis and comparison of the experimental data (dotted lines) with simulations (red lines) yield a reference pulse duration of t_p=9.8 fs and a plasmon dephasing time of τ_{pl}=2.0 fs.

The ultrashort dephasing of our antennae is clearly dominated by radiation damping [25–27]. Note that a damping time of 2 fs corresponds to less than an oscillation cycle of the fundamental plasmon resonance of the antenna of 250 THz. In the context of metal nanoparticles, it has been argued that long plasmon dephasing times are a prerequisite for the occurrence of strong field enhancement and efficient nonlinear optics [26, 27]. Nevertheless, we obtain an increase of the resonant nonlinear emission of more than an order of magnitude as compared to off-resonant driving (see Fig. 1). Obviously, the relevance of long plasmon dephasing times for the efficiency of nonlinear processes is limited when working with optical antennae. In fact, these are devices where radiation energy is collected from a large cross-sectional area and funneled into a small volume near the antenna gap where a strong field enhancement arises. Since electromagnetic coupling works in both directions, optimized antenna geometries will always tend to come with relatively fast radiation damping. Therefore, these systems are naturally compatible with few-femtosecond optical excitation, providing the efficient harmonics generation we find under resonant conditions.

In conclusion, we have studied the nonlinear optical properties of single gold nanoantennae driven with few-cycle laser pulses in the near infrared. Intense third-harmonic emission is obtained when exciting with fundamental spectra below 1.1 eV. At higher photon energies, additional components corresponding to frequency doubling and two-photon induced luminescence are observed together with a fading of THG. We relate these findings to the band structure of bulk Au, especially to a two-photon resonance with the d-band transitions. A plasmon dephasing as fast as 2 fs is directly measured in the time domain. We expect the bright third-harmonic emission to originate from a volume of a few (10 nm)3 close to the feed gap of our antennae. Therefore, they enable selective excitation and sensitive probing of optical properties in nanoscale spatial regions with femtosecond time resolution and ultrabroadband spectral coverage. Consequently, such structures will be attractive tools for new classes of experiments like femtosecond quantum optics with solid-state single-electron systems [28]. Pumped with a few mW from compact fiber lasers, the intensive and coherent nanometer-sized light sources might also find technical applications, e.g., in readout devices for ultrahigh density optical data storage.

REFERENCES

1. P. Mühlschlegel, H.-J. Eisler, O.J.F. Martin, B. Hecht, and D.W. Pohl, Science 308, 1607 (2005).
2. P.J. Schuck, D.P. Fromm, A. Sundaramurthy, G.S. Kino, and W.E. Moerner, Phys. Rev. Lett.94, 017402 (2005).
3. J.N. Farahani, D.W. Pohl, H.-J. Eisler, and B. Hecht, Phys. Rev. Lett. 95, 017402 (2005).
4. T.H. Taminiau, R.J. Moerland, F.B. Segerink, L. Kuipers, and N.F. van Hulst, Nano Lett. 7, 28 (2007).
5. L. Rogobete, F. Kaminski, M. Agio, and V. Sandoghdar, Opt. Lett. 32, 1623 (2007).
6. G.W. Bryant, F.J. Garcia de Abajo, and J. Aizpurua, Nano Lett. 8, 631 (2008).
7. L. Novotny, Phys. Rev. Lett. 98, 266802 (2007).
8. J. Merlein, M. Kahl, A. Zuschlag, A. Sell, A. Halm, J. Boneberg, P. Leiderer, A. Leitenstorfer, and R. Bratschitsch, Nat. Photon. 2, 230 (2008).
9. S. Kim, J. Jin, Y.-J. Kim, I.-Y. Park, Y. Kim, and S.-W. Kim, Nature (London) 453, 757 (2008).
10. P. Ghenuche, S. Cherukulappurath, T.H. Taminiau, N.F. van Hulst, and R. Quidant, Phys. Rev. Lett. 101, 116805 (2008).
11. A. Sell, G. Krauss, R. Scheu, R. Huber, and A. Leitenstorfer, Opt. Express 17, 1070 (2009).
12. See EPAPS Document No. for a real-time video of THG emission of antennas scanned through the excitation focus.
13. J. Aizpurua, G.W. Bryant, L.J. Richter, F.J. Garcia de Abajo, B.K. Kelley, and T. Mallouk,Phys. Rev. B 71, 235420 (2005).
14. M. Lippitz, M.A. van Dijk, and M. Orrit, Nano Lett. 5, 799 (2005).
15. B.K. Canfield, H. Husu, J. Laukkanen, B. Bai, M. Kuittinen, J. Turunen, and M. Kauranen,Nano Lett. 7, 1251 (2007).
16. M.I. Stockman, D.J. Bergman, C. Anceau, S. Brasselet, and J. Zyss, Phys. Rev. Lett. 92, 057402 (2004).
17. J.I. Dadap, J. Shan, K.B. Eisenthal, and T.F. Heinz, Phys. Rev. Lett. 83, 4045 (1999).
18. Y. Zeng, W. Hoyer, J. Liu, S.W. Koch, and J.V. Moloney, Phys. Rev. B 79, 235109 (2009).
19. A. Bouhelier, R. Bachelot, G. Lerondel, S. Kostcheev, P. Royer, and G.P. Wiederrecht, Phys. Rev. Lett. 95, 267405 (2005).
20. T. Tsang, M.A. Krumbügel, K.W. DeLong, D.N. Fittinghoff, and R. Trebino, Opt. Lett. 21, 1381 (1996).
21. G. Stibenz and G. Steinmeyer, Opt. Express 13, 2617 (2005).
22. A. Anderson, G. Steinmeyer, and M.B. Raschke, Conference on Lasers and Electro-Optics/International Quantum Electronics Conference, OSA Technical Digest (CD) (Optical Society of America, Washington, DC, 2009), paper CThK2
23. The FROG error on a grid size of 128 amounts to 0.015.
24. B. Lamprecht, J.R. Krenn, A. Leitner, and F.R. Aussenegg, Phys. Rev. Lett. 83, 4421 (1999).
25. A. Wokaun, J.P. Gordon, and P.F. Liao, Phys. Rev. Lett. 48, 957 (1982).
26. C. Sönnichsen, T. Franzl, T. Wilk, G. von Plessen, J. Feldmann, O. Wilson, and P. Mulvaney,Phys. Rev. Lett. 88, 077402 (2002).
27. T. Klar, M. Perner, S. Grosse, G. von Plessen, W. Spirkl, and J. Feldmann, Phys. Rev. Lett. 80, 4249 (1998).
28. F. Sotier, T. Thomay, T. Hanke, J. Korger, S. Mahapatra, A. Frey, K. Brunner, R. Bratschitsch, and A. Leitenstorfer, Nature Phys. 5, 352 (2009).

INDEX

M

N

O

P

Q

R

S

T

V

W

X